Integrated Management Systems for Construction

Integrated Management Systems for Construction

Quality, environment and safety

Alan Griffith

Routledge
Taylor & Francis Group

LONDON AND NEW YORK

First published 2011 by Pearson Education Limited

First edition published 2011

Published 2013 by Routledge
2 Park Square, Milton Park, Abingdon, Oxon OX14 4RN
711 Third Avenue, New York, NY 10017, USA

Routledge is an imprint of the Taylor & Francis Group, an informa business

ISBN 13: 978-0-273-73065-1 (pbk)

British Library Cataloguing-in-Publication Data
A catalogue record for this book is available from the British Library

Library of Congress Cataloging-in-Publication Data
Griffith, Alan.
  Integrated management systems for construction : quality, environment and safety /
Alan Griffith. – 1st ed.
     p.   cm.
  Includes bibliographical references and index.
  ISBN 978-0-273-73065-1 (pbk.)
  1. Construction industry–Management.   2. Construction industry–Quality control.
3. Construction industry–Standards.   I. Title.
  HD9715.A2G72 2011
  624.068′4—dc22

                                                                              2010032831

Typeset in 10/12.5pt Sabon by 35

Contents

Preface

Integrated Management Systems for Construction focuses on the standards-based management systems used for three key management functions routinely applied within the construction processes: quality, environment and safety. It examines these individually through the concepts, principles and practices associated with each system and in application by the principal contracting organisation. Moreover, it examines the establishment of an integrated management system to embrace these three management functions in dual or triple applications. An integrated management system is an approach which integrates two or more management functions into one coherent and seamless system, enabling the better achievement of the organisation's business activities and outputs. Quality, environment and safety management are key management, assurance and business functions within the principal contracting organisation. However, they are support functions and therefore they must be directed to enhancing the core business, maximising the performance of business processes and their management, and improving the quality of service to the client.

Why are management systems for quality, environment and safety, and integrated management systems, important? And why examine them? Within the construction industry, principal contractors and indeed other participants use management systems to structure and organise their corporate approach to business activities, translating these into operational procedures for application to the management of construction projects. Such systems are process orientated, standards based and document reliant, or what are often termed soft systems, the most prominent and well recognised being the ISO 9001 quality management system. Additional soft systems have since been introduced to guide other management functions such as environmental safeguards and health and safety. This has raised a challenge to find ways of optimising the similarities, compatibilities and potentials of these traditionally separate systems to develop an integrated approach which can deliver greater organisational efficiency and business effectiveness. The fundamental precept of this book is that organisations operating within the construction industry need to be more greatly aware of the evolving concepts, principles and practices within construction which influence their work – management systems and their integration are undoubtedly one such interest.

Alan Griffith

Acknowledgements

I would not profess to having moved the boundaries of research or professional practice within the field of construction management. Research at the truly leading edge in many academic and professional disciplines including construction management is a challenge which is rarely fulfilled. Notwithstanding, I would profess to having engaged in research and its dissemination, at important boundaries of construction management knowledge and practice, and I have done so for many years. Such work has been considerable, in many ways has been significant and has, at times, had good and valuable impact upon construction management as an academic discipline and as a professional practice. Good research is not conducted in isolation and I have been fortunate in working alongside and associating with prominent and indeed some eminent individuals from whom I have received great encouragement to pursue research and scholarship over the years. I would like to take this opportunity to thank these individuals.

First and foremost I acknowledge the late Professor Victor B. Torrance CBE, Past President of the Chartered Institute of Building (CIOB), formerly William Watson Professor of Building at Heriot-Watt University, Professor at Loughborough University and University College London (UCL), and latterly Professor at Universiti Teknologi MARA (UiTM) in Malaysia. Victor was without doubt a great ambassador for the construction industry and a true champion of the construction disciplines within academia. To those whose lives Victor touched he was quite simply an inspiration. I express my very considerable gratitude also to Professor Roger Flanagan, University of Reading. Roger was, in association with the late Professor Bill Biggs OBE, instrumental in acquiring much government funding for my early research and its dissemination. Others to whom I am most grateful for their help, guidance and acquaintance over the years are: Professor Emeritus Norman Fisher, University of Reading; Professor Emeritus Tony Walker, University of Hong Kong; and Professor Li Yin Shen, The Hong Kong Polytechnic University.

In connection with the development of this book I am indebted to Dr Khalid Bhutto, Director of SafeScope Ltd, whose research and professional practice has contributed greatly to the understanding of standards-based management systems and their application to construction. I am also grateful to those organisations which have contributed to the underpinning research information, experiences and guidance included in appropriate parts of the book. More widely, information in the public domain, from authoritative written and Web-based sources, has contributed usefully and meaningfully to the work, and these are, with thanks, referenced as appropriate.

Abbreviations

ABCB	Association of British Certification Bodies
BEM	Business Excellence Model
BMS	Business Management System
BPM	Business Process Model
BQF	British Quality Foundation
BS	British Standard
BSI	British Standards Institution
BSS	British Standards Society
CARES	Certification Authority for Reinforcing Steels
CDM	Construction (Design and Management) Regulations
CEN	European Committee for Standardization
CEO	Chief Executive Officer
CFCs	Chlorofluorocarbons
CIOB	Chartered Institute of Building
CIRIA	Construction Industry Research and Information Association
COSHH	Control of Substances Hazardous to Health Regulations
CSCS	Construction Skills Certification Scheme
CSR	Corporate Social Responsibility
DEFRA	Department for Environment, Food and Rural Affairs
DLCS	Davis Langdon Certification Services
DTI	Department of Trade and Industry
EA	Environment Agency; Environmental Assessment
ECI	European Construction Institute
EFQM	European Foundation for Quality Management
EIA	Environmental Impact Assessment
EMAS	Eco-Management and Audit Scheme
EMS	Environmental Management System
EN	European Normalisation
EQM	Environmental Quality Management
EQS	Environmental Quality System
ESCM	Environmental Supply Chain Management
GD	Guidance Document
H&S	Health and Safety
H&SMS	Health and Safety Management System
HRM	Human Resource Management
HSE	Health and Safety Executive
ICC	Incident Contact Centre
IEMA	Institute of Environmental Management & Assessment
IFAN	International Federation for the Application of Standards

IiP	Investors in People
ILO	International Labour Organization
IMS	Integrated Management System
IMSA	Integrated Management System Assessment
IOSH	Institution of Occupational Safety and Health
IQM	Integrated Quality Management
ISO	International Organization for Standardization
ISO 9001	ISO Standard Requirements for Quality Management Systems
ISO 14001	ISO Standard Requirements for Environmental Management Systems
IWS	Integrated Waste Strategy
JQA	Japan Quality Assurance Organization
MD	Managing Director
OH&S	Occupational Health and Safety
OHSAS 18001	OHSAS Standard Requirements for Health and Safety Management Systems
OHSAS	Occupational Health and Safety Assessment Series
OPSI	Office of Public Sector Information
PAS	Publicly Available Specification
PDCA	Plan–Do–Check–Act
PESTE	Political, Economic, Social, Technological, Ecological
PESTLE	PESTE with Legal aspects
PMP	Project Management Plan
PPE	Personal Protective Equipment
PQP	Project Quality Plan
PR	Public Relations
QMS	Quality Management System
QSCM	Quality Supply Chain Management
QSRMC	Quality Scheme for Ready Mixed Concrete
QUENSH	Quality, Environment, Safety and Health
RIDDOR	Reporting of Injuries, Diseases and Dangerous Occurrences Regulations
SCEEMAS	Small Company Environmental Energy Management Assistance Scheme
SHE	Safety, Health and Environment
SME	Small and Medium Enterprise
SSCM	Safety Supply Chain Management
SWMP	Site Waste Management Plan
SWOT	Strengths, Weaknesses, Opportunities, Threats
TBL	Triple-Bottom Line
TQM	Total Quality Management
TRADA	Timber Research and Development Association
UKAS	United Kingdom Accreditation Service
UNFCCC	United Nations Framework Convention on Climate Change
WBS	Work Breakdown Structure
WRAP	Waste and Resources Action Programme

Introduction

Structure and presentation

Integrated Management Systems for Construction: Quality, environment and safety is structured and presented in five chapters: (1) management systems; (2) quality management systems; (3) environmental management systems; (4) safety management systems; and (5) integrated management systems. These chapters are designed so that they may be read as stand-alone topics in their own right or read as a whole through the integrating theme and focus of management systems. Each chapter examines management systems in application to the processes of construction and their deployment by principal contracting organisations. However, the chapters have also been structured and presented so that the core values of management system standards and the principles and applications of management systems can be appreciated by other users within the construction processes and indeed in other business and industry sectors. Each chapter is divided into distinct sections with a particular emphasis. Again, these sections can be read as individual sub-themes or collectively within the holistic theme.

Management systems in the context of this book are what are often termed soft management systems. In essence, these are document-based organisational approaches to structuring functional management disciplines and are developed and implemented to the requirements of recognised international management system standards. The three most prominent of these systems are examined in this book: quality management systems meeting the standard ISO 9001; environmental management systems meeting the standard ISO 14001; and health and safety management systems meeting the standard OHSAS 18001. These are not the only management systems used within the construction industry, or indeed in other business sectors. Management systems and standards exist for application across a wide range of management disciplines such as auditing, ethical trading and information security, to mention just three.

The adoption and use of any management approach by an organisation is entirely voluntary. Some approaches to carrying out management functions are specifically influenced by legislation and regulation such as construction-phase health and safety planning, and environmental site waste management planning. Organisations use management systems because they can help meet such requirements, but they also apply them because they best place the organisation for meeting a greater range of business influences and demands such as the delivery of better quality to its clients and customers. It is asserted that an organisation will pursue a systems management approach because it presents opportunity and benefit to the operation of the business and is therefore axiomatic. However, management systems inevitably incur costs to the organisation. Costs will be borne in the development and implementation of a management system through structural reorganisation, preparation of documentation, employment of consultants and system assessment, and such costs may be considerable. Within this book, cost is seen as an inevitable consequence of establishing a management system and is therefore not discussed as a pervasive or overriding consideration. The reality is that an organisation will need to balance the business and commercial benefits of adopting a system against the organisational and developmental costs of establishing and maintaining that system.

While international management system standards, such as ISO 9001, present requirements for the establishment of discipline-specific management systems, the application of such systems is not prescriptive. Likewise, in this book, the presentation of material does not purport to be prescriptive. Each chapter presents the relevant management systems standard and its specified requirements; explains the development, implementation and maintenance of the system in the context of an organisation and its business processes and outputs; and examines the system in application to construction, taking the principal contractor as its focus. Management system standards such as those presented by the International Organization for Standardization (ISO) are essentially generic and apply not only to the construction industry but to many different sectors of business. They must therefore be used with great consideration and care. They are 'management tools' to be applied in the context of the specific nature, scope, situation and circumstance of the organisation and its business activities and outputs.

Why look at management systems and their integration? Over the last fifty years, quality management has evolved tremendously to become an embedded and accepted part of normal business activity. As further standards-based systems have been introduced to guide a structured and systematic approach to other key management functions, such as environmental safeguards and health and safety, an emerging challenge is to find effective ways of optimising the similarities and compatibilities of the various different systems through the development of an integrated approach. Organisations at the leading edge of construction practice have already established integrated management systems. While these have been configured to suit the particular business needs of the organisation and operate in ways specific and unique to the organisation, it is clear that integrated management systems in a wider context and application can deliver a range of tangible and important benefits to organisations which utilise them.

The overriding theme of this book is the understanding of management systems, their use, their application to the process of construction, and the potential of systems integration in the development of management systems in the future. The content is grounded in a range of literature which describes and examines management system standards, the establishment of systems and their use. It also draws extensively upon and embraces research and professional practice in the field. This is important because some texts, focusing for example on construction safety management or construction quality management, have tended to be mostly descriptive and without research or industry practice underpinnings. In each part of the book, the explanations of system development and implementation for each management system are founded on research studies and a synthesis of examples from construction industry practices. These present an insight into the lessons learnt together with practical perspectives and comment from construction professionals involved in management system use.

Management systems are founded fundamentally in their documentation. Documents, in the form of system manuals, procedures, plans and instructions, are essential to the appropriate development of any system and its effectiveness in application. Also intrinsic to any system is the use of administrative forms, or pro forma templates, for translating managerial procedures into working guides, monitoring mechanisms and operational records. This is particularly true for construction, where the site-based activities of management and work supervision predominate. Appropriate examples to illustrate system development and implementation are presented in each chapter of this book. However, it is not intended to provide sets of system documents or administrative forms for use. These are available on a contracted commercial basis from private-sector management consultancy practices and in electronic form from Web-based business providers. Moreover, the development and use of system documentation need to satisfy the express requirements of the organisation, so generic documents, while helpful, are not suitable for blind or routine adoption.

The construction processes which take place on any project site are influenced significantly by a plethora of general and industry-specific legislation and regulations. These impinge directly upon the methods used to manage the construction project environment and health, welfare and safety on site. Such legislation and regulation must be taken into account along with a host of customer, contractual and other requirements when adopting any management approach. Where such factors are relevant, then these are referred to within the appropriate chapter of the book. However, it is not intended within the scope of the book to digress into detailed discussions concerning European Directives, statutory instruments, regulations or other influences, other than where these impinge upon the explanation of management systems and their application to construction. Information concerning legislation, industry regulations and forms of contract, in whole or in part, can be obtained from the official documents themselves. Relevant references are provided at the end of each chapter to assist the acquisition of further general and specific information.

To avoid direct repetition in each chapter of the book, some aspects are given emphasis in one chapter and then only outlined or mentioned in other chapters. An example of this will be seen in Chapter 2, where quality management system assessment and certification is detailed. The generic orientation of these aspects applies equally to environmental management systems and safety management systems, so the reader can easily make the connection between generic characteristics and their application to specific systems.

The theme of this book has both a national and worldwide scope and relevance. In the UK, standards for many aspects of industry, business and commerce are set by the British Standards Institution (BSI) and the International Organization for Standardization (ISO). Such standards include those specifying management systems for quality, environment and safety. In Europe, ISO standards to guide management system development predominate, while in many parts of the world ISO standards govern exclusively, or, alternatively, strongly influence the development of locally evolving standards including those relevant to management systems. Collectively, over 160 national standards-making bodies are in the worldwide General Assembly of the ISO. The ISO has developed over 18,000 international standards and associated technical and guidance documents, and in the specific fields of quality, environment and safety it has introduced over 700 standards (ISO, 2009). Throughout the construction sector worldwide, many companies are supporting the application of management systems within their declared corporate social responsibilities (CSRs) of business operation. Moreover, quality, environment and safety can form parts of greater integrated corporate management initiatives focusing on, for example, business ethics, valuing people, engagement with community stakeholders and sustainable development. Some of the world's largest, most progressive and successful companies are at the forefront of such developments and these include: Balfour Beatty (UK); Hochtief (Germany); Bechtel (USA); Skanska (Sweden); Vinci (France); the Taisei Corporation (Japan); and Leighton Holdings (Australia). It is clear, therefore, that the relevance of this book for many companies and organisations which operate within the construction sector both in the UK and indeed worldwide is unequivocal.

Chapter 1: Management systems

Chapter 1 focuses on *management systems*, so what is a management system? A management system is, in practice, a set of procedures for guiding the management of the organisation's activities that need to be undertaken to deliver successfully the service or product being provided. In more simple terms, management systems are the organisation's way of doing things without which the organisation would likely be chaotic, unmanageable and dysfunctional. Management systems bring order, structure and stability to the many functional management disciplines which are used in the day-to-day operations of almost all organisations.

The management of any organisation calls for clear objectives, direction, leadership and control. Organisations establish documented, or soft, management systems to establish the necessary framework of policy, command, communications, processes, procedures and resources that enable the organisation to achieve its business objectives. Within the construction industry, client organisations, principal contracting companies, contractors and other participants all use management systems to structure their corporate approach to business activities, translating these into operational procedures for application on the construction projects they contribute to by way of their services and products.

Development, implementation and maintenance of a management system must permeate the whole organisation and guide all its business processes and resources. The system must be embedded within organisational objectives and policies and within procedures and practices. Moreover, the system needs to become an intrinsic part of the organisation's culture, its thinking and everyday operation. Therefore, people are central to the way in which the organisation configures and establishes its management systems, as they lie at the heart of making management systems work effectively. Moreover, the success of any system implementation depends crucially on commitment from individuals and groups at all levels within the organisation, and in particular from executive, or top, management from which organisational culture, vision and drive must materialise.

Management systems are regarded in some organisations merely as administrative, or support, services. Indeed, they are support services, but when seen in such a narrow way they can lose focus, become self-administering and self-perpetuating, and their value to the organisation is diminished, redirected or even lost. There is no point in establishing management systems for their own sake. Systems need to service and support the organisation, they need to focus on and contribute to the organisation's core business, and they need to add value to the wider and holistic vision and business of the organisation. Seen in this way, management systems can give active support, assurance and business service.

Most systems are 'standards based' – management systems which are developed, implemented and maintained to a recognised framework and structure can be the most effective. Organisations with a quality management system certified to the ISO 9001 quality system standards are recognised for their uniformity, consistency, reliability and customer focus. Their quality management system assures the quality of the business processes by which they deliver their services and products. Most systems are also 'process orientated' – management systems focus on the processes used to deliver the organisation's business outputs. An effective management system, with a focus on its own individual management discipline and function, directs its attention and attributes to supporting the management of those processes which deliver the core business.

Management and management systems have their developmental roots in systems theory. Systems theory is founded on the concept of developing synergy – the whole adds up to be greater than the sum of the parts. This is fundamental to all organisations since indeed organisations comprise a great many different parts in the forms of people and resources. The axiom is that the organisation must bring together all the contributing parts with a view to developing a greater whole organisation. This holds true for management systems also. A management system must in its own right and in association with other systems generate synergy and add value to the holistic entity that is the organisation.

Chapter 1 examines: management theories, the organisation and business processes; definition of a management system and its types and features; the application of management systems; and research into and experiences of management systems in use. Chapter 1 provides a fundamental and important introduction to the prominent approaches to establishing management systems and their application to construction.

Chapter 2: Quality management systems

Chapter 2 focuses on *quality management systems*. Effective quality management is a prerequisite to the business of any organisation. Quality, in business today, is simply a necessity, a given, a requirement, a demand made of the provider by the customer. The quality of a product or service and performance of the provider are, probably, the most significant differentiating characteristics among providers in the eyes of the customer. A successful provider such as a principal contracting organisation, which consistently meets the requirements of its customers, will have a competitive advantage in the commercial marketplace.

A structured and systematic approach to the management of quality is needed to address the demands of customers for better quality outputs and business performance of providers. A quality management system (QMS) can help those organisations which provide products and services to improve the quality of their provision and enhance customer satisfaction. A quality management system focuses on delivering better managerial, supervisory and operational performance of its organisational resources, more efficient execution of its business processes, and the highest quality of business output. There are, therefore, good reasons for an organisation to adopt a formal and systematic approach to quality management and establish an effective quality management system.

ISO 9001 specifies a framework for an appropriate and effective quality management system. ISO 9001 is not a quality assurance system but rather, in practice, presents a set of guidelines to assist organisations to manage quality in the context of their business activities. It provides a common-sense approach to quality management systems which an organisation can use to augment the management of its business processes and outputs. The certification of a quality management system offered by ISO 9001 is valuable as organisations certified are known and recognised for their business performance and commitment to customer service.

The benefits of establishing a recognised quality management system are tangible and clear. Rewards include: increased customer satisfaction and the likelihood of repeat business; greater access to public-sector and private-sector contracts; business processes and procedures become uniform, standardised and consistent; operations can become streamlined and more efficient; management and workforce better understand the objectives of the organisation and their contributions to it; and the reputation and standing of the organisation are enhanced through quality management system certification. It can be seen that the benefits are many, diverse and powerful.

Principal contractors, as larger organisations within the construction industry, are well aware of the commercial advantage associated with using quality management systems. Indeed, all public-sector contract letting involves some element of pre-qualification. Many clients in the private sector also view performance in quality management as one of many criteria for contract selection or tendering pre-qualification. Small and medium enterprises, or SMEs, will also find quality management systems helpful to their businesses. Within construction, as throughout industry, SMEs that do not reflect an ISO 9001 approach to quality management can find themselves overlooked by prospective clients. An important point to remember is that quality is not a given, it is not automatic and is never guaranteed – quality has to be actively pursued and consciously managed. Those organisations which use a formal quality management system are best placed to deliver good quality successfully to the customer.

Chapter 2 examines: fundamentals of quality management and quality management systems; the quality management system standards and requirements; developing and implementing a quality management system; independent assessment of systems; and system application to construction.

Chapter 3: Environmental management systems

Chapter 3 focuses on *environmental management systems*. Organisations throughout the industrial, manufacturing and commercial sectors are under increasing pressure to make their activities and outputs more environmentally acceptable. Good business practices which deliver contracted services and products while preserving the environment are a high priority for both customers and providers. Within the construction industry, clients in the public and private sectors, consultants, contractors and suppliers of equipment and materials all have a greater awareness of and responsibilities for safeguarding the environment. Construction project sites in particular are frequently criticised for the disruption and damage that they cause to the environment and the adverse effects that they have on local inhabitants. Construction industry organisations generally, and contractors specifically, have much to gain from improving their environmental performance. In addition to meeting head-on the challenges from rapidly evolving legislation, increasingly stringent regulation, more demanding clients and a more discerning public, there are organisational and commercial benefits to being eco-conscious. Being green and being seen to be green is fast becoming a staple of business practice.

Within almost all economies, larger companies generally perpetuate larger environmental impacts, and many of the larger companies are construction contracting organisations. It is imperative that such organisations take the lead in conducting and promoting environmental management. One approach to appropriate and effective environmental management is to adopt an environmental management system (EMS). This is a structured framework for planning, managing and monitoring the environmental performance of the organisation and systematically ensuring that the organisation complies with environmental legislation and regulation. Depending upon an organisation's status, situation and circumstances, it may establish an individual and bespoke intra-organisational system, use BS 8555 as a guide for staged systems development, or follow the requirements of the international standards-based system ISO 14001.

Environmental management is influenced by the requirements of environmental legislation. For example, current legislation requires that, for construction projects which meet specified criteria, an environmental impact assessment, or EIA, is carried out. While the responsibility to identify and carry out such assessment is placed upon the client, the information collected must be taken into account when the principal contractor develops any project-specific implementation plan as part of its environmental management system. Development and implementation of the system may also be influenced by any environment-related contractual requirements set by the client.

A principal contractor who establishes an effective environmental management system can reap substantial benefits. This is true of organisations both large and small and which operate throughout many different business sectors. First and foremost, an effective system can help the organisation to satisfy onerous legal and regulatory environmental responsibilities and client requirements. Second, financial savings can be made through more efficient organisational practices such as reducing energy use and waste materials. Third, good environmental management can improve the organisation's reputation and standing in the commercial marketplace with clients and business stakeholders.

Chapter 3 examines: environmental management and organisational awareness; system standards; system development and implementation; assessment and certification; the environmental supply chain; and environmental application to construction.

Chapter 4: Safety management systems

Chapter 4 focuses on *safety management systems*. The management of occupational health, safety and welfare is relevant to all sectors of industry, business and commerce. It is arguably the most important function of management in any organisation. It is particularly important for organisations in construction as the industry remains one of the most hazardous within which to work. Construction is inherently dangerous, with accidents to people and workplace incidents an ever-present threat. Therefore, safety management remains a real and continuing challenge for construction management.

Perhaps the overwhelming challenge for any organisation is to establish and embed an effective health and safety management approach within the day-to-day business such that it becomes an intrinsic and virtually unconscious component of organisational culture and operations. This often requires a real step change in organisational recognition of the safety management function.

The pertinent and key approach to achieving a safe and healthy work environment is to ensure that all safety-related matters are considered, planned, organised, controlled, monitored, recorded, audited and reviewed in a structured, systematic and consistent way. The most appropriate way is for the organisation to establish an effective occupational health and safety management system (H&SMS). An appropriate system will establish mechanisms for achieving effective health and safety throughout the organisation and will also add value to the organisation by embedding a meaningful and recognised internal culture, so enhancing the holistic capability to improve the core business.

The Occupational Health and Safety Assessment Series (OHSAS) Standard 18001: Occupational health and safety management systems – Requirements was introduced to meet the demand from business organisations for a standard by which they could assess their systems. OHSAS 18001 presents a structured framework around which an organisation can develop an effective system in the context of organisation-based needs and customer requirements. This is further emphasised by the International Labour Organization (ILO), which has long championed the use of appropriate structured and systematic approaches to management and workforce supervision within the construction processes through its publications, and in particular its recent occupational safety and health training packages (ILO, 2010).

The principal contracting organisation is at the sharp end of health and safety management system deployment within construction. It must manage the day-after-day, project-after-project, hazardous and risk-laden construction phase on site. This requires a systematic and robust approach. An effective health and safety management system can provide the principal contractor with a range of benefits including: the minimisation of health and safety incidents, accidents and associated financial and resource loss; the recognition that shortcomings in health and safety performance are failings which can be addressed through better management; awareness that health and safety can be embedded within the core business support services to add value to the organisation; systematic identification of occupational hazards and levels of risk; and better and safer resource utilisation. In addition, there are tangible financial, commercial and marketplace benefits for an organisation which proactively supports good health, safety and welfare management. Perhaps more fundamental than these benefits is the potential, through an effective system, to limit hazards, reduce occupational risk and save lives.

Chapter 4 examines: the principles of safety management; health and safety standards and systems; development and implementation of effective health and safety management systems; and system application to construction.

Chapter 5: Integrated management systems

Chapter 5 focuses on *integrated management systems*. An IMS is a management system which integrates the various management functions necessary to supporting the operation of the business into one comprehensive and coherent management system to enable the achievement of the organisation's business. Two caveats are paramount to such an approach: (1) the IMS must actually integrate those systems which would ordinarily operate as separate and dedicated systems; and (2) the IMS must integrate the systems within the structure, elements, activities and operation of the organisation's core business and its outputs. In essence, the approach must bring together the business practices of the organisation into one complete, seamless and effective management system. Clearly, this presents a tremendous challenge for any organisation.

Any system which has an influence on the core business of the organisation should be considered carefully and meaningfully for integration. The organisation may consider integration of all its activities and management functions within a unified management structure or may choose to integrate strategically selected parts. Where a system has a significant impact on the operation and performance of the business, then it should be integrated for best effect and value to the organisation. All management systems can be integrated including certificated systems and wholly intra-organisational-based systems. An overriding benefit of standards-based certificated management systems is that that they are developed and applied in a uniform, consistent and robust way and, moreover, in a way that is independently approved, recognised and respected.

Why would the organisation want to integrate its management systems? After all, some people see systems as overly bureaucratic, grossly inefficient, operationally cost ineffective and organisationally divisive. Furthermore, as more management systems are introduced, the greater the organisational complexity that might be created. In some organisations this may indeed be true. However, for other organisations management system integration can be both positive and beneficial. Integration can: simplify routine organisational procedures throughout managerial functions; reduce duplication of procedures and paperwork, so reducing costs; harmonise and optimise business practices; balance the demands and conflicts of many separate systems; give greater focus on core business objectives; improve communications, co-operation and teamwork; and develop improved organisational culture. So, there are tangible and good reasons to pursue management system integration. Even more fundamental than this is the overriding expectation within the business marketplace that service providers can manage multiple deliverables simultaneously. This is exemplified within the construction process where contractors are expected to deliver projects on time, to budget and to the highest standards of quality while also maintaining occupational safety and environmental safeguards.

Perhaps the best route to management system integration is to adopt the business process approach. This involves examining the organisation holistically, establishing clear and directed policy, objectives, procedures and practices based around the processes of the organisation and its business outputs. Further, the organisation can capitalise on the use of its standards-based systems to bring together its practices and apply the key elements of all management system approaches – policy, organisation, documentation, monitoring, audit and review. In this way, the organisation can establish an effective IMS which intrinsically pursues the principal tenet of all management systems – continual improvement.

Chapter 5 examines: management standards and specifications for integrated management systems; management systems and the business; developing and implementing an integrated management system; and system certification.

References

ISO (2009). *General Information.* International Organization for Standardization, Geneva. www.iso.org

ILO (2010). *ILO Construction OS&H.* International Labour Organization, Geneva. www.ilo.org

CHAPTER 1

Management systems

Introduction

Chapter 1 focuses on *management systems*. It examines management theories, the nature of organisation, and the processes involved in delivering business outputs. It defines what a management system is and describes and explains the types, features and characteristics of such systems. It proceeds to examine the application of management systems to construction and the processes involved, together with research reports and experiences of systems in use ascertained from construction professionals involved in management system development and implementation. Chapter 1 provides a fundamental and important introduction to the prominent approaches used to establish management systems for quality, environment and safety within construction.

Management: theories, the organisation and business processes

This section examines: the nature of management and the influence of systems concepts upon a construction-related organisation; prominent management theories including systems theory and the influence of quality management theories; management and organisation, focusing on the open system and holism; organisational management and the different types and characteristics of business processes; and the application of the systems approach and the challenge of continuous improvement in organisational activities and outputs.

Management

Management is not a single discipline but one which crosses the boundaries of many disciplines – it is both multidisciplinary and interdisciplinary in application. Management and its association with systems applications centres on the relationship between the company and those functions of management that need to be structured and supported in delivering the processes associated with the company's business. Within the context of the construction industry, management systems can be a part of the structure and organisation of any company associated with the construction processes. Many client organisations, consultants, contractors and materials and equipment suppliers can and indeed do implement management systems in the delivery of their inputs. More prominent perhaps, in the context of construction projects, management systems tend to be associated generally with those organisations which deliver the construction phase on site during construction projects – principal contractors and contractors. The use of management systems by supply and service organisations is driven predominantly by the requirements of clients. This is exemplified in the view of Davis Langdon and its subsidiary company Davis Langdon Certification Services (DLCS), a certification body operating globally which assesses quality, environmental, occupational health and safety, and integrated management systems across a wide range of industries including construction: 'Market forces are increasing the demand for organisations to implement management and product systems that operate in accordance with recognised standards' (DLCS, www.davislangdon.com, 2010).

Systems management concepts have permeated management and organisational thinking for well over two centuries. A systems approach to configuring the organisation of business activities within a company is well advised as systems thinking can meet a number of the most fundamental and important organisational needs. A management system is, simply put, a way of doing things. Systems arrange, develop and apply protocols and sets of procedures which bring structure, order and stability to the processes of running any business. Without such basic arrangements being in place, a company would be subject to general dysfunction, possibly even chaos. When configured as holistic, transparent and open, a system is one which not just satisfies the intra-organisational requirements of the company's business processes, but extends to interact with and address the external and wider requirements of the business environment.

Any management system should be simple but not simplistic. It must be easy to understand, interpret and apply by those personnel in the company who work with and around the system. A system must give focus and shape to the many functions of management which are used to oversee the undertaking of the processes involved in delivering the business, whether the outputs are services, products or projects. Because business activities generally flow continuously or in cycles, management systems must give repeatable, consistent and reliable outcomes. Systems must be able to arrange those functions of management key to the core business in and throughout the corporate organisation. Moreover, the systems which focus and drive the corporate organisation must be capable of being translated into sensible and appropriate procedures and tasks which actually deliver the business outputs – the services or products of the company. This point is especially important within the construction industry, where the services and products which input companies provide are fundamentally task based and delivered at the project site.

In many organisations, the structure and organisation of management systems is relatively straightforward. This is true because many businesses are based on a single and central corporate management location from which their business outputs are distributed. Construction organisations differ from the single-location scenario as the corporate organisation delivers its business outputs via project sites. Project delivery within the construction industry more often than not involves geographically and sometimes widely dispersed sites. Such sites are temporary in nature with a transient state of inputs and resources. These inputs are usually interdisciplinary, with provision from a variety of contributing companies and organisations. Furthermore, companies offering services and projects within the construction industry handle multiple projects simultaneously, and many of these differ in a wide variety of type, content and complexity.

The construction industry comprises a vast number of small building companies and small to medium enterprises (SMEs). For such organisations, the provision of their services and products is relatively simple, with the management of their corporate and project activities centralised in a single or restricted location. For larger-sized companies such as principal contracting organisations, however, management of their provision may require systems which have one ethos and focus but two levels of practical management – the corporate and the project. Establishing systems which satisfy the requisite functions of management seamlessly throughout the entire organisation is a prerequisite to the use of such systems and paramount to the efficiency and effectiveness of the company's business.

For a management system for any functional management discipline within construction, the principal goal of the 'corporate organisation' is to conceive, develop and perpetuate the company's ethos, policies, aims and objectives which drive its core business and allow it to survive and prosper. The principal goal of the 'project organisation' is to undertake effectively those projects which the company undertakes, delivering its services and products on time, to budget and with quality while making a profit. Both organisational levels need to support each other, operate holistically and capitalise on the synergy that their combined resources can create and maintain.

A strong framework of organisation is needed and this must cascade throughout the entire company. The lead for this must come from the corporate, or executive, level through to the directive, or management, level to the operational, or supervisory and task, level. Policies, aims and objectives for the business determined by the executive level will need to be translated into actions throughout the entire managerial hierarchy; they will need to be written down with clarity and communicated effectively to all within the company. This is achieved by establishing documents which describe the company, the processes involved in delivering its business, and how these processes are to be managed. For any given management function, these documents identify, define and describe: the structure and organisation of the company; the procedures by which business processes will be managed; plans to describe organisation and management procedures in specific situations; and working instructions to guide the undertaking of tasks at the point of delivery of the business outputs.

The knowledge and capability to plan and organise and undertake activities in seeking to achieve the needs and wants of individuals, groups and organisations is absolutely essential to their being, their welfare and their perpetuation. Management is the broad function of putting in place the necessary mechanisms to do all that is needed to deliver the outputs of these needs and wants. Companies of all types require management to guide their business activities and execute the processes associated with delivering their business services or products to their customers. Management pervades throughout any company, helping to shape vision and culture and business goals at the corporate level, while at the operational level the supervision of processes and tasks is essential to delivering outputs to time, budget and quality. Therefore, management is an activity that is fundamental to all that human beings do and essential to the configuration and operation of any organisation.

'Management' is, however, a term that can be ambiguous. It does not have a robust meaning in the minds of many managers and employees. It can be a loose term giving rise to misrepresentation in use and misunderstanding by perception. This is because each user of the term management tends to have in mind a discrete application in the context of the user's work activities and with little appreciation for the frame of reference of the receiver. Management is in reality a term with broad practical application. It can range from the handling of a limited, simple and routine organisational activity to the planning, resourcing and undertaking of extensive technologically complex projects. Management is not a single discipline confined to an individual but one which extends across the boundaries between many different disciplines and can involve the activities of many people. It is multidisciplinary in nature and interdisciplinary in application. Within any organisational setting management goes on day by day, handling a range of issues from the small and trivial to the large and vital. All in their own way contribute to the collective activity of what constitutes the term management.

In the context of organisations and a systems approach to management, the term is used to emphasise the application of a multitude of specific functional management concepts, principles and practice. Management is the collection of practices used throughout a company and its organisation to apply the functional concepts, principles and practices to all its activities. Many elements of management practice will be generic with widespread use throughout a company while other elements will be specific, governed by the precise nature of the company's business activities. Furthermore, generic elements are transferable from organisation to organisation while others will appear quite specific, even unique to the business activities of other organisations. Management practice may be generic throughout companies across many different business sectors but be specific in their application within their sectors as corporate management is translated and applied to particular business outputs – services, products or projects. Management therefore takes on a configuration appropriate to the nature of a company, its activities and its environment. So, while there are common threads to the nature, form and application of management, there is no single prescription for its arrangement within any organisation. Likewise, the management systems established to assist in the organisational delivery of each and every functional management discipline will be configured around sound generic elements coupled with specific elements essential to meeting the particular needs of the organisation and its business.

Management theories

The concepts, principles and practices of what is conventionally termed management have been influenced significantly throughout the centuries by the thinking and philosophies of prominent individuals and collective schools of thought. These different and sometimes conflicting perspectives have been important as they have changed the ways in which the work of people is seen and have changed the ways in which organisations are shaped. Moreover, such views have influenced the ways in which the organisation of human endeavour has evolved towards efficiency, effectiveness, adaptability and holism. Over the years, many similar and also many very different theories on appropriate ways to manage have emerged, yet in their own way and collectively all have contributed to modern management thinking. Moreover, they have contributed greatly to the ways in which companies today structure, guide and operate their business activities.

Of greatest prominence are perhaps the *modern management theories*, or those which have emerged since the end of the Second World War. These moved management thinking on considerably from those theories which preceded them – the *scientific management school* and the *human relations school*. Both of these subsequently drew in a host of managerial thinkers who theorised concepts and principles which have significantly influenced the development and application of management systems and procedures which informed today's organisational practices.

The scientific management school

'Scientific management' propounded by F.W. Taylor (1911) focused on industrial engineering and the principles of production and, in particular, its organisation. Deemed the 'Father of Scientific Management', Taylor is credited with the early attempts at determining the best way of undertaking human-based work by simplifying its component parts and tasks. Unfortunately, Taylor's contribution to management thinking was, at the time, somewhat misinterpreted. Organisations used his ideas simply to invoke more work from employees at less cost rather than seeking greater productivity through improved and more humane working practices. Nonetheless, Taylor's work was followed by a number of related contributions, the most prominent being that of Frank Gilbreth (1911), associated with time and motion studies, Henry Gantt (1919), noted for the long-standing work-charting method known as the Gantt, or bar, chart, and Henry Ford (Ford and Crowther, 1922), renowned for innovation in assembly line production techniques. These management perspectives were followed by the *classical school* and its most prominent supporter Henri Fayol (1949), the French industrialist and entrepreneur famous for his treatise on the processes of management. Fayol's work gave rise to the identification and embedding of the well-recognised 'five processes of management' – forecasting, organising, commanding, co-ordinating and controlling – which have in various guises permeated management thinking and practice for well over the last half century.

The human relations school

The second of the two principal schools of management thinking – human relations – emerged in the 1920s and focused on productivity and the link to human welfare and motivation. The principal advocate of the approach was Elton Mayo (1949). Mayo became famous for the 'Hawthorne Experiments' at the Western Electric Hawthorne Plant in the USA. His theory suggested that there was a direct relationship between the sense of belonging to a working group by an individual and their level of motivation and productive outputs. An important finding of his work was that informal structures within any organisation are as significant, if not more so, as the formal structures and give rise to positive effects upon communication, teamwork, motivation and morale.

Both the scientific management school and the human relations school have identified many positive aspects to the management of the modern organisation. Conceptually, they have laid the foundations upon which many theories have emerged over the last century. These have addressed contemporary issues surrounding organisational structure, resource deployment, supervisory approaches, human resource management, production processes, quality improvement and the like. The most prominent and influential theories of the post-1945 period have been *systems theory*, *contingency theory* and an array of theories surrounding *quality management*. All of these have proved important and influential to organisational management thinking, with quality management theories of particular importance to the development and implementation of management systems.

Systems theory

'Systems theory' (Bertalanffy, 1950, 1955, 1968; Checkland, 1981) is generally attributed to the German philosopher G.W.F. Hegel (1770–1831). Hegel's work focused on creating a systems map to explain the natural workings of organic biology. He suggested that in a system of any kind the whole adds up to be greater than the sum of the parts, or produces synergy; the whole determines the characteristics of the parts; the parts cannot be fully understood if seen in isolation from the whole; and the parts are interrelated and therefore interdependent. These attributes of systems are used in the configuration of almost all organisations. Furthermore, such a systems perspective enables an organisation to develop: philosophy – its way of thinking; management – the design and operation of the organisation as a whole; its method of analysis – problem-solving techniques; and systematic thinking – logical and regular consideration (Hamilton, 1997). So, systems theory can provide a framework for the managed direction of organisational activity through the provision of all its managed parts with a focus on and benefit to the core business it undertakes – it creates synergy and a holistic perspective.

Contingency theory

'Contingency theory' (Burns and Stalker, 1961; Fiedler, 1967: Lawrence and Lorsch, 1967; Kast and Rosenzweig, 1973) focuses on the need for any organisation to become and remain adaptable to change. Management theorists are consistent in their view that over time an organisation will and must evolve to accommodate change within both the organisation itself and the environment within which the organisation exists and operates. Contingency theory has some common threads and overlaps with systems theory. Organisations exist to prosper in a changing political, regulatory and financial marketplace where prevailing circumstances must be grasped as catalysts for change and actions responsive to the needs of change. Essentially, the tenets of contingency theory are that: there is no one right or best way to manage; the configuration of an organisation must be commensurate with its environment; effective organisations have effective sub-elements, or systems; and the operation of the organisation is closely linked to the management style adopted.

Quality management theories

A range of management theories have evolved from the scientific, human and classical schools of thought over the last century linked directly to 'quality management' techniques. These theories are pertinent to the evolution, development and implementation of management systems. The theories underpinning quality management are especially important as specific elements of the theories have influenced systems developments and continue to form component parts of systems applications. The work of W.E. Deming (1950; 1960; 1982) is perhaps the most prominent of noted management theorists. His work is of direct application and significance to the development of management system understanding and development. Deming propounded the view of quality management within a cycle of plan–do–check–act, or PDCA, a key attribute within standards-based management systems advocated today.

This cyclic and total view of quality was supported by the work of A.V. Feigenbaum (1951; 1961), the originator of 'total quality control' – 'an effective system for integrating quality development, quality maintenance and quality improvement efforts of the various groups within an organisation, so as to enable production and service at the most economical levels that allow full customer satisfaction'. Feigenbaum went further to describe the 'total quality system' as:

> the agreed company-wide and plant-wide operating work structure, documented in effective, integrated technical and managerial procedures, for guiding the co-ordinated actions of people, machines and the information of the company and plant in the best and most practical ways to assure customer quality satisfaction and economical costs of quality.

The work of Feigenbaum was augmented by that of Joseph Juran (1951; 1955; 1964), who advocated quality management based on the related and contributing elements of planning, control and improvement. Moreover, he propounded the need for these elements to be linked directly to the perspective of quality held by the organisation's executive level of management with direction coming from the top.

More recent theories have emerged from theorists and practitioners including Crosby (1979; 1984), who propounded the 'zero-defects approach', Ishikawa

(1984), who suggested the application of 'quality circles', and Imai (1997), noted for the 'Keizen', or continual improvement techniques. All of these and other quality management theories have contributed to the awareness and understanding of quality as a concept to be managed. Moreover, transferable lessons from the field of quality management have added to the evolution of management approaches used in other fields of organisational management including construction. This will be seen subsequently in relation to the development and application of management techniques and systems for environmental management and safety management.

Management and organisation

Management processes

It will be seen subsequently that management is the handling and regulation of a transformation process. That is, inputs are brought together and managed through a process of conversion to become outputs. Management consists of all those aspects which are organised and applied in delivering this working process. Construction works in this way with many inputs, contributing to a conversion process to deliver outputs in the form of services, products and projects on a daily basis. Management therefore involves all of the five processes of management mentioned earlier – forecasting, organising, commanding, co-ordinating and controlling. Each of these must be applied effectively throughout the undertaking of any business process, but this requires structure and organisation. The system applied to achieve this provides the necessary arrangement of the management processes in direct relation to the management function being implemented. So, quality management is a system which, for the implementation of quality, directs forecasting, organising, commanding, co-ordinating and controlling towards the delivery of quality. However, any one function is easy to handle but a system must also be capable of dealing with the set of relationships among multiple parts of an organisation and make them come together and work as a whole. A systems approach is most helpful in this regard as it underpins the tenet of holism and the focus on the positive interrelationship and interdependencies of the parts. Furthermore, successfully undertaking any process once is insufficient as it must deliver the management function to the input, conversion, output process again and again. It must be continuous and consistent in operation. Again, a systems approach is there to do just this.

The open system

A company or organisation does not exist in isolation. It exists and functions within a living environment. An organisation is surrounded by so many factors of influence upon its being and working that it can only be regarded as operating within a continued state of flux. If it thinks that it does not, then effectively that organisation is static within its environment and simply cannot survive. All organisations are influenced by both internal and external variables, which all react to one another and are in a constant state of change. Construction projects, because of their technical, resourcing and managerial complexity, are a case in point. Construction projects are complex as they usually, although not always, have: multiple resource inputs; intricate technical conversion processes; and exacting multiple output performance requirements – time, cost, quality (value for money), environment and safety, and more. Furthermore, the site-based processes are conducted within a surrounding technical environment of external resourcing: legislation and regulation. Further again, all of this takes place in a variable, often volatile, political and financial marketplace. This environment can be likened to concentric circles of influence with boundaries radiating outwards from the project with all, in isolation and as a collective, impinging upon the scenario.

Because of this, organisations must operate within an 'open system', a system which interacts with its environment. Systems theory recognises the need for the organisation to be open in its outlook and operation and use open management systems. Moreover, the open system adopts 'wholeness' rather than 'fragmentation'. Traditionally, the management of organisations has tended to favour the fragmentation, or segmental, approach. In this approach, management functions in application are understood by taking sections of the organisation apart, understanding the working of the parts and then understanding the whole. The problem with this approach is that functions are seen in isolation and with no reference to other parts or the wider workings of the organisation. The wholeness approach sees the wider organisational picture with a focus on seeing functions in relationship to other functions. Whereas a segmental view is narrow, static, overwhelming and cannot accommodate organisational change, a wholeness view integrates thinking and is readily amenable to change.

Wholeness and structure

The concept of and level of commitment given to wholeness, or holism, is fundamental to the structure of any organisation. The structure, in turn, influences all that is configured within the organisation. An organisation which clearly sees its business activities and the management functions which deliver the business is generally proactive, dynamic and forward thinking – the organisation intrinsically becomes a morphogenic entity. An organisation which does not appreciate wholeness clearly becomes staid and unresponsive to its business environment – the organisation intrinsically becomes a morphostatic entity.

Companies need to create a structure within which the open system predominates, perpetuating creative thinking, knowledge development and exchange and problem-solving capabilities. Structure creates organisational direction and focus by which individuals and groups can see their place within the arrangement of the company and their contribution to the organisational parts and the whole. Human relations are assisted where there is potential for harmony and less likelihood of organisational dysfunction. Certainly, change is effected more easily where strong and directed structure prevails. Perhaps the most important feature of a holistic perspective on structure is the provision of a stable and certain working environment. An employee who feels a part of the organisation, sees their part in it and can see their future within the evolution of the company is likely to be a more willing and productive worker. Management systems can more easily be developed and implemented in this type of organisation as the structure naturally supports systems attributes and use.

Vertical and horizontal structure

It will be seen in detail in a later section that companies often support management system development through the establishment of vertical structures. This is understandable from a segmented, closed-system perspective. In such a configuration, each management function and therefore management system is treated as separate, operating in parallel with other systems and with no sharing of knowledge base, information and experience. While such systems work very well in isolation, in the modern company, where business holism and organisational synergy is demanded, the vertical system is simply outdated. Of far greater value to any company is the horizontal structure of functions and systems, where cross-functional management is propounded. Such an arrangement encourages the sharing of information with and across multiple systems, so creating synergy and a whole-organisation perspective.

Management and the business

For any company to operate in an efficient and effective way, it must be managed in a systematic manner where clear direction and control are intrinsic to everything that it does. Its activities are governed primarily by the methods used to handle the plethora of management functions that need to be applied in carrying out the business. This can be fulfilled though the establishment of formal, or documented, management systems which perpetuate the business and continually improve its performance while also meeting the needs of its customers and marketplace. To steer an organisation to improved performance across the range of management functions that are conducted in almost all organisations, a number of fundamental principles need to be met:

- *Customer focus* – An organisation must understand the needs of its customers, as it is dependent on upon them for its very existence.
- *Management leadership* – An organisation requires managers who are leaders to provide purpose and direction to the business and who can structure, resource and drive it to meet its objectives.
- *Involvement of people* – An organisation involves people who provide the wherewithal to make the business happen, as without them the business could not function.
- *Business processes* – An organisation's activities and resources need to support those processes which deliver its products and/or services.
- *Management system approach* – An organisation should relate its business processes to management systems such that they are carried out systematically and consistently to standards of performance.

- *Continual improvement* – An organisation should seek to ensure that its performance is continually monitored, reviewed and improved, and this should be reflected in its processes and systems.

- *Evidence-based evaluation* – An organisation should evaluate its performance based on evidence from analytical data provided from its systems implementation.

When any company looks at its business and the way it intends to manage its activities, management systems can present prominent advantages to an organisation because they:

- align management functions with company policies and objectives;
- introduce structure and organisation to management functions;
- allow management functions to be conducted systematically;
- provide management procedures which are consistent in application;
- specify roles and responsibilities of those involved with management;
- clarify the boundaries between different management functions;
- communicate to employees how management functions are implemented.

Requirements of management systems and alignment with business outputs

The requirements for management system development and implementation are specified in international standards. The standard for any one kind of management system, for example a QMS, specifies requirements which are generic and therefore applicable to organisations in any industry or commercial business sector. Standards are not product, service or project related. A standard specifies the requirements for the management system. The system is then applied to manage the processes involved in delivering the product or service.

The requirements for products and services and the situation within which they are provided, for example a project, are specified by a client or customer in a form of contract between supplier and purchaser. In addition, requirements can be specified by the organisation as it sees itself meeting the anticipated marketplace. The requirements for products and services may also be specified in product/service standards, for example those concerned with quality, and process standards and regulations, for example those concerned with safety. The important aspect is that the management systems used must be carefully aligned with the processes which they manage and linked to the output of the processes as expected by the customer. Only in this way is the business delivered efficiently and effectively. This can be seen in the key considerations needed in the general approach to establishing a management system.

Key business considerations for management systems

A general approach to planning, delivering and implementing any management system consists of the following key considerations:

- the needs of the customer and other stakeholders;
- the policies and objectives of the organisation;
- the organisational processes necessary to fulfil the policies and objectives;
- the assignment of responsibilities to manage processes towards the objectives;
- the provision of resources to attain the objectives;
- the establishment of procedures and instructions to manage the processes;
- the monitoring of processes to determine their efficiency and effectiveness;
- the identification and elimination of non-conformities in the processes;
- the encouragement of continual improvement in management of the processes;
- the audit and review of systems to improve the overall management approach;
- the feedback on performance to improve provision to customers through improved policies and objectives.

An organisation which addresses all of the above key considerations will have gained a thorough understanding of its business, the processes involved and its systems of management to provide a sound basis for review and continual improvement. This is essential to achieving overall business success through meeting the expectations of stakeholders and customers.

Management and business processes

The process model

The concept of the 'process model' is fundamental to structuring an organisation, its activities and their management. A process is an activity, or set of activities, that uses resources to convert inputs to outputs and lies at the heart of delivering all products and services. Processes may be undertaken sequentially or concurrently but all usually impact upon another somewhere in the delivery of the business. Standards encourage a process approach in the development and implementation of management systems. Therefore, in order for an organisation to function efficiently and effectively, it has to identify and understand all the processes that interrelate in providing a product or service.

Products, projects and some, but not all, services follow a process approach, or the process model of development. This is certainly true within the construction industry. The process model is one where inputs go through a process of conversion to become outputs, as shown in Figure 1.1. When systems are used to manage a process, feedback from that process is used as a basis of control and so a management loop is created where one learns from experience and modifies subsequent action accordingly, as shown in Figure 1.2. Fundamental features of all management systems are: continual analysis; measurement; review; and improvement to the management functions as they are applied to a process.

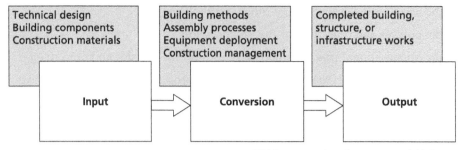

Within construction, the process model brings together design, components and materials to be converted through construction methods and management to become the completed product

Figure 1.1 The process model

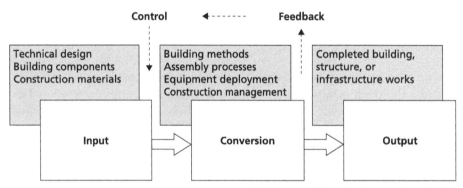

Within construction, a basic system based on the process model utilises continual feedback on performance of the conversion phase to inform the ongoing input phase

Figure 1.2 A basic system

Characteristics of processes

The outputs of a company, in particular where these are products, follow a process of: the acquisition of materials, components and resources; manufacture and/or assembly; and distribution and delivery to a customer. This follows the convention of the 'process model'. Where services are provided to a client within a project arrangement, again this relates to a process model, as a project is itself a combination of processes leading to an end output. So, products and services are delivered by a process or collection of processes, hence the term process model. The model is characterised by inputs which undergo conversion to become outputs. To ensure conformance to predetermined standards, the inputs, conversion and the outputs are planned, monitored and controlled; that is, the process becomes managed. Therefore, links are established between the process model and management systems as shown in Figure 1.3.

This is simplifying what can be a complex arrangement of activities resulting from the many inputs, direct and indirect, which contribute to a process. Any

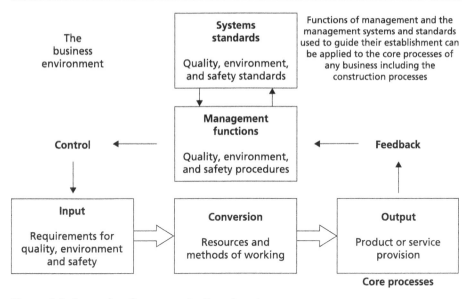

Figure 1.3 Example of an organisational system

process is configured within the corporate and operating practices of the company, which itself exists within the wider business and commercial environment and which itself is subject to independent controls and influences. Figure 1.4 shows a range of factors which impinge upon any process or project activity. These can include the implications of: input constraints specified by customers such as time and cost; the methods used in the process, such as operations and sequence; the

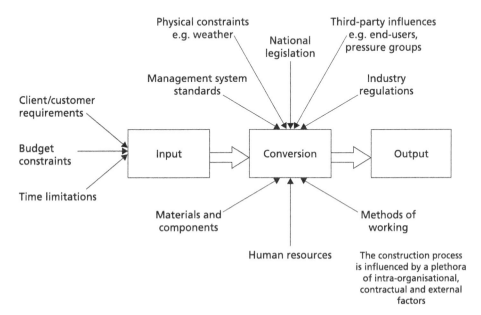

Figure 1.4 Factors impinging upon processes

resources deployed, such as competence and availability; the equipment and materials used, such as training in use and waste control; the standards applicable to activities, such as work regulations and legislation; and others created by any specific process being undertaken.

Key aspects of managing processes

Three key aspects are essential to delivering efficiency and effectiveness when managing processes and must be reflected in the management approach and systems used. These are:

1 the clear identification of the processes needed to deliver the business;

2 the arrangement of management responsibilities;

3 the deployment of knowledge, resources and skills.

Types of process

To achieve the above requirements most companies use three types of processes to carry out their business, as shown in Figure 1.5. The three types of processes, all seen clearly within the total construction process, are as follows:

1 *Core processes* – processes directly involved in producing the product or delivering the service, that is, those that are essential to the perpetuation of the business because they form the essence of the company, what it seeks to achieve and why it exists.

2 *Support processes* – processes carried out to ensure that the core processes work efficiently and effectively, that is, they are essential to supporting the core business.

3 *Assurance processes* – processes carried out to ensure that the company is fulfilling its business policies and responsibilities and conducting its organisational and operational procedures appropriately.

Functions of management can be directed to any and all of the above groups of processes. The three functions of quality, environment, and health and safety are essentially support processes. They are not directly involved in the core processes but are essential support functions to them. In addition, they are assurance processes because they involve ensuring that the company meets its business objectives to internal and external standards of conduct and operation.

To optimise the business, a company should focus and concentrate on maintaining and improving its core processes first and foremost. This is, after all, the very reason why the company exists. The core processes drive the products or services that it provides to meet the requirements of the market and customer base. This presents implications for both support and assurance processes. They are essential to helping the core processes deliver the business but are not directly income deriving and therefore not income critical as are the core processes. In almost all companies the costs incurred in maintaining these processes

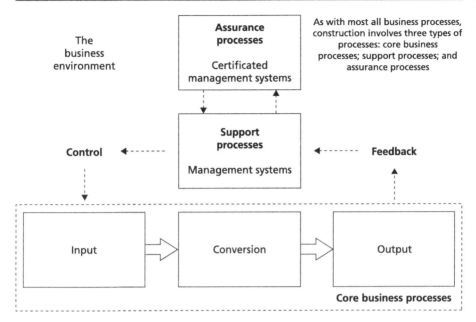

Figure 1.5 Types of processes

are necessary yet expensive operational overheads. They may be regarded as extravagant and even superfluous where they are not supporting the core processes effectively. This means that support processes and assurance processes must be structured, organised and resourced with the greatest of efficiency and implemented with the best effectiveness.

Deming methodology

A system will develop policies, goals and strategies to ensure the effectiveness of management, and experience of implementation will allow lessons learnt to be fed back into future policies, goals and strategies. This is commonly termed the 'continuous improvement loop' of thinking, planning, doing and measuring or the 'Deming' methodology involving PDCA (Griffith *et al.*, 2000; Griffith and Gibson, 2007). This is shown in Figure 1.6. It is also associated with the 'Kaizen', or small steps, approach to performance improvements through constant review. For efficiency and effectiveness in the long term, the system must be able to go full circle, evaluating, or auditing, all aspects of and inputs to its implementation with a view to enhancing management, providing better outputs and adding greater value to the organisation.

Systems theory in detail

Systems theory can help explain the strong links between the process model and management systems. Systems theory, like all management theories, presents a way of understanding organisations and how they are structured and operate.

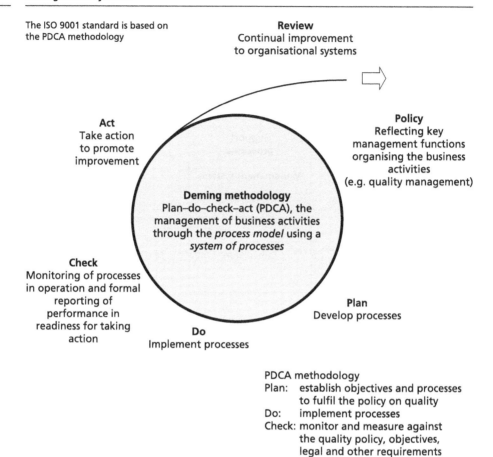

The ISO 9001 standard is based on the PDCA methodology

Review
Continual improvement
to organisational systems

Act
Take action
to promote
improvement

Policy
Reflecting key
management functions
organising the business
activities
(e.g. quality management)

Deming methodology
Plan–do–check–act (PDCA), the
management of business activities
through the *process model* using a
system of processes

Check
Monitoring of processes
in operation and formal
reporting of
performance in
readiness for taking
action

Plan
Develop processes

Do
Implement processes

PDCA methodology
Plan: establish objectives and processes
        to fulfil the policy on quality
Do: implement processes
Check: monitor and measure against
        the quality policy, objectives,
        legal and other requirements
Act: take actions to continually improve
        quality of performance and outputs

Figure 1.6 'Deming methodology': plan–do–check–act (PDCA)

Definition

A system is an arrangement and set of relationships among multiple parts operating as a whole.

(Hamilton, 1997)

Features

Systems theory is generally attributed to the German philosopher Hegel (1770–1831). Hegel identified distinctive features which, today, are characteristic of the ways in which almost all companies structure their organisation and resources:

- The whole adds up to become greater than the sum of the parts, or the production of synergy.
- The whole determines the characteristics of the parts.

- The parts cannot be fully understood if they are seen in isolation from the whole.
- The parts are interrelated and therefore interdependent.

Principles

A 'system' is 'the whole' made up from 'the parts'. Hamilton (1997) superimposes philosophical reasoning to present an important set of principles for the theory. These are:

- The whole is primary and the parts secondary;
- Integration is the condition of the inter-relatedness of the many parts within one;
- The parts so constitute an indissoluble whole that no part can be affected without affecting all other parts;
- Parts play their role in light of the purpose for which the whole exists;
- The nature of the part and its function is derived from its position in the whole and its behaviour is regulated by the whole-to-part relationship;
- The whole is any system or complex or configuration of energy which, no matter how complex, behaves like a single piece;
- Everything should start with the whole as a premise and the parts and their relationship should evolve.

In organisational terms 'the whole' is 'the system' and 'the parts' are 'elements' that combine to form that system. It is these parts which are applied to the process model to manage the inputs and their conversion to become outputs.

The paramount function of any management system is to counteract the risk of problems occurring throughout the undertaking of a business process. This requires a clear understanding of the process, the specific functions needed to manage that process, and those activities required to ensure that the inputs consistently lead to the desired outputs. To achieve this, three aspects need to considered:

1 *Definition of the system* – what it has to do, what its operational boundaries are, what parts of the organisation and its operations come within its remit.

2 *Description of the system's component parts* – inputs, conversion, outputs and feedback loops.

3 *Determination of the system's environment* – interaction of the component parts with the external dimensions of the company and its business.

Systems theory applied to organisational processes

If this is applied to the management of quality, environment and safety, an example of the configuration created is as shown in Figure 1.7. In the case of providing a product, the core processes would involve inputs in the form of, for example, customer specifications for quality, environment and safety of materials and components; and conversion using, for example, safe and environmentally considerate working methods to produce a finished output. Support processes will be provided in the form of management functions applied to monitor and control the quality, environment and safety of the core processes to plan.

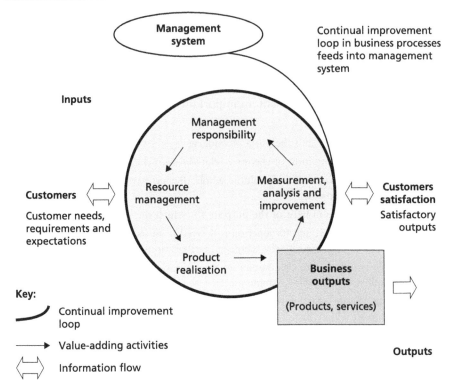

Figure 1.7 Continual improvement loop and its link to the management system
Source: Adapted from ISO 9001

Assurance processes will ensure that quality, environment and safety management will comply with external legislation, regulation and standards, so linking the management functions with the external business environment.

This illustration is perhaps simplistic in that it does not convey the complexity of inputs and multitude of management functions and external dimensions that need to be satisfied in many business processes. Nor does it represent in any way the uncertainty and variability of the human inputs which have a great influence over how efficiently and effectively processes are managed. This is particularly pertinent within construction, which is subject to considerable variations in types of work, production techniques, resourcing and management approaches, all of which can differ from organisation to organisation and from project to project. Construction is truly unique in its range of scope, in its transient operating environment and in its degree of uncertainty.

Management systems: definition, types and features

This section examines: the definition of a system and its various configurations, or types, focusing on the 'company-wide' application and the 'project-based' application within construction; the advantages of the various systems; and the key features and general features which contribute to the development of a management system.

Definition

A management system must focus and co-ordinate the act of control of those elements within an organisation which determine and fulfil its company policies and business objectives. An appropriate definition is:

> A management system sets out and describes, for a particular management function, the organisation's policies, strategies, structures, resources and procedures used, within the company, or corporate, organisation and any sub-organisation, to manage the processes that delivers its products or services.
>
> (Griffith, 1999)

Therefore, for a particular management function such as 'quality', the system describes the company's approach to the management of quality throughout its organisation and operations. A system develops a framework of protocols and sets of procedures and instructions which give structure, order and stability to the particular function being managed:

> A management system is, more simply put, a structured and coherent way of doing things.
>
> (Griffith *et al.*, 2000)

Types of system

There are essentially two directions which a company can take when seeking to establish a management system:

1 A *'company-wide' management system* – implemented within the entire organisation under a corporate-based system umbrella which is applied to all company activities and processes.

2 A *'project-based' management system* – which provides a system for specific organisational activities and processes.

The application of management systems within the construction industry can sometimes appear confusing as both types of system can be and are used. Smaller contractors, for example, may not have the need or capability to utilise a fully developed company-wide management system but need particular elements and attributes of a systems management approach. They would therefore adopt a project-based system to manage explicit aspects of their service or product delivery. Conversely, a larger company such as a principal contracting organisation will likely have the need to apply systems-based management throughout its corporate organisation and with application to all construction projects. Therefore, a company-wide system would be established. Presenting further confusion, some organisations refer to their project-based management systems as 'plans' rather than systems. Also, plans form an important element of systems development as a plan is the sub-component which accommodates the situation-specific dimension of a system. One therefore has to be careful in defining and describing any management system. It is best to think in terms of the two types previously mentioned – the 'company-wide system' and the 'project-based system'.

The configuration of a management system must be determined by a company based upon its own individual and sometimes quite unique requirements, its circumstances and its position in the marketplace. Both the company-wide system and the project-based system have advantages and disadvantages for individual companies in individual situations.

Advantages of a company-wide management system

Some of the prominent advantages of the company-wide management system are as follows:

- The policy guiding the system is applied to the whole organisation.

- A uniform approach to the particular function of management is applied throughout the organisation.

- A set of determined management procedures and working instructions is applied to business processes.

- One management system standard, if applicable, is used throughout the operation of the system.

- All employees become familiar with and are clear about the management system approach.

- Management commitment cascades throughout the company to the service, product or project delivery points.

Advantages of a project-based management system

The advantages listed above should not be taken to imply that there are no advantages to a project-based management system. Indeed there are situations where a project-based approach can be advantageous, as follows:

- There can be a dedicated focus on the delivery points for services, products or projects where a larger management system might impede this.

- Management procedures and working instructions can be tailored and directed to an individual situation more readily than a company-wide system.

- The ownership of the system is close to those delivering the service, project or project.

- Decision making in relation to business processes and the management system which delivers them is more direct than the company-wide system.

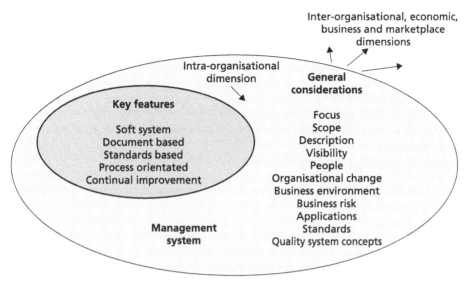

A management system must take on a holistic perspective bringing wider
and greater efficiency and better management effectiveness while
considering the intra-organisational, inter-organisational and external
dimensions to the business – it should 'add value' to the whole organisation

Figure 1.8 Key features of and general considerations for management systems

Features of systems

Key features of systems

There are a number of key features of any management system (see Figure 1.8),
as follows.

Soft systems

Management systems are often referred to as 'soft' systems. This means that
they are paper-based systems which describe designated management functions
through written documentation.

Document based

Management systems are 'document based'. This means that the management
of business processes is described through a set of written documents, usually
taking the form of a system manual, management procedures, implementation
plan and working instructions.

Standards based

Management systems can be 'standards based'. This means that specifications
to guide system development and implementation are available, such as those
provided by the ISO, examples being ISO 9001, ISO 14001 and OHSAS 18001.

Process orientated

Management systems tend to be 'process orientated'. This means that the under-taking of a company's activities tends to focus on the management of processes which deliver the business outputs.

Continual improvement

Management systems seek 'continual improvement'. This means that systems adopt the Deming methodology of PDCA or a continual circle of system improvement and refinement.

General features of systems

There are some important general features of all management systems, as follows.

Focus

A system must have a clear focus for its business activities and a clear structure to the ways in which its business activities are to be carried out. Moreover, its focus and structure must have a clear perspective of the way in which all parts of the organisation are to operate within the environment that surrounds it. This is achieved by adopting:

> a sensible and workable framework to its activities given its business environment, robust management procedures describing the actions to be taken in key areas of organisa-tional activity, clear working instructions to guide operations and tasks and, where appro-priate, pertinent situation-specific plans to describe the provision of products and services in differing circumstances.
>
> (Griffith, 1999)

Given these requirements, systems must be directly in tune with the organisation's business. They must encapsulate its business ethos, shape its culture, determine policies and goals, and develop processes and procedures, all with the intent of delivering optimum services and products. Management systems do much more than provide a framework within which the compliance with procedures and instructions is required. They provide a culture within which activities, resources and people become focused, through implementation of the management func-tion, towards supporting the holistic business of the company. Management systems are, therefore, not merely structural in concept and context but also socio-technical.

Scope

The system, or systems, used in managing a company will be characterised by generic elements of the management function in question, and these elements are common to both the corporate organisation and any sub-organisations that it employs. These are augmented by specific elements of the management function to meet the particular requirements of the sub-organisation delivering the service or product. The configuration of the corporate organisation and output sub-organisations is determined by the scope of the company's business activities, how many there are and where they take place. The scope may include business oper-ations exclusively from a head office location, as seen with small engineering companies, or extend to multiple geographically dispersed operations as seen in the configuration of retail-chain outlets. Scope may also range from a one-site business activity such as that of an independently run hotel to an extensive complex of process manufacturing facilities. In addition to guiding the delivery of products or services, systems can also be applied to project situations. This scope of operation is seen extensively within construction, where large contracting companies are directed from a head office and deliver their services through construction projects carried out in many locations. The management systems used guide the company's management functions at head office level while also guiding management at the points of business delivery.

Description, explanation and understanding

Any system must be set out, described and explained, hence the need for system documentation. Documentation must be amenable to translation from the corporate business perspective to working practices conducted at the delivery point of products, services or projects. This is particularly important as the corporate system is essentially management function driven while the delivery system for product or service supply is operationally driven and task based. Furthermore, a system should be simple, although not simplistic, such that it can be easily understood by those persons who implement the system at the various levels of the organisation and those who must work within the operational working instructions that the system establishes.

Visibility

An organisation may develop widely recognised systems such as those specified by national and international standards for management systems or develop its own in-company approach based around the processes and procedures it uses to meet the requirements of its clients or marketplace. Systems are visible and easier to understand where a company's structure, organisation, procedures and practice follow the requirements of formal standards, such as ISO 9001 for quality, and associated recognition, or certification, schemes. In contrast, systems can be less easy to appreciate where company procedures are essentially internal and operate informally. That is not to say that internal and informal systems are of little benefit as such approaches often serve a company well within the sphere of its particular business activities and clientele. The visibility of management systems in both structure and use are absolutely essential to their effectiveness and maintenance. This is influential upon the intra-organisational perspectives of employees who must operate and support the systems and also in promoting the appropriate external profile where the perceptions of clients and customers can be crucial to business success.

People

Systems embrace the social elements of any company and its organisation. A company must consider the human dimension and those people who implement management systems and make them work. People – individuals, groups, employees

– form the core of companies, lie at the heart of any organisation and make things happen. They are central to shaping the organisation, to getting the work done, and driving and maintaining the company's business. Within the business processes, management procedures and operational tasks, it is the human resources that come together, combine, deliver and maintain the business on a day-to-day basis. The success of implementation relies upon people at all levels in the organisation supporting and upholding the systems they use. It is all too easy to view companies and their organisations and systems in isolation from the people that work with them. Effective systems will establish management procedures and working instructions that explain how the company works, what part people play within them and, moreover, facilitate participation and provide support. Without people, management systems cannot be implemented and the business simply cannot deliver to survive and prosper. In establishing its environmental management system, the construction company Carillion has become a:

> beacon of good green practice . . . employees think the company does a lot for the environment . . . they think Carillion is getting better at protecting the environment . . . staff are aware of the environmental implications of the work they do, know that how they do their job has a bearing on the environment and are kept fully informed of eco issues and campaigns.
>
> (*Sunday Times*, 2009)

Clearly, this is an excellent example of the need for company managers to embed employee support for environmental initiatives and the willingness and commitment of the workforce to embrace them.

Organisational change

Perhaps one of the important aspects, if not the most important aspect, of systems development and implementation is that of organisational change and its management. The introduction of any management system will necessitate a degree of change to those organisational procedures used to manage the business processes. Changes will need to be made to the modes of operation of employees, to the ways in which they interface with processes as a result of systems approaches and in the ways they interact and interrelate to others in the organisation. All of these may make the people who will manage and operate the new system feel quite uncomfortable, uncertain and even threatened by the organisational changes needed to establish the management system. While change in any form, even if minor in reality, will create unwanted organisational unease, it must be appreciated that change is an aspect to be managed, can be managed and should be managed to ensure that any disruption to processes, procedures and people is minimised as far as is practicable. While some disruption from change is probably inevitable, the way in which change is managed is crucial. Many problems do not stem from the idea, recognition and acceptance of change, but are due to the organisational pace of change and the type and amount of communication required to make personnel aware and cognisant of what is happening within the organisation and how it will affect them. Developing a style of management which supports change and gives a good understanding of organisational policy, objectives, structure and organisation such that a shared sense of purpose is achieved is absolutely essential within any company.

A good example of eliciting support within the company is the work of Envirowise (2010), a managed government programme providing practical advice to companies on environmental matters, which has advised many contractors on the need for positive organisational change when establishing formal management systems. It suggests that successful contractors can capitalise on initiatives centred on gaining the support of staff and workforce. This is achieved by establishing senior management buy-in to concepts and principles of using a formal environmental management system, gaining input from the wider workforce and providing relevant training to increase employee awareness, understanding and appreciation of the need for change to the benefit of the company and all its employees.

Business environment

Companies do not exist in isolation but rather they operate within a marketplace and a wider business world. They must meet the demands of customers, often in competition with other providers. They must be managed efficiently and effectively if their products or services are to maintain currency and remain desirable to their customer base. This requires that an organisation has clear business aims, objectives and policies in line with the expectations and outputs that customers want. The business environment therefore plays a key role in shaping the configuration of the management systems used to manage the business. Beyond the requirements of customers, businesses of all kinds are influenced by factors emanating from their wider environment, for example legislation of their particular business sector and marketplace, regulations which govern the provision of their outputs and standards of organisational and business operation. Furthermore, companies exist in an environment where clients, customers and the public are increasingly interested in and seek to influence them, for example in the ethical dimensions of business operation. At a further level, companies conduct their business within, at times, an inconsistent political and often volatile commercial environment. All of these aspects influence how a company goes about its business and shape the management systems which are used in configuring its response to its business environment and marketplace.

Business risk

Davis Langdon, a large construction service company, propounds that 'A management system that has been properly established, developed and implemented will assist in managing risk and mitigating liability for yourself and your business' (DLCS, www.davislangdon.com, 2010). The level of risk to which a company is exposed as a result of its business activities is an important influence upon which functional management systems will be established and the ways in which the systems will be used. The consideration of risk is a key element of any and all systems. All companies are susceptible to risk from their operations, some more than others. This is determined primarily by the nature of the business being undertaken. Construction has a high level of risk throughout all its activities. Construction is human resource intensive and it takes place in an active working environment. Moreover, many construction duties and tasks are complex and hazardous. They give rise to environmental and safety dangers which must be considered and mitigated as part of the everyday operations which take place on a construction site.

Risk management is a key element of construction-related management systems. It involves the processes of risk identification, risk assessment and quantification, risk responsiveness and risk control. Within construction, risk management will form a part of the arrangement of company organisation at the corporate level where business risk is considered. It will form a major element in the consideration of project management where a company's services and products are delivered. Management systems for quality, environment and safety all incorporate risk management as a key element to inform management procedures, implementation plans and, in particular, the work instructions used on site.

Applications

Management systems can be applied to almost any aspect of a company's business and operations. This can include procurement, planning, finance and human resources, to name but a few. The most common applications of systems management include those featuring in this book – quality, environment, and health and safety. This tends to be because these particular management functions are absolutely essential to delivering the business of almost all companies. Also, these functions are at the forefront of ensuring that companies meet the requirements of current legislation and those regulations which impinge upon the provision of products and services.

Standards

International standards present a model for specific management systems where compliance with a recognised standard assures customers that the provider of a product or service has the management systems in place to deliver to a consistent level of provision. This assurance forms an important part of the business relationship between the supplier and purchaser. The QMS has been at the heart of business across almost all industrial sectors for over half a century. As key aspects of modern business management have evolved beyond merely the provision of quality, international standards have further perpetuated and driven the development of standards-based systems in other functional management areas. These are recognised through the most well-known application, that of quality systems meeting ISO 9001 (ISO, 2005) and its evolutionary derivatives, and also in application to the EMS under ISO 14001 (ISO, 2004) and H&SMS to OHSAS 18001 (BSI, 2007).

Quality system concepts

The drive for better quality of product or service has seen the international standard for quality systems ISO 9001 become the norm within business. While this standard is dedicated to quality provision, and guides organisations explicitly in their implementation of a QMS, elements of its constitution are common to any management system. A QMS therefore shares characteristics with an EMS and H&SMS. It is these links that enable organisations to use quality system concepts when considering the establishment of other management systems. It is also these links that provide the basis for bringing essentially separate management systems together towards configuring an IMS.

Holistic perspective

The implementation of a management system has to do more than merely control the particular management function; it should bring wider efficiency and organisational effectiveness, adding value to the company and its business. Management systems have, traditionally, been regarded as support systems to the core business processes. Although vital to delivering the appropriate level of product or service to satisfy a customer, they are regarded as non-core income stream activities. Another way to look at management systems is to see them as central rather than peripheral. At the core business of the organisation is 'the system' with management functions acting as highly focused input sub-systems structured to support directly that core business. In this way, a management system assumes a much greater status as one where it exists, first to be optimised in application, and, second, that focuses not only on itself but also for the greater contribution that it can make to the organisation. As such, it takes on a holistic, or whole-organisation, perspective. The holistic perspective of management system application has much in common with 'systems theory' (Checkland, 1981). Systems theory propounds that the whole becomes greater than the sum of its parts, or produces synergy. By looking at management systems in this way, they too can be configured to produce synergistic effects which can add greater value to the company.

Management system application

This section examines: what a management system is in application to the organisation; the key elements which make up a management system; the requirements for and types of documentation to formalise an operational system for application; systems applications to construction organisations, appreciating the individual, often unique, characteristics of the construction process; and the diffusion of management system application.

What is a management system?

Having outlined some of the key and general features of management systems, it is appropriate to ask and explain: what is a management system in practice?

A management system is:

- an arrangement of organisational policies, goals, strategies, structures, processes, procedures and resources needed to implement a given management function;
- a hierarchy of documents – policies, manuals, plans, procedures, instructions and forms – which defines, describes and explains the application of the management function within the organisation;
- a set of management arrangements that facilitate the provision of the organisation's business outputs – products and/or services;
- a way of managing the process model of providing products and/or services;
- a method of providing continual analysis and enhancement of the management function though an improvement loop of thinking, planning, doing and measuring – linking activities back to policies, goals and strategies;
- as comprehensive as required to meet the specific organisational objectives of the management function being carried out, but also able to conform to recognised and registered, or certified, system standards;
- simple to understand for those involved with system implementation, and visible such that workers can associate with and support it;
- reliant upon the contribution of the people, at all levels in the company, who implement it;
- configured around the intra-organisational dimension but also accommodating the requirements of the inter-organisational dimension and wider business environment;
- applicable to any function of management, although quality, environmental, and health and safety management are international standards/guideline-based and are therefore most recognised and accepted widely in most industry sectors;
- a way of doing things that not only optimises the management function but also contributes to the holistic activities and business of the company.

Because organisations are often extensive and sometimes complex in their arrangement of business activities, processes and resourcing, a management system must establish an effective organisational framework of responsibilities at various organisational levels. There are, typically, three levels of management:

1 Company, or corporate (executive management) level;

2 Operational (management and work supervisory) level;

3 Implementation (operative/worker) level.

These three levels need to permeate the configuration and description of the system. This is achieved by, first, developing a structure within which each management level takes responsibility for specific elements of the system. Second, each element within the system needs to be described by written documents.

Management system key elements

A management system must establish an effective structure of command and documents to guide the application of the system in specific areas of organisational activity. This is also essential to meeting the requirements of those standards for management system development and implementation. The principal areas are:

- policy making (company executive);
- organisation (executive/management);
- risk assessment (management) – sometimes subsumed within the planning element;
- planning (management);
- implementation (supervisory/operative–worker);
- auditing and review (management).

A schematic of key system elements is shown in Figure 1.9. It is not proposed at this stage to describe and discuss the details, complexities and implications of these areas of activity. These will be covered comprehensively in later parts of this book where specific applications are described and discussed.

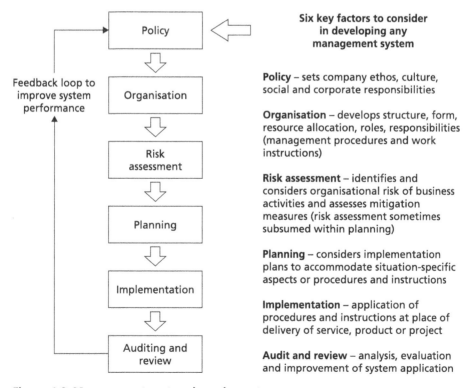

Six key factors to consider in developing any management system

Policy – sets company ethos, culture, social and corporate responsibilities

Organisation – develops structure, form, resource allocation, roles, responsibilities (management procedures and work instructions)

Risk assessment – identifies and considers organisational risk of business activities and assesses mitigation measures (risk assessment sometimes subsumed within planning)

Planning – considers implementation plans to accommodate situation-specific aspects or procedures and instructions

Implementation – application of procedures and instructions at place of delivery of service, product or project

Audit and review – analysis, evaluation and improvement of system application

Figure 1.9 Management system key elements

Documentation

The formalisation of management systems is achieved through establishing documentation. Documentation can embrace any written or illustrative information defining, describing, specifying, reporting or certifying organisational activities and/or business requirements, processes, procedures, results and reviews. Documentation enables an organisation to conduct its activities with purpose, direction, order and consistency.

The use of documentation contributes to:

- the achievement of organisational activities in meeting customer requirements;
- the satisfaction of organisational policies and objectives;
- the fulfilment of activities to predetermined standards of performance;
- the determination of effectiveness in managing organisational processes;
- the provision of consistency, repeatability and traceability in organisational activities;
- the collection of objective evidence on organisational activities;
- the review of organisational efficiency and effectiveness based on evidence rather than conjecture.

System documents

An organisation requires the following types of documents to be used in its management systems:

- Documents that provide information about its commitment to the particular function of management around which the system is configured – *policy*.
- Documents that provide information about the management system – *manuals*.
- Documents that provide information about particular requirements – *specifications*.
- Documents that provide information about recommendations for practice – *procedures*.
- Documents that provide information about the management system in application to a specific product, service or project – *implementation plans*.
- Documents that provide information on how to carry out activities concerned with processes – *work instructions*.
- Documents that provide information on the evidence gathered from activities – *records*.

It is not appropriate to be overly prescriptive with regard to the structure outlined. Each organisation must make up its own mind as to the nature, extent and detail of documentation required. This will depend on the type and size of the organisation, the requirements of its business processes, the nature of its products/services, the demands of its customers, the legislation and regulations that govern its operations and the requirements of its industry sector. In most management systems, four levels of documentation tend to be adopted:

1 A management system manual
2 A set of management procedures
3 An implementation plan
4 A set of work instructions

Management system manual

This document defines, describes and explains in the context of the particular management function, such as quality management, the company's policy, organisation structure, resources, business processes and personnel responsibilities for delivering the management function throughout the company. The manual describes the overall framework of organisation and procedures by which the company operates with regard to legislation, regulation, management standards and the wider world. In simple terms the manual explains the organisation of the organisation.

Management procedures

These describe the procedures involved in managing the company's business processes to meet the policy set and in line with the organisation structure, resources and responsibilities described in the system manual. A single procedure is a document that specifies and describes how a business activity is to be carried out. Management procedures give the organisation directives on standard methods of operation for corporate and project activities which, ordinarily, remain constant and uniform irrespective of the types of work being undertaken. In some organisations, management procedures will be termed company standing procedures or company standing instructions. The procedures may be contained within the management system manual or can be contained in a separate set or sets of documents, with this depending upon the scope and extent of the total system documentation.

Implementation plan

This is a plan which describes the management system manual and management procedures in application to a specific situation such as a project. It describes the characteristics of that project, taking into account aspects such as applicable legislation, regulation and forms of contractual arrangements for its undertaking. Similarly, a plan might cover the delivery of a particular service or product which has situation-specific characteristics. While the system manual and management procedures will apply to all organisational activities, the implementation plan will be individual to a particular business application such as a project.

Work instructions

These describe instructions needed to undertake the processes which actually deliver the service, product or project by those carrying out the processes. Some work instructions will be generic and apply to all business processes while others, influenced by particular characteristics described in the implementation plan, may be specific in nature.

The document pyramid

Systems are formalised by presenting the company's management approach in a hierarchy of documentation. This usually takes the form of a company handbook or manual embracing the particular function being managed, for example a quality management handbook or a quality systems manual. Specific documentation will depict management activities at different levels of the organisation. Typically, there will be a set, or compendium, of systems documents for a particular management function, like quality, describing the corporate policy and framework of operation, management procedures, working instructions and administrative forms. In addition, there will be documents describing specific systems necessary in providing a product or service in any given circumstance, for example within a contract or project procured by a client. These are not always required, but where they are they normally take the form of an implementation plan, sometimes referred to as a contract or project plan. This is necessary because the provision of an output to a particular customer in a particular location and to particular requirements will, more often than not, differ from that of another. These documents are situation specific.

The full set of documents embraces both the generic and specific elements of carrying out the management function throughout the whole organisation. The hierarchy of documents is sometimes termed the pyramid structure of documentation, because the policy is sharply focused at the top, broadening out through procedures and working instructions to administrative documents at the base. Also, there is greater opportunity at the base of the pyramid for record forms to overlap, with modification to use, whereas at the top of the pyramid policies are individually focused and not so readily amenable to overlap and integration. The document pyramid is shown in Figure 1.10.

Although the conventional and most logical approach is to use four levels of system documents in the order outlined, the sequencing of the levels within the document hierarchy is often varied. Some organisations reverse the implementation plan with the work instructions while others swap the plan round with the manual. Some organisations use different terminology to describe the sets of documents they implement. Such variations, which can appear to defy straightforward thinking, in fact serve the particular organisation better because it suits the specific understanding, needs and applications. So, it is important not to dwell too much on the finite detail of what may or may not be the perfect sequencing of documents but rather to configure documents as the organisation best sees and uses them. The important points are that the organisation:

• configures the management system document pyramid to meet the organisation's particular requirements;

• defines and describes the documents in the context of the organisation's understanding and application;

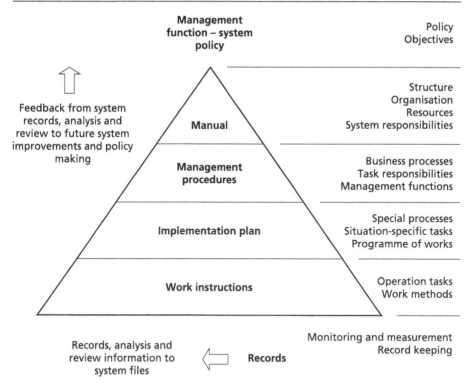

Figure 1.10 The management system document pyramid

- arranges the set of documents in a sequence which best suits the organisation's application.

Application of systems to specific management functions

Management systems may be used to configure a range of organisational management functions. A number of these management functions are standards based, such as quality, environment and safety, while others are intra-organisationally based, such as systems used for planning and procurement. Some systems stand outside those influenced by international systems standards, such as Investors in People (IiP), which applies to the management of human resources, and where national-based independent recognition and certification are applied. Figure 1.11 illustrates some examples of management functions together with applicable management system standards, drivers and performance indicators.

Optimising system effectiveness

A company must seek to ensure that any management system it establishes operates at optimum effectiveness. The essence of optimising system effectiveness is to ensure open communication, transparency and pragmatism. In addition, a conscious effort must be given to putting in place management mechanisms which

Management function	System type	System standard	Key driver	Performance indicator
Project planning	Company	None	Software	Business excellence
Cost planning	Company	None	Modelling software	Business excellence
Human resources	Standards based	Investors in People	IiP model	Certification
Procurement	Company	None	Databases and software	Business excellence
Quality	Standards based	ISO 9001 EFQM	Management system specification	Certification
Environment	Standards based	ISO 14001 EMAS	Management system specification	Certification
Safety	Standards based	OHSAS 18001	Management system specification	Certification

Figure 1.11 Examples of organisational and standards-based management systems

continually monitor, audit, review and refine the system. The most obvious and sensible way to achieve this is the adoption of Deming's methodology of PDCA. Moreover, this can be allied to the process model of business activity with its continual improvement loop linking management responsibility to resources, outputs and analysis. These aspects were described and illustrated earlier in this chapter.

The promotion of system effectiveness requires particular undertakings at both a company level and a project level. These undertakings must be directly relevant to the particular management function handled by the system.

At company level:

• executive and management commitment to company policy and strategy;

• clear statement of company policy, aims, objectives and goals throughout the entire company;

• employee ownership of the system through involvement in their development and implementation at all levels;

• setting of challenging yet achievable and measurable goals for system performance;

• adequate resources to perpetuate, maintain and improve the system;

- appropriate training for managers in system supervision and workforce in the methods of working;
- mechanisms to ensure monitoring, auditing, review and system improvement.

At project level:

- identification of key matters applicable to any project, or service/product delivery point;
- risk assessment within the boundaries of management functions;
- development of action plans to meet identified needs;
- distribution of good practice management procedures and work instructions;
- briefing to all employees on business processes and system applications;
- training of all personnel on system implementation;
- provision of practice guides and notes to assist system operation;
- clear reference on executive and management support to system operations;
- determination of audit procedures to check and test system operation;
- mechanisms to encourage reporting of systems deficiencies for rectification;
- procedures for system analysis, review and improvement;
- feedback channels from systems application to company business and process evaluations.

Systems application to construction organisations

It has been seen that organisations within the construction industry differ in terms of their structure and operational characteristics from organisations in others sectors of business. A large principal contracting company, for example, tends to adopt a decentralised structure to accommodate construction projects taking place in many and often geographically dispersed locations. Although a large proportion of construction activities and tasks are the same from project to project, some aspects are project specific as they are unique to a particular project. Multiple projects are often undertaken simultaneously, and each may have a wide range of interdisciplinary elements. Given these characteristics, a construction company must seek to develop and apply strong and robust corporate structure, organisation and management from which its management approach can cascade to all its project sites.

Given the required framework for effective systems establishment, management system applications, for the purpose of managing construction work at the project level, require the following:

1 *Corporate, or company-wide, structure* – to establish 'organisation' within which the management of the company *and* its project sites can be configured.

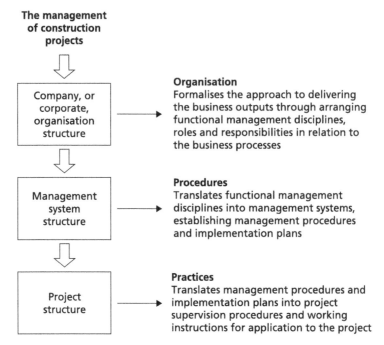

**The management
of construction
projects**

**Company, or
corporate,
organisation
structure**

Organisation
Formalises the approach to delivering
the business outputs through arranging
functional management disciplines,
roles and responsibilities in relation to
the business processes

**Management
system
structure**

Procedures
Translates functional management
disciplines into management systems,
establishing management procedures
and implementation plans

**Project
structure**

Practices
Translates management procedures and
implementation plans into project
supervision procedures and working
instructions for application to the project

**Figure 1.12 Key elements of structure for the management of construction
projects**

2 *Management systems structure* – to establish 'procedures' by which the
company's construction projects can be successfully delivered.

3 *Project structure* – to establish 'practices' by which the construction works will
be undertaken at the project site.

Such an arrangement can be seen in Figure 1.12.

The need for organisational structure

The key common thread in the three requirements is that of 'structure'. Structure
is absolutely essential to any construction-related organisation. An overarching
corporate structure must be sufficiently rigid to provide the necessary formality
for the management of the company's business activities while remaining
sufficiently flexible to accommodate change in the company's business marketplace. This is perhaps the chief reason why any organisation should be morphogenic,
or outward looking. The organisation must maintain a strong sense of its holistic business within a constantly changing environment while keeping a close eye
on the delivery of its business outputs. A sensible and well-conceived corporate
structure, supported by clear and well-defined management systems and project
practices, will place a company in the position of being able to remain aware of

both its external and intra-organisational environments to deliver its business outputs successfully to clients and customers.

Organisational structure and specialisation

A tradition within construction is that of the 'specialisation' of inputs. In fact, specialisation is true of most companies and organisations. Specialisation of activities is driven principally by the core business of the company and the need to develop knowledge and capabilities to deliver the core business through the type of people and resources that it employs and the ways in which it deploys these throughout the organisation. While it is true that other configurations of structure, organisation and resourcing can be used to reduce or even remove the reliance on specialisation, the construction industry tends to be heavily dependent on specialisation in some form because of the technological disposition of its output products and the activities needed to deliver them.

Within construction, specialisation forms the fundamental basis of a company's structure. Vertical specialisation gives rise to the hierarchy of authority and accountability while horizontal specialisation provides the differentiation of the functions of management at various levels within the overall hierarchy. Together, vertical and horizontal specialisation form the basis of the holistic organisational structure of the company within which communication, command, resources, processes, procedures and practices are established. An organisation where there is overt vertical specialisation with strong control is termed a *mechanistic* structure. In such a structure, company policies, processes, procedures and practices are highly organised, well formalised and documented and have a closely defined division of labour and resources. An organisation with a prominent horizontal, or flat, structure is said to be an *organic* structure. In this structure, control is predominantly founded on less formality and has looser arrangements of people and resources. It is not implied that either of the two structures is good or bad as both have positive and negative attributes and elements to them. In practice, almost all companies within the construction industry combine aspects of both the mechanistic and organic structure within their organisation. Moreover, companies understand the overwhelming need to provide a certain degree of mechanistic approach to establish a strong, robust and continuing overarching corporate management structure while allowing the organic approach to develop and embed the informal and natural interaction of employees. Both are essential to a modern and healthy organisation operating within the construction industry today.

People

It was mentioned earlier that people are absolutely essential to the operation of any management system, and therefore the consideration of people is a prerequisite to the framing of any structure and organisation. People are at the heart of any organisation; they make the organisation work. People carry out the functional roles of management and also make management systems work.

This is particularly true within construction as the contributing work activities are so human resource dependent and labour intensive. So, the structure required at both corporate and project levels must be arranged around the careful consideration of the company's staff and worker base. It is not so much a case of identifying individuals within the structure, although this can be useful, but rather of clearly identifying personnel roles and their accompanying responsibilities. This should embrace the whole organisation if the needs of business holism are to be accommodated.

Configuring structure

Like most organisations, construction companies of all types tend to reflect their configuration of structure in 'organisation charts'. Such charts identify the functional management disciplines, roles, responsibilities, relationships and routes of communication. A set of organisation charts may well be needed to reflect the complexities of a larger company or to illustrate management arrangements at different levels in a company. This is often the case within, for example, a principal contracting organisation, where a set of charts may be needed to depict the different structures at corporate management level and project site management level. Because management systems, if so designed, pervade the entire company, there will need to be clear links shown between the management functions in each organisation chart. For a principal contracting company it is not uncommon to find the whole-organisation structure reflected in two or more organisation charts with management function and system links included in their format and presentation.

Organisation charts will likely be required to reflect the differing structures in different types of companies. Construction companies can be small and simple in structure or, conversely, large and complex. This differs according to a wide range of influences as shown in Figure 1.13. A small contracting company, and there are many such companies within the construction industry, may well adopt the simplest of organisation charts. Small companies tend to be centralised in one location, operate to a narrow geographic spread of business activity and employ a limited resource base. Therefore the representation of their whole-business activity can be achieved in one simple organisation chart. A medium to large organisation may well have to reflect business activities undertaken nationwide, with corporate management from a head office location and outputs delivered at multiple and geographically dispersed project sites. This may require more than one chart to reflect head office structure and project structures. Further, a large organisation may operate nationally and internationally and this may need to be illustrated in a set of organisation charts to capture the corporate-base structure together with location-related sub-structures and project structures.

It is important that the organisation chart reflects the various levels of management within the wider structure of the company and links this to the management systems which are put in place to handle the functional management disciplines. As a result of these needs the charts will likely indicate the three broad

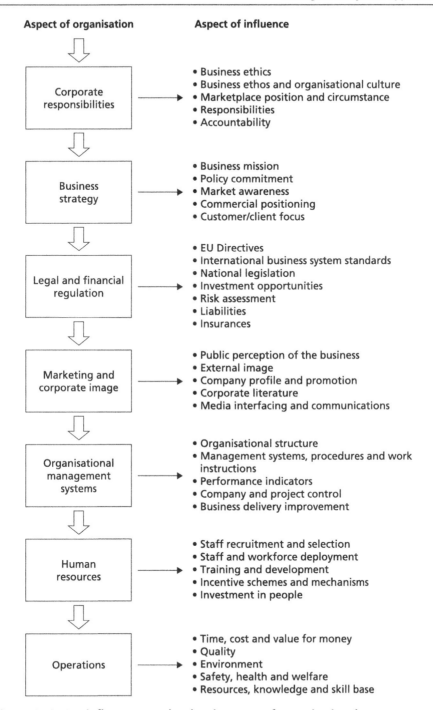

Figure 1.13 Key influences on the development of organisational structure

groupings of management – executive, directive and operational – and link these with the key management systems – quality, environment and safety – in whatever arrangement the systems are configured.

The structures adopted must give a clear perspective on the positions of the management functions within the organisations depicted. The awareness and understanding of and support and commitment given to managerial functions are dependent upon individuals knowing their place within the structure and the contributions that they make to both their management function and the wider organisation. Establishing and maintaining support for the structure and organisation throughout the staff and workforce base is essential. Effective organisational structure positively promotes teamwork, again essential to the wellbeing of the company's operation and its outputs. Teamwork is a catalyst for the shared understanding of management system development and implementation, the creation of synergies from resource inputs and the perpetuation of benefits for the wider organisation. People management is therefore a prerequisite within the arrangement of management structure adopted. Moreover, the company must ensure that its project structures encourage the cascading of commitment from the corporate structure, and that it is translated into a positive and productive working environment on the construction site.

It must be emphasised that there is not one ideal or perfect approach to establishing the corporate and project structures within a company. Companies will adopt an approach which encapsulates many generic elements of good organisational structure augmented by company-specific elements and apply these within the context of each company's particular business environment and circumstance. Notwithstanding, there are common schematics of organisation charts to depict the general arrangement of functional management disciplines, the various levels of management and management system links used by construction organisations. These are shown in Figures 1.14 and 1.15.

Diffusion of management system applications

The importance of understanding the development and implementation of management systems is supported by the large number of business organisations implementing standards-based systems for quality, safety and environmental management. There has been a growing understanding and general theme of diffusion in systems applications over the last decade and throughout all business sectors, with over a million organisations in 175 countries worldwide becoming certified to ISO 9001 and approaching 150,000 organisations with an EMS certified to ISO 14001. While the number of organisations certified to OHSAS 18001 for the same date is unknown, it is likely that the number runs to the tens of thousands. Moreover, thousands of organisations worldwide utilise management systems for their QMS, EMS and H&SMS which have yet to be certified to international standards (Griffith and Bhutto, 2008).

With the evolution of management system standards there has been a general diffusion of management system developments from the earliest applications of

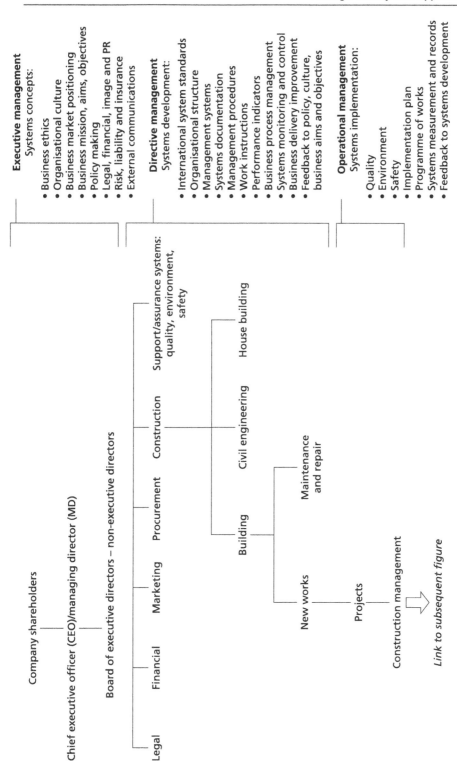

Executive management
Systems concepts:

- Business ethics
- Organisational culture
- Business market positioning
- Business mission, aims, objectives
- Policy making
- Legal, financial, image and PR
- Risk, liability and insurance
- External communications

Directive management
Systems development:

- International system standards
- Organisational structure
- Management systems
- Systems documentation
- Management procedures
- Work instructions
- Performance indicators
- Business process management
- Systems monitoring and control
- Business delivery improvement
- Feedback to policy, culture,
  business aims and objectives

Operational management
Systems implementation:

- Quality
- Environment
- Safety
- Implementation plan
- Programme of works
- Systems measurement and records
- Feedback to systems development

Company shareholders

Chief executive officer (CEO)/managing director (MD)

Board of executive directors – non-executive directors

Legal   Financial   Marketing   Procurement   Construction   Support/assurance systems: quality, environment, safety

Building   Civil engineering   House building

New works   Maintenance and repair

Projects

Construction management

Link to subsequent figure

Figure 1.14 Organisation chart showing company structure for medium–large contracting organisation, the levels of management and systems establishment

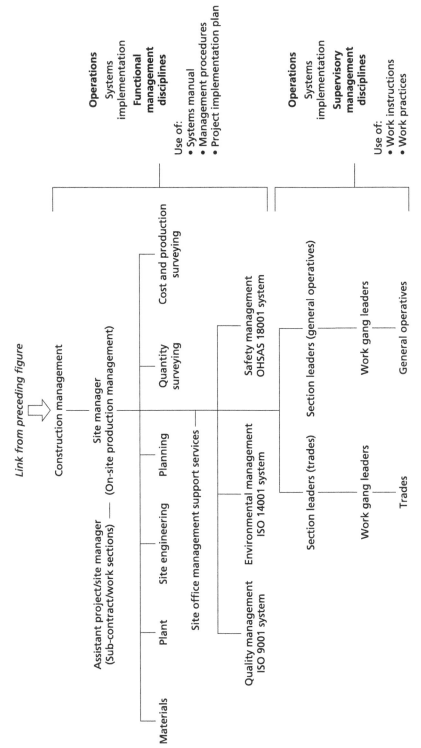

Figure 1.15 Organisation chart showing project structure for medium–large contracting organisation and systems implementation at functional management and supervisory management levels

the QMS in the 1970s. This has seen the blurring of the boundaries between the QMS and EMS to form the EQS (Environmental Quality System) and more recently the absorption of H&SMS into early and rudimentary merged and integrated management systems. Of course, as these three more established management standards-based systems become embedded into everyday practice, so their concepts and principles of approach are influencing the evolution of standards and systems in other fields of business, including: business auditing; information security; innovation; and supply chain management. A prominent example of new and similar management system standards is that of IiP embracing human resources management.

Management system research and experience

This section examines empirical and first-hand research undertaken to understand better the nature and application of management systems within construction, focusing on the experiences of management systems and their use by construction professionals operating within major contracting organisations.

Research and experience

Research

Although there has been a strong pattern of adoption of standards-based management systems, practical experiences of QMS, EMS and H&SMS in application have not always been positive. Over a decade ago, research by the New Zealand Foundation for Research, Science and Technology (FRST, 1998) reported that half of all the organisations it surveyed saw little added value from their management systems in use. Most companies appeared to be motivated by the desire to comply with current standards, regulations and certification rather than the will to embrace the wider and holistic opportunities available to their organisations. It appeared that contractors were tending to use paper-based management systems for compliance purposes without giving appropriate attention to the quality assurance of the core business outputs. Also, problems appeared to be occurring within client and customer organisations which were demanding certificated services and products without fully understanding the practical limitations of delivery.

In relation to the construction industry specifically, the Construction Industry Research and Information Association (CIRIA, 2000) aired some concerns at the apparent rapid upsurge in management system development and use. It suggested that this might be due to little more than market positioning by organisations where their real focus might be the pursuit of improvement to business and project effectiveness. In addition, management systems can be applied to particular parts of an organisation rather than the whole enterprise, and the usefulness of this has been questioned (Hoyle, 1998). Applications of this type are often seen both within and outside a company as inefficient, cost ineffective, bureaucratic and divisive. Furthermore, these applications often perpetuate substantially negative impacts on organisational stakeholders and, in particular, those who implement the systems (Beckmerhagen *et al.*, 2003).

Townsend (1999) highlighted the leading-edge developments in the use of QMSs by the manufacturing and engineering industries. Their applications recognised the advantages of bringing systems together as integrated quality management, or IQM, in the late 1990s. The 'QUENSH' (QUality, ENvironment, Safety and Health) approach is a prominent example (Renfrew and Muir, 1998; Winch, 2002). Subsequently and likewise, standards- and systems-based quality management within the construction industry has become an established and accepted function over the last 25 years and continues today most prominently. Pressures from clients, legislation, industry-governing bodies and professional institutions have emphasised the overwhelming need not just for high-quality end-products but also for construction outputs that are delivered using safe and environmentally friendly methods. The 'SHE' (Safety, Health and Environment) integrated project management approach used extensively within the construction industry reflects the real recognition of such needs and pressures (ECI, 1995).

As contracting companies with the construction industry have gained valuable experience from establishing multiple management systems and learned from the early lessons of the manufacturing and engineering sectors, so too they have challenged the conventional wisdom of using separate and dedicated systems. IMS are seen by principal and large contractors as a way of addressing some of the difficulties of separate systems and pursuing greater efficiency, effectiveness and business performance (Karapetrovic and Wilborn, 1998; Thompson, 1999). This is strongly reinforced by the British Standards Institution (BSI) support for 'Integrated Management System Assessment' (IMSA). This facilitates the integrated dual and triple certification QMS, H&SMS and EMS in whatever combination a company would like to present its systems.

Throughout the last decade much research has shed light on the evolution and application of management system standards, systems, procedures and practices (Karapetrovic and Wilborn, 1998; Renzi and Cappelli, 2000; Wilkinson and Dale, 2001; Matias and Coelho, 2002; Mendel, 2002; Karapetrovic and Jonker, 2003; Zeng *et al.*, 2005; Zutshi and Sohal, 2005; Marimon *et al.*, 2006). While the opportunities that exist to maximise organisation and management through the implementation of management systems have been well identified, the hindrances to effective use have also been highlighted. The key issue within construction and for the future of IMSs across all business sectors appears to be to find a practical and successful way to interweave the various, often disparate, systems into one all-embracing management system which meets the real needs of the contractor while satisfying the real requirements of the other participants in the total construction process. The dilemma is not so much the conceptualisation of the IMS, but finding simple and intelligible ways to make sure the system operates in practice with genuine effectiveness and benefit to the project and the wider organisation.

The experiences of construction industry contractors

Research in relation to IMSs in construction by Griffith and Bhutto (2008) examined the experiences of construction industry contractors, issues surrounding IMS development and use, and the key features of systems integration and implementation frameworks. These are summarised as follows.

Contractors' experiences of traditional standards-based management system application

- The decision to implement management systems is predominantly influenced by legislation, client pressure, and the need by the contracting organisation to avoid perceived competitive disadvantage in the construction marketplace.

- QMSs meeting ISO 9001 have successfully been implemented by contractors over the last 25 years, formerly as BS 5750 and in recent years by the ISO series.

- There is growing emphasis on the implementation of EMSs meeting ISO 14001.

- Most contractors are not certified to OHSAS 18001, relying upon first-party and second-party (non-independent) recognition of their H&SMSs for project application.

- Current management system structure tends to be vertical and separate for each system, making systems operationally independent and disparate in terms of shared communication and information.

- Requirements for documentation of management systems are, sometimes, regarded as onerous, bureaucratic, inefficient, ineffective and divisive by managers and system users.

- While there is acceptance of a certain workable degree of formal procedure and documentation, staff often regard systems as a burden and a hindrance to 'getting the job done'.

- Site staff, in particular, do not fully understand management systems so that real and holistic benefits to both the project and the company are lost to a simplistic compliance and checklist culture and mentality.

- Management system implementation tends to be let down by lacking intra-organisational culture change, commitment, education and support.

The key points highlighted by the contracting organisations show that traditional standards-based management systems continue to create considerable uncertainty in application. This is made worse by a distinct lack of corporate understanding of systems needs at the project implementation level. Moreover, the perception that systems implementation is little more than a burdensome compliance reaction means that the potential benefits of an effective and holistic systems approach will, likely, not be realised.

Contractors' perceptions of systems integration and its potential through recent IMS development and applications

- There appears to be no clear understanding of or consensus for effective IMS structure, with contractors interpreting management system development and implementation in their own way.
- There is considerable duplication of effort and tasks in current management system applications.
- There is agreement that current standards-based systems have many similarities which should be seen as complementary to reduce duplication and effort in systems development and use.
- Large-scale integration is seen as possible but only through a simple, generic and recognised IMS framework, or model, for system development and implementation.
- QMSs and EMSs have been integrated with some success by contractors, yet H&SMSs are seen as more inflexible and based predominantly on compliance with health and safety legislation rather than wide-ranging systems standards.
- Recent approaches to the IMS have focused on 'merging' systems documentation for specific management functions with an existing QMS.
- Practical implementation of IMSs relies on the development of simple integrated manuals, management procedures and work instructions supported by well-defined and described site-specific plans linked to site processes.
- Implementation plans, or project management plans (PMPs) in construction-related terminology, were seen as the most important tool for effective IMS implementation on projects, as these must translate corporate policies into comprehendible and useable operational procedures.
- IMS development requires considerable intra-organisational culture change, focused vision and strong corporate and management commitment.

While contractors recognised the similarities in the separate standards-based management systems which could be integrated to reduce duplication and effort in fulfilling key management functions, the lack of understanding of how best to integrate systems may have been holding back potential developments. Applications had tended to focus on 'merging' rather than truly 'integrating' the systems. The perception that integration can appear overly complex called for clear project management plans and simpler procedures and documentation configured around more clearly defined site processes.

Key features of systems integration and implementation frameworks

- It is recognised that there has been some predictable early-stage cultural resistance to IMS implementation on projects, and this needs to be overcome.
- Differences in scope of the various systems must be recognised, with the core system, usually a QMS, used as a vehicle for perpetuation towards the IMS.

- An IMS will only be effective and successful if it is developed in a way that managers and operations-based personnel can easily understand and, in so doing, embrace its intentions, spirit and opportunities.

- Rapid identification and alleviation of hindrances to an IMS such as duplication of system tasks and documents, bureaucracy and inefficiency are essential to acquiring early support and longer-term commitment.

- Creating 'visible' benefits of an IMS in implementation is vital to gaining wider acceptance by system users and receivers.

- The BSI Integrated Management System Assessment (IMSA) can be a useful focus for IMS dual and triple certification system developments.

- An IMS has strong similarities with 'business process models' as these are configured around the holistic vision and objectives of the business and, therefore, a business management system (BMS) orientation might be best adopted.

- A BMS approach is, in concept and application, compatible with conventional construction processes and typical contractors' corporate and project activities.

- Risk assessment is an essential element of all management systems and should form the central feature of an IMS and be reflected prominently in any implementation, or project, plan.

The key point highlighted is that an IMS might be best configured around a business process model which is clearly compatible with the conventional arrangement of construction processes. The long-standing, tried and tested base management system is the QMS, and this should be used as the integrating vehicle. Linking systems to processes and, in particular, risk assessment – a crucial element of all construction projects – should be the central feature of the project management plan. As a prerequisite to such a configuration, the IMS must engage with the hearts and minds of all system users and receivers if the detrimental perceptions highlighted earlier are to be minimised and the positive benefits are to be encouraged and embedded.

Establishment of a framework or general approach to an IMS

The establishment of a framework or general approach to an IMS is not without difficulty, as identified clearly in the foregoing observations. Five stages of best practice development and implementation were identified in the research as essential: (1) *organisational mapping* – to configure functional management in direct relation to processes within the context of the holistic perspective of the organisation; (2) *statement of vision, policy and objectives* – to be clearly defined, communicated and accessible to all employees; (3) *commitment to IMS/BMS* – to deliver 'integration' rather than piecemeal approach; (4) *structure* – to ensure homogeneous and consistent application across functions and throughout the organisation; (5) *evolution* – to ensure continual improvement and achieve organisational betterment.

   A traditional and major difficulty highlighted was employee resistance to organisational change and the need to 'capture hearts and minds' to secure support for system implementation. Communication and involvement were identified as vital to this. While it was admitted that systems initiatives are essentially top-management driven, and in business terms perhaps rightly so, there has to be a facility for grassroots involvement and bottom-up communication.

An essential aspect of the research was to verify the findings and suggestions for best practice with practitioners from the leading-edge contractors involved in IMS developments. In this regard a number of senior managers scrutinised the findings and provided useful views and comments. Key suggestions for IMS development and application included: (1) a clear differentiation of project-specific information contained in any implementation plan from the generic information contained in the system management manual, as the plan changes with each project; (2) companies with particular activities devolved from their corporate centre may require a homogeneous approach with modified or bespoke procedures; (3) an IMS must cater for a variety of procurement methods and subcontract elements; (4) system development must be accompanied by IT support to facilitate intranet-based documentation and ease of site implementation; (5) for smaller contractors, it may be advantageous first to consider site practice and work back through procedures to the development of a formal documented system.

Summary

Research has illustrated that organisations in many commercial sectors of business including construction have spent vast sums of money on developing and implementing management systems for quality, safety and environment, while, collectively, the construction sector spends millions of pounds on these initiatives each year. However, it was highlighted that many organisations have reported dissatisfaction with traditional systems applications. Therefore, the challenge is to question convention and seek an alternative and more effective approach to management system configuration. Construction has accepted this challenge, with a number of international building and civil engineering contractors at the very forefront of IMS development. Today, the notion of integration within construction is melding into a greater vision of contracting as a holistic business process guided by a single multi-dimensional system of management.

Management systems: key points, overview and references

This section presents: a summary of key points from the collective sections of Chapter 1; an overview of Chapter 1; and a list of references used in the compilation of Chapter 1.

Key points

- 'Management' is not a single discipline but one which crosses the boundaries of many disciplines – it is multidisciplinary and interdisciplinary in application.

- In the context of organisations and a systems approach to management, 'management' is the term used to emphasise the application of a multitude of specific functional management concepts, principles and practice.

- The concepts, principles and practices of management have been influenced over the centuries by the thinking and philosophies of prominent individuals and collective schools of thought.

- 'Modern management theories' – the 'scientific management school' and the 'human relations school' – moved management thinking on considerably from those procedures which preceded them.

- A management system is, simply put, a way of doing things – such systems arrange, develop and apply protocols and sets of procedures which bring structure, order and stability to the processes of running a business.

- A management system must be an open system that works for the company's business activities in a proactive, dynamic and forward-thinking way – it should be morphogenic (responsive) rather than morphostatic (unresponsive).

- A management system should relate its business processes to the functions of management such that they are carried out systematically and consistently to set standards of performance.

- The concept of the 'process model' is fundamental to structuring an organisation, its activities and its management, as almost all businesses use the model of input–conversion–output in delivering their services, products or projects.

- There are three types of processes which companies generally use in support of their business activities: core processes; support processes; and assurance processes.

- The construction industry and its processes have a close affinity with the process model and to the core, support and assurance processes which influence its management.

- Systems theory is central to company organisation as it links the process model of delivering business outputs with the establishment of management systems to deliver the functions of management.

- Systems theory provides a framework for the managed direction of organisational activity through the better management of its parts to the benefit of the whole business – it creates synergy and holism.

- A management system is normally one of two types: a 'company-wide' system covering the activities of the entire organisation under a corporate-based umbrella; or a 'project-based' system applied to specific organisational activities and processes.

- The key features of any management system are that they are: soft systems based; document based; standards based; process orientated; and that they pursue continual improvement.

- Management systems used by companies within construction generally adopt the company-wide umbrella approach translated into project-based application, and are standards and document based, process orientated and seek continual improvement to a customer-focused agenda.

- The key elements of organisational activity which management systems must include are: policy; organisation; risk assessment; planning; implementation; and auditing and review.

- Documents describing a management system are normally arranged in a hierarchy termed the document pyramid, and this includes: a management system manual; a set of management procedures; an implementation plan; and a set of work instructions.

- System effectiveness in implementation must be promoted at two levels: the company, or corporate, level; and the project level.

- A management system should establish three structures within an organisation: a company-wide structure; a management system structure; and a project structure – together these embrace the system in operation throughout the whole organisation.

- There has been a general diffusion of management system standards and systems across all industry sectors, and within the construction industry sector particularly, since the early applications of quality management in the 1970s.

- QMS applications have been joined by the development and applications of EMSs and H&SMSs in recent decades as new international management standards have emerged, so perpetuating greater use of management system applications.

- Research experience of management systems in use across business sectors including construction has not always been positive and a range of inhibiting factors have emerged which present challenges to be addressed in the pursuit of effective applications.

Overview

A well-formulated, carefully developed and effectively implemented management system is vitally important in successfully delivering any construction management function. Consideration must be given at corporate, directive and operational levels if both the company-wide and project application dimensions are to be ensured. Top management are charged with setting the right tone within the organisation for a system to be successful in application. They must develop, communicate and embed the holistic mission, policy, organisation and resource structure which drives the management system. Directive and supervisory management must be able to uphold the holistic vision through the application of management procedures which perpetuate the effective handling of business processes. Moreover, employees working with the processes must use the system efficiently and effectively to deliver the outputs of the business – the services, products or projects. Essential to well-configured management systems is clear and intelligible written systems documentation which describes why the system exists, what it seeks to achieve and how it is to be implemented. Key to this are well-formulated and well-written system manuals, management procedures, implementation plans and work instructions to guide system application throughout the whole organisation. In this way, the system will be understood by all employees such that greater support can be embedded throughout the entire company. Paramount to system perpetuation is the ethos to evolve the system by learning from what has taken place with the view to improving future operation. Therefore, the system must be seen as dynamic rather than static, with the continual commitment to monitor, analyse, audit and review the system. In this way, a system becomes a living part of the company's business, its processes and its management.

References

Beckmerhagen, I.A., Berg, H.P., Karapetrovic, S.V. and Wilborn, W.O. (2003). Integration of management systems: focus on safety in the nuclear industry. *International Journal of Quality and Reliability Management*, 20, (2), 210–228.

Bertalanffy, L. (1950). An outline of general systems theory. *British Journal for the Philosophy of Science*, 1, (2), 134–165.

Bertalanffy, L. (1955). An essay on the relativity of categories. *Philosophy of Science*, 22, (4), 243–263.

Bertalanffy, L. (1968). *General System Theory: Foundations, Development, Applications*. Braziller, New York.

BSI (2007). *OHSAS 18001: Occupational Health and Safety Assessment Series*. British Standards Institution, London.

Burns, T. and Stalker, G.M. (1961). *The Management of Innovation*. Tavistock, London.

Checkland, P. (1981). *Systems Thinking, Systems Practice*. Wiley, New York.

Construction Industry Research and Information Association (CIRIA) (2000). *Integrating Safety, Quality and Environmental Management*. Report C509, CIRIA, London.

Crosby, P. (1979). *Quality is Free*. McGraw-Hill, New York.

Crosby, P. (1984). *Quality Without Tears: The Art of Hassle Free Management*. McGraw-Hill, New York.

Deming, W.E. (1950). *Some Theory of Sampling*. Wiley, New York.

Deming, W.E. (1960). *Sample Designs in Business Research*. Wiley, New York.

Deming, W.E. (1982). *Quality, Productivity, and Competitive Position*. MIT Press, Cambridge, MA.

Envirowise (2010). Eco management the easy way. *Construction Manager*, March edition. The Chartered Institute of Building (CIOB), Ascot.

European Construction Institute (ECI) (1995). *Total Project Management of Construction Safety, Health and Environment*. Thomas Telford, London.

Fayol, H. (1949). *General and Industrial Management*. Pitman, London.

Feigenbaum, A.V. (1951). *Quality Control: Principles and Practice*. McGraw-Hill, New York.

Feigenbaum, A.V. (1961). *Total Quality Control*. McGraw-Hill, New York.

Fiedler, F.E. (1967). *A Theory of Leadership Effectiveness*. McGraw-Hill, New York.

Ford, H. and Crowther, S. (1922). *My Life and Work*. Garden City, New York.

Foundation for Research, Science and Technology (FRST) (1998). *Integrate: integrate project survey*. FRST, Wellington.

Gantt, H. (1919). *Organising for Work*. Harcourt, Brace and Howe, New York.

Gilbreth, F. (1911). *Motion Study*. Van Nostrand, New York.

Griffith, A. (1999). Developing an integrated quality, safety and environmental management system. *Construction Papers*, No. 108, *Construction Information Quarterly*, **1**, (3), 6–18.

Griffith, A. and Bhutto, K. (2008). Contractors' experiences of integrated management systems. *Proceedings of the Institution of Civil Engineers (ICE): Management, Procurement and Law*, **161**, (3), 93–98.

Griffith, A. and Gibson, D. (2007). *Quality Management: Quality Standards-Based Integrated Management and Systems*. Association of Building Engineers (ABE), Northampton.

Griffith, A., Stephenson, P. and Watson, P. (2000). *Management Systems for Construction*. Longman, Harlow.

Hamilton, A. (1997). *Management by Projects: Achieving Success in a Changing World*. Thomas Telford, London.

Hoyle, D. (1998). *ISO 9000 Quality Systems Handbook*. Butterworth-Heinemann, Oxford.

Imai, M. (1997). *Gemba Kaizen: A Commonsense, Low-Cost Approach to Management*. McGraw-Hill, New York.

Ishikawa, K. (1984). *Guide to Quality Control*. Asian Productivity Organisation, Japan.

ISO (2004). *ISO 14001:2004 Environmental management systems – Requirements with guidance for use*. International Organization for Standardization, Geneva.

ISO (2005). *ISO 9001:2005 Quality management systems – Requirements*. International Organization for Standardization, Geneva.

Juran, J. (1951). *Quality Control Handbook*. McGraw-Hill, New York.

Juran, J. (1955). *Case Studies in Industrial Management*. McGraw-Hill, New York.

Juran, J. (1964). *Managerial Breakthrough*. McGraw-Hill, New York.

Karapetrovic, S.V. and Jonker, J. (2003). Integration of standardized management systems: searching for recipe and ingredients. *Total Quality Management*, **14**, (4), 451–459.

Karapetrovic, S.V. and Wilborn, W.O. (1998). Integration of quality and environmental management systems. *The TQM Magazine*, **10**, (3), 204–213.

Kast, F. and Rosenzweig, J. (1973). *Contingency Views of Organisation and Management*. Science Research Associates, Chicago.

Lawrence, P.R. and Lorsch, J.W. (1967). *Organisation and Environment*. Harvard University Press, Cambridge, MA.

Marimon, F., Casadesus, M. and Heras, I. (2006). ISO 9000 and ISO 14000 standards: an international diffusion model. *International Journal of Operations and Production Management*, **26**, (2), 141–165.

Matias, J.C.O. and Coelho, D.A. (2002). The integration of the standards systems of quality management, environmental management and occupational health and safety management. *International Journal of Production Research*, **40**, (15), 3857–3866.

Mayo, E. (1949). *Hawthorne and the Western Electric Company: The Social Problems of Industrial Civilisation*. Routledge, New York.

Mendel, P.J. (2002). International standardization and global governance: the spread of quality and environmental management standards. In Hoffman, A.J. and Ventresca, M.J. (Eds), *Organizations, Policy and the Natural Environment: Institutional and Strategic Perspectives*. Stanford University Press, Stanford, CA.

Renfrew, D. and Muir, G. (1998). Quenching the thirst for integration. *Quality World*, **24**, (8), 10–13.

Renzi, M.F. and Capelli, L. (2000). Integration between ISO 9000 and ISO 14000: opportunities and limits. *Total Quality Management*, **11**, (4/5/6), 849–856.

Taylor, F.W. (1911). *The Principles of Scientific Management*. Harper & Row, New York.

Thompson, D.A. (1999). A system approach to TQM. *Manufacturing Engineering*, 78, (3), 104–106.

Townsend, P.F. (1999). Integrated management system – a key to rethinking construction? An overview of integrated management systems. In *Construction Productivity Network (CPN) Workshop Report No. E9080*. Construction Industry Research and Information Association (CIRIA), London.

Wilkinson, G. and Dale, B.G. (2001). Integrated management systems: a model based on a total quality approach. *Managing Service Quality*, 11, (5), 318–330.

Winch, G. (2002). *Managing Construction Projects: An information processing approach*. Blackwell Science, Malden, MA.

Zeng, S.X., Tian, P. and Shi, J.P. (2005). Implementing integration of ISO 9001 and ISO 14001 for construction. *Managerial Auditing Journal*, 20, (4), 394–407.

Zutshi, A. and Sohal, A.S. (2005). Integrated management system: experience of three Australian organisations. *Journal of Manufacturing Technology Management*, 16, (2), 211–232.

CHAPTER 2

Quality management systems

Introduction

Chapter 2 focuses on *quality management systems*, or QMSs. Effective quality management is a prerequisite to the business success of any construction organisation by providing products and services which directly meet the requirements of clients and customers. Chapter 2 examines the fundamentals of quality, the requirements for its control, and the systems which can be used to help and perpetuate an effective management approach. It progresses to look at international system standards and requirements, in particular ISO 9001, and explains what an organisation has to do to establish an effective quality system to support its business activities. It details the development and implementation of an appropriate QMS, discusses the assessment of the quality system by independent accreditation bodies, and looks at system applications within construction.

Fundamentals of quality management and QMSs

This section examines: the need for QMSs in relation to meeting the set of principles for ensuring effective quality management; the relevance of the systems approach to quality management; QMS standards; and complementary management approaches including the total quality management approach and business excellence models.

The need and rationale for quality management and QMSs

To survive and prosper in the business marketplace, a provider of products and services must, first and foremost, ensure that it meets the needs and expectations of its customers. Perhaps the most prominent and demanding of all business criteria is to provide products and services which truly and consistently meet their fitness for purpose. The performance of the provider and the quality of its business outputs are the most distinguished differentiating characteristics of products and services in the commercial marketplace. A successful provider is one which recognises the importance of quality to its business activities, understands the need for the proactive management of quality and puts in place the mechanisms to ensure that quality management is undertaken systematically, rigorously and continuously.

Customers want products – goods and services – which meet their exacting needs and ever-increasing expectations. These needs and expectations are defined and described in product specifications, or what are more commonly termed customer requirements. Customer requirements are normally specified in the arrangement, or contract, for the provision of products between the provider and the customer, but they can also be determined by the provider itself. As customer needs and expectations evolve as the business changes, so too do the demands placed upon the provider. Providers are constantly challenged to improve the quality of their outputs, and in so doing their organisations are driven to find better ways of undertaking their business processes.

A carefully considered, structured and systematic approach to the management of quality is a prerequisite to addressing the demands of customers for better organisational performance and quality of business outputs. A QMS can help those companies which provide goods and services to improve the quality of their provision and enhance customer satisfaction. A QMS places a company in a position to more readily identify, consider and respond to customer requirements. An effective QMS focuses on delivering better managerial, supervisory and operational performance of its organisational resources, the more efficient execution of its business processes and the highest quality for its business outputs. A QMS can best place a company in the position of seeking continual improvement to its business performance. Within a company the application of an effective QMS can engender awareness of the holistic dimension to its business activities, while outside the company a QMS can convey to the business marketplace its commitment to customer-focused quality of outputs. Therefore, there are good business reasons for any company to adopt a reasoned quality management approach, and support this with the establishment of a robust QMS.

Within the construction industry, quality is a key performance indicator by which the construction output – the final product – is judged. Quality is the judge of many dimensions within construction. It is also an indicator by which the quality of service provision to the client from consultants and contractors and the standard of products provided by suppliers to the construction project are assessed. Building designers or principal contractors, for example, can differentiate themselves from their competitors in the eyes of those clients who might employ them by the standard of service they provide.

The construction industry is diverse in so many ways, from its range of end-products to the methods by which construction work is carried out. While this diversity makes the industry so interesting and challenging, this same diversity can blunt a sharp focus of what might represent good quality. Clients expect their construction projects to be delivered within the specified timeframe and budget but they also expect a high level of quality in the finished product. Moreover, they expect an optimum balance between time, cost and quality – value for money. The intrinsic quality of a building product can be in its latent characteristics such as reliability and durability, and it can also be in overt features of function and form. Quality may be appreciated in the ease of use of a building or it can be seen in the aesthetic shape or finish of a building. In essence, quality is an ethereal sense of 'what is right' in a building or structure from the perspective of the owner or user.

Notwithstanding, quality is the mark of performance because it can be assessed against tried and tested standards which are well recognised and accepted by construction clients:

> Quality is the goodness or excellence of any product, process, structure or other thing that an organisation consists of or creates. It is assessed against accepted standards of merit for such things and against the interests/needs of products, consumers and other stakeholders.
>
> (Smith, 1993)

All organisations operating within the construction sector need to meet the quality requirements of clients, customers and stakeholders. For this reason alone, construction organisations need structures and protocols which arrange and drive their business towards meeting these needs. They need to be focused on quality, organised to perpetuate quality and managed to deliver quality. The provision of quality requires many things, but first and foremost it requires a principled and systematic approach to principles and practice.

Quality management principles

Although an aspiration, the business attribute of quality can never be guaranteed. Quality is a characteristic of product or service delivery that an organisation must desire, strive for and commit to. Quality is a facet of a company's business activities, its organisation and its resources that must be directed and managed. Quality is more than simply delivering inherent quality with a product or service. It is about demonstrating appropriate performance of the core business within the context of customer expectations. This is therefore dependent upon the whole organisation focusing on the quality agenda and all personnel contributing to its achievement. Only a holistic view of the business can embed an ethos and culture for quality, and the perceptions, attitudes and commitment of the workforce follows from this.

The achievement of quality relies upon focused and effective management, and this relies upon effective managerial leadership and control. Successful quality management can only result from clear messages on the importance of quality to the company's business and its operations. Such messages must come from top management and cascade throughout the organisation so that they permeate all its activities and its people. It must be remembered that quality is only one of a number of fundamental management functions which are required to deliver any business successfully. The achievement of quality is not so much bound up in managing quality per se, but rather in putting in place sound principles of good business and organisation such that quality becomes an intrinsic product of good management.

For effective quality management to be realised, top, or senior, management must actively and purposefully lead a company towards better organisational

Figure 2.1 Quality management principles

performance. ISO 9001 propounds a set of eight *quality management principles* which can assist management to achieve this (see Figure 2.1):

1 *Customer focus* – as organisations need customers to sustain their business, they must understand current and future customer needs and requirements and focus on exceeding the customers' expectations.

2 *Leadership* – as leaders give direction and purpose to an organisation, it is their responsibility to establish an internal environment in which people can play their part in achieving the objectives of the business.

3 *Involvement of people* – as people lie at the heart of any organisation, they should be encouraged to engage in the business and use their knowledge, skills and abilities for the betterment of the organisation.

4 *Process approach* – as results are achieved more effectively when managed as a process, so this approach should be supported for its value to the organisation.

5 *Systems approach to management* – as a system can contribute to the efficiency and effectiveness of achieving organisational objectives, so a systems approach should be adopted and actively applied to business and managerial processes.

6 *Continual improvement* – as continual improvement to the organisation's performance is a key facet of enhancing its business, it should be a permanent objective and activity.

7 *Factual approach to decision making* – as effective decisions are based on objective analysis of information, so the gathering and use of such data should be promoted.

8 *Mutually beneficial supplier relationships* – as providers and customers depend upon each other, the establishment of a mutually beneficial relationship should be encouraged to enhance the ability of both parties to create value.

This set of quality management principles underpins the ISO 9000 family of standards and forms the basis for the QMS requirements contained within ISO 9001.

Adoption of a QMS

The establishment of a QMS is a strategic business decision of the company. The fundamental basis for that decision will be made in the light of a great many business-related and organisational influences, its current circumstances and its future positioning. The ways in which a company considers, configures and applies its QMS are influenced by the following key aspects:

- the particular nature of the company's business activities and associated ethos, policy, goals and objectives;
- the size of the organisation and its workforce;
- the organisational structure and location of its facilities;
- the products, or goods and services, that the company provides;
- the mode of provision of the company's output (for example services can be provided through a project arrangement);
- the business processes that the company uses in the provision of its outputs;
- the business, organisational and wider environment including financial, political, marketplace, legislative and regulatory influences;
- the evolving business needs of the organisation.

A company that develops and implements and maintains a QMS is in a good position, strategically and operationally, to deliver the function of quality management. Moreover, a company that establishes a QMS which is independently assessed to ISO 9001 requirements will be recognised for its ability to provide its business outputs to known and consistent standards for performance and quality. Such an approach can clearly assist a provider of products and services to focus best on the needs and expectations of customers, which is crucial to maintaining market positioning. A company whose core business does not meet the requirements of its customer base and which does not provide outputs that meet the key requirement of fitness for purpose will not retain its customers – it will simply go out of business. A company which actively supports high quality of provision based on business competence, customer focus, market differentiation and value for money will, likely, gain a competitive advantage within its business sector. A QMS is an important endeavour in placing a company in the appropriate and fruitful position to serve the marketplace and prosper.

A QMS approach

For a company to develop and implement an effective QMS it must give thorough and careful consideration to key aspects of its configuration. These are:

- the needs, expectations and likely requirements of customers, clients and business stakeholders;
- the ethos, culture, policy, objectives and goals for quality within the organisation;
- the quality standard of the core business product – goods or services – supplied;
- the processes and responsibilities necessary in realising the business outputs;
- the provision of human, physical and organisational resources needed to support the quality of processes and outputs;
- the establishment and application of management procedures to plan, monitor and control business processes;
- the implementation of mechanisms to determine the efficiency and effectiveness of business processes and the performance of their management;
- the prevention of non-conformities in business output and, where they do occur, measures of detection, investigation and correction;
- the collection of data, information and records to evidence the performance of business processes and their management;
- the application of continual improvement to the system used to manage the organisation's business activities and outputs.

The process approach to systems management

It was identified in Chapter 1 that an activity or a set of activities that involves resources in its undertaking to convert inputs to outputs is termed a process model. The process model can be applied appropriately and effectively to the configuration of a QMS. The model is important to QMS development as all organisations have to handle a plethora of interrelated and interacting business processes. Many products have multiple inputs, or components, to their realisation, and many services have multiple input contributions from individuals. The output from one process often impinges upon the subsequent process and so on through to product realisation. This process approach is central to the provision of almost all goods and services, so the development and implementation of a QMS must respect its fundamental concept. The main advantage of the process approach is the continual control that it gives to the link between organisational processes and their interaction – it allows them to be managed in a coherent and holistic way. It also allows them to be closely managed to a remit of customer-focused requirements while meeting a range of organisational business requirements (see Figure 2.2).

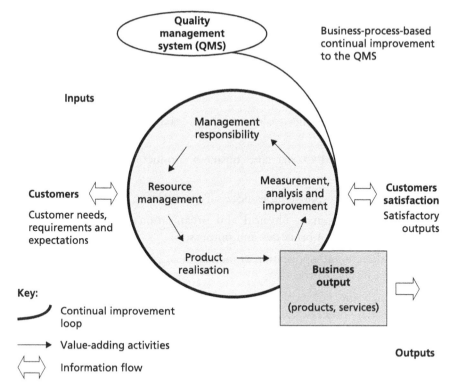

Figure 2.2 The model for a process-based quality management system (QMS)
Source: Adapted from ISO 9001

QMS and standards

A QMS may be recognised by a number of assessment, or approval, schemes depending upon its nature and application and the standards against which the system is assessed. The most prominent of all standards for a QMS is the ISO 9000 family of standards. These standards have been developed to assist companies of any nature, type and size to develop, implement, operate and maintain an effective QMS and support the continual improvement of that QMS.

The ISO family of standards (ISO, 2008) includes the following:

- ISO 9000 – this describes the fundamentals of a QMS and specifies the terminology for these systems.
- ISO 9001 – this specifies the requirements for a QMS where an organisation needs to show and evidence its capabilities to provide its product – goods or services – to meet customer requirements and enhance customer satisfaction while also meeting regulatory requirements.
- ISO 9004 – this provides guidelines that consider the efficiency and effectiveness of the QMS, aiming to improve organisational performance and the satisfaction of customers and stakeholders.
- ISO 19011 – this provides guidance on auditing quality and the QMS.

The standard is applicable to and of interest to:

- companies seeking business advantage through the application of a QMS;
- companies seeking recognition from their customers that the goods and services provided meet their needs and expectations consistently and accurately;
- companies which purchase and use products to assure them that the standard of goods and services meets their business requirements and expectations;
- any stakeholder who has an interest in, is associated with, or interacts with the quality requirements of business products;
- parties who have an involvement in the internal or external auditing of a QMS in relation to the requirements of ISO 9001;
- those who have an involvement in the assessment of a QMS in relation to the requirements of ISO 9001;
- providers and recipients of education and training for QMS development and application;
- companies which may be extending existing or establishing ISO-related management systems within their organisation.

As a collection of documents associated with the function of quality management and the systems application of quality management, they present a comprehensive and coherent set of system standards which are recognised in both the national and international business arenas. The guidance which they provide is widely recognised by companies, organisations and institutions throughout all sectors of industry, business and commerce. ISO 9001 is not a quality assurance system. It sets out standards to help a company manage quality within its organisation. It is, therefore, a management tool to be applied sensibly and creatively to addressing the challenging demands for better quality within business.

Requirements for QMS and requirements for business outputs

The requirements for a QMS and the requirements specified for the outputs, or products, of a business are different. ISO 9001 differentiates between these two requirements. It specifies the requirements for the configuration of a QMS. These requirements are generic in nature and can be applied to an organisation operating within any industrial or commercial business sector irrespective of the business outputs. Companies can be the provider of goods or services or both. Some may just act as a supplier of products while others may supply and also provide post-supply services, for example maintenance and servicing of products. ISO does not present any requirements relating to a product or a service as these will be determined by the provider and, where appropriate, other business stakeholders.

ISO 9001 relationship with ISO 9004

ISO 9001 (the requirements for a QMS) and ISO 9004 (guidelines for improving quality-related effectiveness, organisational performance and customer satisfaction) are QMS standards designed to be complementary but may also be applied independently. ISO 9001 focuses on the QMS and on meeting customer requirements. It can be used directly for putting in place the contractual arrangements between providers and customers or in relation with certification schemes where a company requires independent recognition of its QMS. ISO 9004 focuses on management for achieving long-term organisational success. It has a wider remit than ISO 9001, embracing the attributes of system effectiveness together with the performance of the organisation and the pursuit of continual improvement. ISO 9004 is not intended to be used in conjunction with business contracts, regulatory bodies or certification schemes.

Compatibility with other ISO management systems

ISO 9001 does not include requirements specific to any other management system. Therefore, environmental management systems (EMSs), health and safety management systems (H&SMSs) and indeed other systems are governed by the requirements of separate standards. Notwithstanding, ISO 9001 does facilitate the alignment of a QMS with related management system requirements. This allows an organisation, with some ease, to adapt its existing management systems to develop and implement a QMS that complies with the standard. As seen throughout this book, there are a great many similarities between management systems directed to quality, environment and safety.

Target organisations for ISO 9001 certification

A question central to the considerations surrounding the adoption of a standards-based QMS is: why would a company seek to become ISO 9001 certified? There are a good number of benefits and rewards to be secured through management system certification and these are mentioned below. The simple fact is that an ISO 9001 based QMS has become a staple part of business over the last half century and is considered to be virtually prerequisite in acquiring and conducting business – so much so that many companies, in particular the larger ones throughout all business sectors, simply cannot choose to ignore the powerful influence of quality management and the systems approach.

A company has good reason to seek ISO 9001 QMS certification if it needs to:

- be responsive to its customer base;
- avoid competitive disadvantage in its marketplace;
- secure tendering opportunities for public- and private-sector contracts;
- compete directly with ISO 9001 certified comparator organisations;
- improve organisational performance and quality of business outputs.

Benefits of ISO 9001 application

An organisation that establishes, maintains and seeks to improve its QMS continually can gain very significant benefits in the business marketplace. Moreover, deeper and valuable rewards may be accrued both intra-organisationally and externally through the association with ISO 9001, in particular where independent certification of an ISO 9001 standards-based QMS has been achieved. Some of the prominent benefits are as follows:

- *Intra-organisational*:
  - Defines policy, business objectives and goals towards quality.
  - Provides structure and organisation for quality management.
  - Facilitates operations and performance to a known quality standard.
  - Business processes and management procedures are standardised and uniform.
  - Organisational resources are directed to a defined function of management.
  - Operations gain greater efficiency and effectiveness.
  - Employees have greater awareness and understanding of the core values of the business, its processes and its outputs.
- *External*:
  - Evidence that business management and its outputs meets known and accepted standards.
  - Customer satisfaction in the product or service provision increases.
  - Enhanced reputation within a commercial sector for quality of outputs and business performance.
  - Access to public-sector contracts is assisted.
  - Pre-qualification success for private-sector contracts is improved.
  - Recognition of the company increases following certification.
  - Competitive advantage is enhanced.
  - A wider national/international business market becomes available.

It can be seen from the above points that an effective, recognised and respected QMS has many benefits to offer to an organisation.

Complementary management approaches

An ISO 9001 based QMS is the most prominent and recognised approach to quality management that an organisation can take. Notwithstanding the popularity of its use worldwide, there are alternative approaches to quality management which a company could pursue. The most well-known and recognised approach is that of *total quality management (TQM)* – a business approach focusing on the management of quality in all aspects of the organisation. Within the conceptual approach of TQM a prominent model for its deployment is *The European Foundation for Quality Management (EFQM) Excellence Model* – a framework for the development and implementation of organisational management systems to create capacity for better business performance and quality of outputs.

A description and discussion of TQM and the EFQM Excellence Model follow subsequently. An important dimension of TQM and the EFQM framework is that particular facets and aspects of their application are complementary to the establishment of an ISO 9001 based QMS. In broad terms the direction and aspirations of TQM marry with ISO 9001 such that aspects of good practice within each are useful and interchangeable. The significant difference is that TQM is a management concept and the EFQM framework an application tool, whereas ISO 9001 is a standard against which a QMS can be developed and implemented and then independently assessed and certificated.

TQM approach

TQM is a business approach focusing on the management of quality in all aspects of an organisation. TQM is not a management system and it does not postulate a business or organisational process or even present a management tool – TQM is a philosophy (Griffith *et al.*, 2000). For a company, TQM suggests a way of thinking about how a company develops and implements objectives, organisation, business processes and human resources. The focus is on how these elements, and many more, can be configured and applied to improve organisational performance and standards of quality in a holistic way. TQM takes a whole-organisation perspective. It is a management approach focusing on perpetuating quality based on sound business processes, the contribution of people, and the use of management systems, all the while closely meeting customer expectations (see Figure 2.3).

The evolution of management thoughts and theories including those underpinning the philosophy of TQM was examined in Chapter 1 of this book. Management theorists including Deming (1950), Juran (1964), Crosby (1979), Feigenbaum (1979), Ishikawa (1984) and Imai (1997) have all been influential in postulating advances in the approach to managing quality. The family of ISO 9000 standards considered and distilled the thoughts and philosophies of these management thinkers into the key principles of quality management described earlier: customer-focused organisation; leadership; involvement of people; process approach; system approach to management; continual improvement; factual approach to decision making; and mutually beneficial supplier relationships.

Government support

Government support in the UK and, indeed, in other parts of the world has encouraged the application of TQM generally within economic and business sectors. Initiatives have seen the growing application of TQM where standards of service, in particular those associated with public services, have improved without a corresponding cost to the taxpayer. Support for TQM has assumed considerable prominence within UK central government departments associated with business and commerce over recent years, an example being the Department of Trade and Industry (DTI) promotion *From Quality to Excellence* (DTI, 2009).

TQM and business advantage

The adoption of an effective TQM perspective can achieve the following:

- Encourage an organisational culture of quality and performance.
- Make a company more competitive.

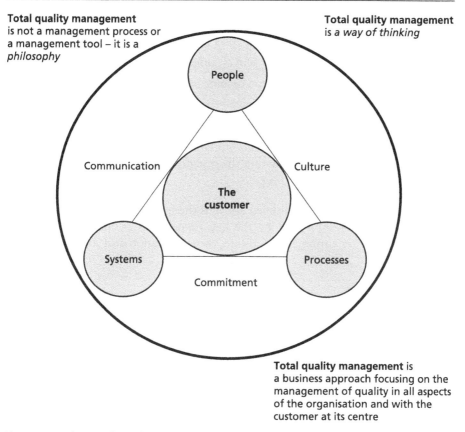

Total quality management
is not a management process or
a management tool – it is a
philosophy

Total quality management
is *a way of thinking*

People

Communication

Culture

The
customer

Systems

Processes

Commitment

Total quality management is
a business approach focusing on the
management of quality in all aspects
of the organisation and with the
customer at its centre

Figure 2.3 The total quality management (TQM) approach
Source: Adapted from DTI (2009)

- Help a company survive, grow and prosper.
- Make a company more outward, commercial and customer focused.
- Build business links with clients, customers and supply chain stakeholders.
- Perpetuate a positive working environment where success is championed naturally.
- Ensure better business processes, lower costs and higher-quality outputs.
- Focus management procedures on continually improving quality standards.
- Reduce non-conformity of outputs to requirements and reduce wasteful practices.

Business drivers

TQM can drive a company to improving its business approach in the following ways:

- The organisation can gain a long-term commitment to continual improvement.
- The philosophy of zero defects, or right-first-time, can be embedded.

- Employees understand the importance of the provider–customer relationship.
- Procurement and purchase to total costs and added value is encouraged.
- The realisation that systems need active, competent and dedicated management.
- Education, training and supervision become intrinsic to organisational activities.
- Departmental barriers can be eliminated through effective process management, good communication and teamwork.
- Developing actions based on analysis of process and facts gained from evidence.
- The development of experts within the organisation.
- A systematic approach to TQM.

The list of drivers for TQM outlined above is the foundation of its development and implementation supported by the key elements of people, processes and systems within an environment of culture, communication and commitment (see Figure 2.4).

Implementing TQM

For almost all companies, TQM will necessitate change, perhaps a radical paradigm shift in organisational thinking and culture. The required volume and depth of change can be less threatening and less dysfunctional if there is a distinctly identifiable vehicle for the deployment of the TQM philosophy – this can be a QMS. The QMS focus allows organisational change to take place and embed in planned and measurable steps over a developmental timeframe. Within this, employees can, in an informed way, embrace the various stages of system implementation and see and play their part in those stages.

Quality initiatives can be less than successful where organisational culture is not given due respect and consideration. The culture for TQM, as with any management system development, is founded in the beliefs and values of the company set by top management. It is therefore the role of executive and senior

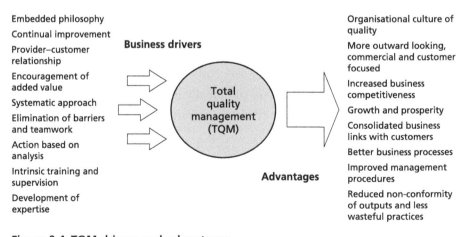

Figure 2.4 TQM drivers and advantages

management to give leadership to TQM establishment and engender the support of all their employees.

The fundamental building blocks for TQM are the business processes and the process model of their constitution. In this respect, TQM development is no different from the ISO 9001 QMS requirements discussed in this part of the book.

To embark on the road to TQM establishment, the company should consider the following key questions:

- What is the company's core business?
- What is the ethos and vision for the business?
- Where is the organisation now and where does it want to get to and why?
- What is the business mission?
- What factors are influential to achieving the mission?
- What aspects might impede the mission?
- What are the company's values, objectives, goals and targets?
- What resources are available to deploy to the task?
- How is the company going to structure and organise for the task?
- What is the timeframe for completion of the mission?

The crucial issue in implementing TQM is that it is not a quick fix for the problems surrounding quality management experienced by many companies. Since TQM is a philosophy, it is not time constrained in the same way that a process can be. Therefore, the developmental process is as short or as long as it needs to be to fulfil the task. The establishment of a QMS within the TQM approach can be helpful, however, as the QMS can represent a defined achievement milestone.

TQM and business excellence models

Models such as the EFQM approach are ostensibly applied to determine business excellence among participating companies and organisations. Business excellence is a result of applying the TQM philosophy, and outstanding achievement in this respect is recognised by the receipt of an award for quality. Therefore, excellence models measure performance rather than providing a philosophy. That said, the EFQM framework is a good and well-recognised vehicle for the deployment of the TQM philosophy. It can be applied purely to deploy TQM, with the award-related aspect assuming a secondary and optional dimension. Much of the EFQM Excellence Model can be used to address the pursuit of TQM, and specific aspects may assist an organisation in its preparations for ISO 9001 QMS establishment.

EFQM Excellence Model

The EFQM Excellence Model is a framework for the development and implementation of organisational management systems devised by the European

Foundation for Quality Management (EFQM). The framework is designed to assist companies to enhance their organisational capabilities to deliver better quality of output provision and encourage their drive towards being more business capable and competitive. The purpose of the model is to provide a self-assessment framework for measuring the strengths of organisational activities throughout a company and highlighting areas for improvement. The term excellence is used as the model focuses on the core business of a company and seeks excellence in the provision of its business outputs – services and products. The application of the model allows a company to identify ways to make significant and lasting improvement to how it delivers its provision to the commercial marketplace. This can lead to greater clarity of policy and business purpose, more efficiency in the way organisational activities are undertaken and better quality of service to clients and customers. The success of the model also lies in the way that it can make companies think about the ways in which it does things with a view on wholeness and its holistic perspective to organisational structure, organisation and operation (BQF, 2009).

The EFQM Excellence Model was introduced in the early 1990s as a framework for assessing applications for the UK Excellence Award. It is one of the most widely used organisational development frameworks in Europe and forms the basis for many national and regional business excellence awards. Evolution of the model has changed the focus of its application over the last decade. While the pursuit of excellence awards to support business profile and activity has been the focus of some companies, for others the use of the model has been to support the self-assessment of organisational activity as part of the development, implementation and maintenance of organisational management systems.

Self-assessment of business activities, the processes involved in their undertaking and their management applies to companies large and small throughout and across many industrial and commercial sectors. Although the origins of the model lie in the private sector, any organisation within the private, public and voluntary sectors can apply its concepts and principles to good effect. The model is not prescriptive and it does not follow any set standard or rules of application. Rather, the model provides a generic, structured and sensible set of expectations about what constitutes an effective organisation and its management. Since the model is not prescriptive, a company can use it in its own way with a focus on its own activities and how they can be managed with greater effectiveness as part of organisational continual improvement.

The model is primarily intra-organisational, helping a company to improve its own business activities. The model is therefore not an external evaluator. Notwithstanding, the practice of self-assessment makes an excellent contribution to companies seeking to develop certificated standards-based management systems. The model represents a practical tool to assist the development of management systems. It allows an organisation to assess and measure where it is on the track to organisational and management excellence. It helps the organisation to understand existing gaps and weaknesses in what it does and promotes the consideration of an appropriate and effective response.

The model is not a standards-based approach in the way that quality systems are related to ISO 9001. An organisation does not have to meet set specifications and clauses influencing its practices. The model provides a tool for the strategic and operational mapping of the organisation leading to a quantitative assessment of performance, quality and excellence. Many organisations simply do not know how well they perform in organisational and business terms. Many do not know how their performance stands next to that of their competitor or comparator companies in the marketplace. Poor or underperforming companies often have management frameworks and systems which are inadequate or inappropriate to the business that they undertake. The model gives such organisations the opportunity to assess their own situation and also their frame of operation against other organisations.

High-performing companies have organisational structures, management frameworks and management systems which clearly define and describe what they do and why they do it. These are the types of companies that have adopted the tenet of the EFQM approach, have applied the model and are reaping the benefits from it. Furthermore, many such companies have used the EFQM framework as the basis of preparation for establishing ISO 9001 quality systems and associated standards-based management systems.

Use of the EFQM Excellence Model

The model is used by companies in many sectors, including construction, to assist in the following:

- The application to self-assessment of organisational strategy and operations in relation to quality of outputs and business competitiveness.
- The identification of organisational areas for improvement.
- The establishment of structure to support the development and implementation of management systems appropriate to intra-organisational applications and in preparation for the establishment of standards-based certificated management systems.
- The provision of a framework for inter-organisation comparison of performance and business excellence.

In addition, the focus on continuous improvement within the model encourages a range of broader perspectives by the company on the way in which it undertakes its activities. These include:

- The encouragement of whole-organisation thinking and holistic management.
- A greater clarity between business strategy, organisational processes and their management and the resulting outputs.
- A stronger focus on measuring performance, evaluating and improving on results.
- The importance given to interface management with clients and customers.
- The encouragement of employee engagement and involvement to drive improvement.

- The drive in organisational development towards achieving holistic business success.

As the use of the model is not prescriptive it allows a company to:

- map the model to the specific needs of the organisation;
- develop a holistic approach to organisation and management which can be incorporated with standards-based management systems such as ISO 9001;
- collect organisational information for self-assessment in a systematic, thorough, efficient and cost-effective way;
- establish a customised and dedicated framework of organisational performance measurement that encourages continuous improvement towards excellence;
- recognise the demands that the organisation will likely face in the future.

The model can be applied in four key ways:

1 As a framework which a company can use to develop its business vision, structure, approach and management by determining aims, objectives and goals in an intelligible and measurable way.

2 As a framework which a company can use to help recognise and understand the systemic and holistic nature of its business and the relationship to its environment and other organisations.

3 As a diagnostic tool for determining the existing position and health of the organisation with a view to improving its business goals, priorities and resources.

4 As a basis for the EFQM Excellence Award, allowing the best practice of Europe's most successful companies to be shared and promoted.

The model is continually updated by the EFQM, which is based in Brussels. The focus of the EFQM establishment is: to ensure that companies across Europe maintain an up-to-date awareness of the model and its application; to promote appropriate and widespread use of the model; and to ensure that there are consistent organisational practices used throughout Europe. These supporting attributes of the model are a prerequisite to self-assessment and benchmarking, and are important where companies apply for certification of their organisational management systems to accredited recognition and certification schemes.

Essence of the EFQM Excellence Model

The essence of the model is that it presents a non-prescriptive framework based on nine criteria. Five of these criteria are considered to be 'enablers' – *what the organisation does*, while four criteria are considered to be 'results' – *what the organisation achieves*. The results are influenced by the enablers, and feedback from the results helps to improve the enablers. In this way continuous improvement in quality and standard of organisational performance is achievable. The model recognises that there are different ways to achieve excellence in all aspects

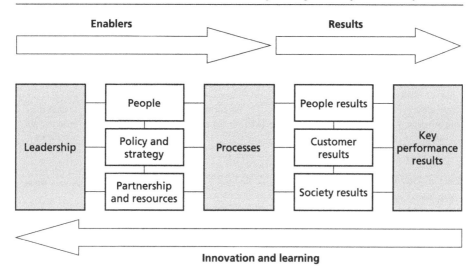

Figure 2.5 The European Foundation for Quality Management (EFQM) Excellence Model

Source: Adapted from the British Quality Foundation (BQF) EFQM Excellence Model: www.bqf.org.uk

of organisational performance. Notwithstanding, it does propound that excellent results are based directly on actively driving the enablers (see Figure 2.5).

Self-assessment of the nine criteria leads to a measurable grade of performance for each enabler and results criterion which when combined provide an overall quantitative score of organisational excellence. The higher the overall score (up to the maximum of 1000), the higher the level of excellence achieved. The overall score represents a European-wide indicator of performance such that comparison with other organisations can be made. Because the model is metric, or measurement, based, it is amenable to use alongside similarly structured models, tools and standards such as the *Balanced Scorecard*, the UK government's *Charter Mark* and *the Investors in People (IiP) standard* (Kaplan and Norton, 1996; IiP, 2009; UK Government Cabinet Office, 2009). See Table 2.1.

ISO 9001 Quality Management Systems – Requirements

This section examines the requirements, or set of clauses, for the development, implementation and maintenance of a QMS.

The requirements

The *requirements* for the development, implementation and maintenance of a QMS meeting the ISO 9001 standard are presented in this section. Requirements are presented within the standard as a set of clauses, with sub-clauses, accompanied by two informative annexes and a bibliography. For the precise wording, terminology, description and explanation of the requirements the reader is directed to

Table 2.1 Simple application of EFQM performance scoring

EFQM (weighted) performance criteria		
	Enablers:	**Results:**
Enablers:	1 Leadership (10%)	6 Customer results (20%)	**Results:**
	2 Policy and strategy (8%)	7 People results (9%)	
What the	3 People (9%)	8 Society results (6%)	What the
organisation	4 Partnerships and resources (9%)	9 Key performance results (15%)	organisation
does	5 Processes (14%)		achieves

- The nine performance criteria are those against which the organisation's progress towards excellence is assessed
- Each criterion is supported by a number of sub-criteria which pose questions that need to be considered in the organisation's assessment of its business activities
- Each of the sub-criteria are assessed for performance (in %) within a five-band scale ranging from 0% (no evidence of excellence) to 100% (comprehensive evidence)
- The results for each of the sub-criteria are added together and divided by the number of sub-criteria to award a 'score' for each of the nine performance criteria

The score for the sub-criteria and hence the criterion (e.g. leadership) are calculated as follows:

1 Leadership

Sub-criterion 1a (Involvement with quality mission)	45%
Sub-criterion 1b (Involvement with system development)	40%
Sub-criterion 1c (Involvement with customers)	45%
Sub-criterion 1d (Involvement with employees)	50%
Total	180
÷ 4 (4 being the number of sub-criteria)	
= Score for the criterion 'Leadership' of:	45

The scores for the nine criteria are weighted (to the EFQM method, e.g. leadership 10%) and 'points' awarded in a table to produce the 'Total Points' awarded as follows:

1 Leadership	45	× 1.0	=	45
2 Policy and strategy	50	× 0.8		40
3 People	55	× 0.9		50
4 Partnerships and resources	55	× 0.9		50
5 Processes	55	× 1.4		77
6 Customer results	60	× 2.0		120
7 People results	60	× 0.9		54
8 Society results	70	× 0.6		42
9 Key performance results	70	× 1.5		105
Total points awarded:				578

With a points total of 578 the organisation would fall short of the EFQM threshold for excellence of 750, but this performance demonstrates that the organisation is well on the way to achieving excellence in its quality management approach and would likely meet the desired level of performance in the future.

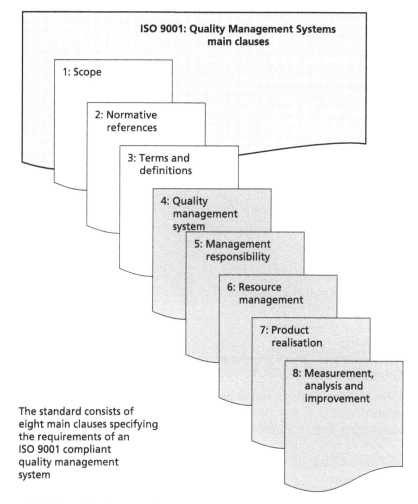

Figure 2.6 The main clauses of ISO 9001

the standard itself. In this section, reference is made to the requirements in the form of statements for use and considerations and questions which are posed to an organisation when seeking to establish a QMS in accordance with ISO 9001.

The contents of the requirements are as shown below. For clarity and consistency of reading, understanding and interpretation, the enumeration of clauses is the same at that used within the standard (see Figure 2.6). The key, or main, clause headings are highlighted in bold.

1   Scope
1.1   General
1.2   Application

2   Normative references

3   Terms and definitions

1 Scope

1.1 General

ISO 9001 specifies the requirements for a QMS where a company seeks to:

I. Demonstrate its ability to provide its business outputs, or *product*, that meets customer, statutory and regulatory requirements.

II. Enhance the satisfaction of its customers through the effective implementation of a QMS and perpetuate continual improvement of that system together with the assurance of conformity to customer, statutory and regulatory requirements.

The standard uses the term *product* to describe the outputs from the business of a company where the product is intended for a customer and is an output from a product realisation process. In practice, a product includes a service, products, or goods, and/or both where they are provided within the arrangement of a project.

1.2 Application

The requirements of ISO 9001 are generic and therefore can be applied to all organisations, irrespective of the type of business outputs and the type, size and configuration of the organisation. As the nature and business of companies can be different and wide ranging, there may be parts of the standard which are not wholly applicable. Clause 7: Product Realisation is the principal section of the standard where this is likely to occur. In the case of Clause 7 specific requirements may be excluded from the needs for compliance when considering the development of the QMS. The reasons why exclusion can apply are discussed subsequently in the next section.

2 Normative references

Normative references are references to documents associated with the main document – ISO 9001 QMS Requirements – such as those explaining the fundamentals of system development and implementation and standard vocabulary used within the documents.

3 Terms and definitions

This clause confirms that the terms and definitions used in the standard apply throughout the whole text. It also confirms that the term *product* can also mean *service*.

4 Quality management system

4.1 General requirements

The essence of Clause 4.1 is to identify and describe those processes involved in the QMS. The key questions to be asked are: has the organisation established, documented, implemented and maintained the QMS; has continual improvement given greater effectiveness to the QMS; and is the QMS configured in accordance with the standard? The organisation is required to:

a) identify the processes needed for the QMS and apply them throughout the organisation;

b) determine the order and interrelationships of the processes;

c) consider the criteria and methods involved in the effective operation and control of the processes;

d) allocate and deploy the resources and information required to support the effective operation and control of the processes;

e) undertake monitoring, measurement and analysis of the processes in application;

f) take actions to achieve planned results from the application of processes and ensure continual improvement of the processes.

A company may choose to outsource its business processes. Where this is the case, it must ensure that conformity of output is maintained through applying appropriate quality control mechanisms. Such arrangement should be identified and described in the QMS documents. Consideration should determine the impact of the outsourced process upon the company's ability to provide the output to the requirements, the respective responsibilities where the control of processes is shared by the company and the outsource party, and the capability of achieving required control through the purchasing process.

4.2 Documentation requirements

4.2.1 General

This clause focuses on the requirements placed upon the organisation when considering the documentation for the QMS. A number of key questions are posed – does the QMS have:

a) a quality policy and set of objectives;

b) a quality manual;

c) the documented procedures required by the ISO 9001 international standard;

d) the documents required by the organisation to ensure that planning, operation and control of the quality processes are effective and that records are maintained?

When configuring and writing the QMS documents, a single document may be used to meet the requirements for more than one procedure. Also, a requirement for a documented procedure may be embraced within a number of documents. Because a QMS will differ according to the company's own needs, the extent and precise nature of the documentation will vary. In this regard, the company should consider the core business of the organisation and its size, the complexity and interactions of its business processes and outputs, and the knowledge and skill bases of its human resources.

4.2.2 Quality manual

The standard requires that the organisation will establish and maintain a quality manual. This is the principal and overarching document of the QMS and forms the base document for all other documents associated with the management system. The manual should contain the following:

a) the scope of the system including details and justifications for any exclusions from the requirements of Clause 7: Product Realisation;

b) the documented procedures or references to other documents containing the procedures;

c) a description of the interaction between the business processes and the QMS.

4.2.3 Control of documents

The salient question that should be answered for this sub-clause is: are the documents which underpin the establishment and maintenance of the QMS controlled? The system should have documented procedures which ensure that:

a) documents are approved for accuracy before they are applied;

b) documents are reviewed and updated, and re-approved before application;

c) changes to documents and their revision status are identified;

d) relevant versions of documents are available for use;

e) documents are legible and identifiable;

f) documents from external origins are identified and their distribution is controlled;

g) obsolete documents are identified, that their use is prohibited, and where they are retained that this is identified.

4.2.4 Control of records

The QMS is required to have records which provide objective evidence that the system conforms to requirements and operates effectively. The key questions are:

• has QMS record-keeping been established and are records controlled;

• is a documented procedure in place to define the controls required for an appropriate records system;

• are records legible, identifiable and retrievable;

• do the records provide evidence of conformity of the system and its effective application?

5 Management responsibility

This set of clauses within the standard has as its focus the commitment of executive and senior management of an organisation to the QMS. It considers how the organisation orientates itself towards meeting customer, legislative and regulatory requirements and how the QMS can benefit from continual improvement.

5.1 Management commitment

The requirement for management commitment asks the organisation if it has:

a) communicated the importance of customer, legislative and regulatory requirements within the company structure and organisation and to its personnel;

b) established the company's quality policy;

c) determined the quality objectives to fulfil the policy;

d) conducted management reviews;

e) provided for the availability of resources.

5.2 Customer focus

The focus of this sub-clause is the consideration of customer requirements. Emphasis should be given to:

- determining customer needs and requirements;
- meeting these expectations;
- delivering overall customer satisfaction.

5.3 Quality policy

The requirement is directed to ensuring that executive and senior management considers the organisation's quality policy and its fitness for purpose. It should ask if the quality policy is:

a) appropriate to the business of the organisation;

b) committed to complying with the requirements and the tenet of continual improvement;

c) supportive to the establishment and review of quality objectives;

d) communicated within the organisation and understood by personnel;

e) reviewed for its currency and continued suitability to application.

5.4 Planning

5.4.1 Quality objectives

The executive and senior management of the organisation are required to ensure that the quality objectives propounded by the company are embedded within the various management functions and at the various management levels within the organisation. The quality objectives determined must be:

- consistent with and underpin the quality policy;
- measurable by objective mechanisms.

5.4.2 Quality management system planning

Senior management are responsible for ensuring that the QMS is appropriately planned such that:

a) the QMS meets the general requirements specified in Clause 4.1 together with the specified quality objectives;

b) the integrity of the QMS is maintained when system changes are planned and applied.

5.5 Responsibility, authority and communication

This clause emphasises the arrangement within the organisation that establishes the devolvement and distribution of responsibilities, lines of authority and routes of communication.

5.5.1 Responsibility and authority

Executive and senior management are charged with ensuring that the assignment of responsibilities and lines of authority are appropriate, clear and communicated within the organisation.

5.5.2 Management representative

A requirement of the QMS is that executive management should appoint a 'management representative' with the assigned authority to:

a) ensure that the processes necessary to the effective functioning of the QMS are established and maintained;

b) report to executive management on the performance of the QMS together with identifying any needs for system improvements;

c) promote awareness of customer requirements throughout the whole organisation.

5.5.3 Internal communication

The company should ask itself if it has:

- established appropriate communication processes within and throughout the organisation;

- communication mechanisms appropriate to determining the effectiveness of the QMS.

5.6 Management review

5.6.1 General

Executive and senior management are required to review the QMS in ways appropriate to its suitability, adequacy and effectiveness. Therefore, the pertinent questions to be asked are:

- are reviews planned and at what frequency are they conducted;

- does a review consider the opportunity for improvement and need for changes;

- does a review embrace the quality policy and quality objectives of the organisation?

5.6.2 Review input

This sub-clause considers the inputs to the management review process and should incorporate information concerning the following:

a) the results from QMS audits;

b) feedback from customers;

c) the performance of processes and the conformity of business outputs;

d) the status of preventive and corrective actions;

e) follow-up actions from preceding management reviews;

f) changes to any aspect that might affect the QMS;

g) management recommendations for improvement to the QMS.

5.6.3 Review output
The output from a management review should consider all decisions and actions in connection with:

a) improvements to the QMS and their effectiveness;

b) improvements to the product related to the requirements of the customer;

c) resource requirements of the QMS.

6 Resource management

6.1 Provision of resources

This clause focuses on the determination and provision of resources needed for supporting an effective QMS. The pertinent question to be asked is: has the organisation confirmed resources commensurate with:

a) implementing, maintaining and continually improving the QMS;

b) enhancing customer satisfaction by meeting their requirements?

6.2 Human resources

6.2.1 General
The conformity of business outputs to requirements can be influenced by the human resources contributing to input tasks of the processes involved. Therefore, all personnel fulfilling tasks within the QMS should be knowledgeable, capable and competent as a result of their education, training, skills and experience.

6.2.2 Competence, training and awareness
This sub-clause requires that the company asks key questions in respect of the qualities of personnel involved with the QMS. It should be asked if personnel have:

a) necessary competence in their work as it impinges upon the requirements for product conformity;

b) had appropriate training to achieve the right level of competence;

c) been evaluated to ensure the effectiveness of the measures taken;

d) been made aware of the importance of their work in relation to underpinning the fulfilment of quality objectives;

e) had their education, training, skills and experience logged by the QMS record-keeping process.

6.3 Infrastructure

It should be asked if management has identified, provided and maintained the physical infrastructure, or facilities, needed to support the QMS and its ability to achieve the conformity of product. Infrastructure includes:

a) buildings and work places, workspace and supporting utilities;

b) process equipment;

c) supporting services, for example transport and communications.

6.4 Work environment

There are requirements placed upon a company to identify, provide and manage the work environment in relation to its capability to achieve conformity of product.

7 Product realisation

The exclusion of the Clause 7 sections of the standard is permitted according to the explicit nature of a company's business, its processes and its product, or outputs. An organisation is required to plan and develop those processes needed to realise the product.

7.1 Planning of product realisation

In planning for the realisation of its business outputs, a company should ask:

a) what are the quality objectives and requirements for the product;

b) are there appropriate processes, documents and resources specific to the product in place and active;

c) how will the organisation verify, validate, monitor, measure, inspect and test those activities explicit to the product, and what are the criteria for acceptance;

d) what records are needed which can provide evidence that the realisation processes and product meet the set requirements?

7.2 Customer-related processes

7.2.1 Determination of requirements related to the product

The emphasis of this clause is to confirm the needs of the customer and other significant influences. To fulfil this, the company should determine the following:

a) the requirements specified by the customer, including those applicable to delivery and post-delivery of the product;

b) requirements which are not stated by the customer but are significant to the use of the product by the customer;

c) all statutory and regulatory requirements that impinge upon the product;

d) any requirements seen to be necessary by the organisation.

7.2.2 Review of requirements related to the product

The organisation should review the requirements relating directly to the product and should undertake this before it committing to its provision. In so doing, it will ask the following questions:

a) have the product requirements been defined;

b) have requirements differing from those previously stated been considered;

c) does the organisation have the ability to fulfil the requirements as defined?

7.2.3 Customer communication

The salient question here is: has the company identified appropriate ways for communicating with its customers? It may need to communicate with customers in respect of the following:

a) information in relation to the product;

b) enquiries, contracts and ordering processes;

c) customer feedback, including complaints from customers and users.

7.3 Design and development

7.3.1 Design and development planning

The requirement of the standard is that the company must control the design and development of the product. It should do this by determining:

a) the stages of design and development;

b) the methods of review, verification and validation appropriate to each stage;

c) the responsibilities and authorities associated with design and development.

7.3.2 Design and development inputs

The following questions should be asked in relation to the inputs to design and development:

a) have the requirements for product function and performance been determined;

b) have applicable statutory and regulatory requirements been identified;

c) has information from similar previous designs been considered;

d) are there any other requirements essential to the product's design and development?

7.3.3 Design and development outputs

The outputs from design and development should:

a) meet the input requirements;

b) provide information appropriate to purchasing, production and service provision;

c) confirm product acceptance criteria;

d) specify characteristics of the product essential to appropriate, proper and safe use or application.

7.3.4 Design and development review

A company should carry out systematic reviews of product design and development. These should be conducted to planned arrangements and scheduled intervals. Such reviews will:

a) evaluate the ability of the results to meet the requirements;

b) identify issues and problems and suggest appropriate actions.

7.3.5 Design and development verification
A verification mechanism needs to be devised and administered to ensure the following:

- that the design and development outputs have met the input requirements;
- that records of the verification and subsequent actions are maintained.

7.3.6 Design and development validation
A validation mechanism also needs to be devised and administered to ensure that:

- the resulting product is capable of meeting the requirements for application or intended application;
- validation takes place prior to delivery or use of the product;
- results of the validation are maintained.

7.3.7 Control of design and development changes
This sub-clause requires that the following questions are asked and considered:

- have any changes been identified and recorded;
- have changes been reviewed, verified and validated before they are implemented;
- has the review included any effects of changes on products already delivered and in use;
- have records of the review of changes and associated actions been maintained?

7.4 Purchasing

7.4.1 Purchasing process
The company has an obligation under this sub-clause to ensure that any purchased products meet with the purchase requirements specified. A control mechanism needs to be established which identifies and chooses suppliers based upon their capability to supply products which conform to the requirements. To achieve this, a set of criteria will be determined against which products can be evaluated. As with other aspects of the QMS, records should be maintained, providing evidence of an effective process.

7.4.2 Purchasing information
Information relating to purchasing should describe the product being procured together with the following:

a) requirements for approval of the product and also procedures, processes and equipment;

b) requirements for the qualification of personnel;

c) requirements for the QMS.

7.4.3 Verification of purchased product

The following questions are pertinent to meeting the requirements for verification of the purchased product:

- has the company put in place procedures for inspection to ensure that the purchased product meets the purchase requirements;
- has the company specified verification procedures to be undertaken at the supplier's premises;
- has the company established formal product release within the verification arrangements?

7.5 Production and service provision

This sub-clause requires the organisation to establish and maintain controlled conditions when undertaking production and service activities.

7.5.1 Control of production and service provision

The requirements for controlled conditions are as follows:

a) making available information that describes the characteristics of the product;

b) making work instructions available;

c) using suitable equipment;

d) making available monitoring and measuring equipment;

e) implementing monitoring and measurement;

f) implementing product release, delivery and post-activity activities.

7.5.2 Validation of processes for production and service provision

The salient questions which should be asked are:

- can the company validate production and service processes where the output cannot be verified by monitoring and measurement;
- if deficiencies are apparent after the product is in use, is there a process for handling such occurrence;
- does validation confirm the effectiveness of processes to give the results as planned?

The company must establish arrangements for these processes as follows:

a) by defining criteria for review and approval;

b) by approving equipment and qualifications of personnel;

c) through the application of specific methods and procedures;

d) by keeping appropriate records;

e) through re-validation.

7.5.3 Identification and traceability

The requirements of this sub-clause are:

- to identify the product throughout the product realisation stage;
- to identify the status of the product throughout the realisation stage using appropriate monitoring and measurement;
- to give unique identification to a product where traceability is needed;
- to maintain records to support traceability.

7.5.4 Customer property

The company is charged with protecting the property of customers while it is under the company's control or being used by the company. The organisation must therefore identify, verify, protect and safeguard such property as it is incorporated into production processes for inclusion in the provision of the product. Should property be damaged, the company must have a mechanism for reporting this to the customer, and appropriate records should be maintained.

7.5.5 Preservation of product

The company has an obligation to preserve the product during processing and delivery to the customer to maintain conformity of the product. Preservation includes identification, handling, packaging, storage and protection activities, and applies to complete and partial products within the arrangement between the provider and the customer.

7.6 Control of monitoring and measuring equipment

This clause requires the company to consider the monitoring and measurement to be carried out and the equipment needed to provide objective evidence of conformity of product to requirements. The organisation must ask if it has established processes to ensure that monitoring and measurement is carried out consistent with the requirements set for those activities.

To give valid outcomes, all equipment should be:

a) calibrated and verified to traceable national/international standards, and where no such standards exist then the basis of calibration and verification should be recorded;

b) adjusted or re-adjusted as required to give correct results;

c) identifiable so its calibration status is known and related to the equipment;

d) safeguarded from adjustment that might invalidate the measurements obtained;

e) protected from damage during handling, maintenance and storage.

Records should be maintained of:

- calibration and verification;
- validity of previous measuring results where equipment is seen not to conform to requirements;
- actions taken in response to non-conformity of equipment.

8 Measurement, analysis and improvement

8.1 General

The requirements under Clause 8 of the standard charge an organisation with planning and implementing the monitoring, measurement, analysis and improvement processes which:

a) demonstrate conformity to product requirements;

b) ensure conformity of the QMS;

c) continually improves the effectiveness of the QMS.

8.2 Monitoring and measurement

8.2.1 Customer satisfaction

The key questions posed from this sub-clause are: does the company monitor information on the perceptions of its customers; has the organisation established appropriate methods for obtaining and utilising this information; and has the data necessary to assessing customer satisfaction been collected?

8.2.2 Internal audit

This sub-clause requires the company to carry out internal audits to determine if the QMS:

a) conforms to arrangements planned by the company, meets the requirements of the standard, and meets any other requirements set by the organisation;

b) is implemented and maintained effectively.

Arrangements for internal auditing pose a number of questions, as follows:

- are audits planned and scheduled appropriately;
- do audits consider the status and importance of the processes and procedures to be audited;
- have the scope, frequency and methods been determined;
- are the audits objective and conducted by persons independent to the processes being audited;
- has an appropriate audit procedure been configured, with responsibilities clearly assigned and communicated within the organisation;
- are the results reported to executive and senior management;
- are records maintained of audits carried out;
- are post-audit actions identified, implemented and documented?

8.2.3 Monitoring and measurement processes

The company is required to implement suitable methods for monitoring and measuring the QMS processes and comparing against planned expectations. Moreover, where planned results are not apparent, the organisation must have procedures to invoke corrective actions.

8.2.4 Monitoring and measurement of product

The pertinent questions to be asked in response to the requirements of this sub-clause are:

- does monitoring verify that product requirements have been met;
- is there documented evidence of conformity to set requirements;
- are records kept of product release to the customer;
- has product release been commensurate with planned activities and approvals?

8.3 Control of non-conforming product

The company must ensure that any product which does not conform to the set requirements is identified and controlled. The focus is upon preventing unintentional delivery or use of the non-conforming product by the customer. A procedure defining such control mechanisms is required, together with the clear assignment of responsibilities and authorities for any such occurrence.

Non-conforming products can be managed by:

a) taking action to eliminate the non-conformity;
b) releasing the product and authorising use under concession by the provider and/or customer;
c) taking action to prevent its use by the customer;
d) taking action to handle the effects of non-conformity after delivery or during use of the product.

8.4 Analysis of data

This clause requires the organisation to identify, gather and analyse data to determine the effectiveness of the QMS in operation. The emphasis is placed upon identifying where continual improvements to the system can be made and what information is needed to support the considerations underlying this. Data and its analysis should be provided in connection with:

a) customer satisfaction;
b) conformity to product;
c) characteristics of processes and products;
d) suppliers.

8.5 Improvement

8.5.1 Continual improvement

Continual improvement is a tenet of the ISO 9001 approach to QMS development and implementation. The company should perpetuate continual improvement to the QMS through use and evolutionary enhancement of:

- quality policy;
- quality objectives;
- audit results;

- analysis of data;
- corrective and preventive actions;
- management review.

8.5.2 Corrective action

This sub-clause requires the organisation to eliminate the causes of product non-conformities and execute corrective actions where they do occur. Documented procedures are required for this as follows:

a) review of non-conformities;

b) determination of cause;

c) evaluating the need for action to prevent re-occurrence;

d) configuring and implementing preventive actions;

e) recording of results from actions implemented;

f) review of effectiveness of actions.

8.5.3 Preventive action

The organisation must also identify potential preventive actions for managing non-conformities of product to requirements. Again, there should be documented procedures within the QMS as follows:

a) determining potential non-conformities and their causes;

b) evaluating the need for action to prevent their occurrence;

c) configuring and applying actions;

d) recording of results from actions implemented;

e) review of effectiveness of actions.

This section of Chapter 2 has presented the requirements of ISO 9001 – the international standard for the development, implementation and maintenance of and improvement to a QMS. The next section looks in detail at the development and implementation of a QMS with reference to the requirements of the standard.

Developing and implementing a QMS

This section examines how a company or organisation goes about developing and implementing a QMS in relation to the main clauses of the international standard ISO 9001.

Key stages

The approach to development and implementation of a QMS, like that for any management system, follows a process of clear and comprehensive organisational thinking and decision making. The key stages of this process must embrace the main clauses of the ISO 9001 QMS requirements which were outlined previously. The stages reflect the clauses as follows:

Clause 1: General requirements and application of the standard – embraced in quality system preparation and preliminary organisational review of quality performance.

Clause 2: References – embraced in quality system preparation and preliminary organisational review of quality performance.

Clause 3: Terms and definitions – embraced in quality system preparation and preliminary organisational review of quality performance.

Clause 4: Quality management system – the requirements for system documentation and their control and control of records.

Clause 5: Management responsibility – the requirements for management commitment, customer focus, policy making, planning, responsibility, authority and communication, and management review.

Clause 6: Resource management – the requirements for human resources, infrastructure and work environment.

Clause 7: Product realisation – the requirements for planning product realisation, customer-related processes, design and development, purchasing, production and service provision, and monitoring and control methods.

Clause 8: Measurement, analysis and improvement – the requirements for monitoring and measurement, control of non-conformity, analysis of data, and actions for improvement.

The eight clauses describe the requirements of a QMS within the scope of the standard so, prima facie, they may not appear sequential in terms of a company's organisational thinking and decision-making processes. Nonetheless, an organisation seeking to establish a QMS will need to consider, where applicable, each clause carefully and determine its approach to the system requirements. Each of the main clauses has associated sub-clauses and again these must be considered where applicable to the organisation's proposed QMS. It is emphasised that this section should be read in conjunction with the actual documents – the standard – to assist understanding and practical application.

Development and implementation of the key stages

Quality system preparation

[Refer to ISO 9001 – Clause 1, sub-clauses 1.1 and 1.2; Clause 2 and Clause 3]

Fundamental aspects

A systems approach meeting the requirements of ISO 9001 necessitates that the company establishes a QMS which will: (1) consistently provide its business outputs – services or products – that meet customer requirements; and (2) enhance customer satisfaction through implementation of an effective system which remains amenable to continual improvement. Focus, commitment and support by a company to these two fundamental aspects of system development are absolutely essential.

The company will need to remember that the requirements of ISO 9001 are generic. This means that the system will need to be tailored to the particular situation and circumstances of the company's structure, organisation, resources and management approach, and all within the context of its operations and business activities. It must be ensured that the QMS is referenced to the current and latest version of the standard. ISO 9001 applies equally to all the potential outputs of a company's business. The standard refers to 'products' but in practice this means the provision of services or products or both if, for example, they are delivered in the context of a project scenario, which is the case with construction industry projects.

Initial decision

The adoption of a systems approach in functional management disciplines such as environment, health and safety presents an important initial choice to be made by the company. The choice is: should the company adopt a company-wide, or corporate, management system which can be applied to each and every company activity; or should it adopt a project-based system developed for specific organisational activities or projects? In the case of a QMS there is no initial decision to make as a QMS is whole-organisation orientated, focused on the holistic business of the company, and seeks to ensure quality in all aspects of its operations and business outputs. The approach will therefore be:

A 'company-wide' QMS focusing on providing a corporate-based umbrella system which embraces all the company's business activities and its business outputs whether they are services, projects or projects.

Advantages of a company-wide QMS
The advantages of a company-wide QMS are:

- a policy on organisational business performance is applied to the whole company;
- a uniform system of quality management is used throughout the entire organisation;
- a set of standard management procedures and operational working instructions are used to address quality-related aspects of the business;
- a single management system standard applies throughout all activities;
- good practice or best practice can be shared throughout the company and to all its activities.

Disadvantages of a company-wide QMS
Despite the obvious advantages of the company-wide system, there can be a number of disadvantages also. These are:

- decisions taken at corporate level need to be cascaded to the point of output delivery, so direction and focus can diminish within the structure of a large organisation;
- separation of company-level staff from those staff responsible at output delivery points – services, products, projects – can lead to lack of communication, support commitment and ownership of business activities;
- a lack of flexibility in a company-wide system where, for example, projects require tailored systems to handle particular requirements at output points.

Preliminary organisational review for quality and performance

[Refer to ISO 9001 – Clause 1, sub-clauses 1.1 and 1.2; Clause 2 and Clause 3]

Definition

A preliminary organisational review for quality involves:

> The detailed consideration of all aspects of an organisation's business processes with regard to their performance in association with the outputs of the business, whether they be services, products or projects, as a basis for developing a quality management system.
>
> (ISO 9001)

Purpose

The purpose of a preliminary organisational review is to determine the current holistic status of the company in preparation for developing and implementing a standards-based QMS. The review should establish an indicative benchmark for the performance of the company's business processes and set the basis for introducing and embedding systematic and procedures-based quality management. The intention of the review is not to compare the current management approach against the standard. The review is a systematic and comprehensive exercise to determine the position of the organisation and the ways it carries out its business processes in readiness for establishing a QMS. The review should incorporate the activities of the whole organisation from the company's structure through to the delivery of its services, products or projects.

Key activities

A preliminary organisational review should include the following key activities:

- an examination to assess, in measurable terms, current organisational performance and output quality;
- a detailed examination of existing policies, processes, management procedures and working practices in relation to quality to identify weaknesses and gaps;
- the consideration of ISO 9001 requirements as they affect the quality management and development of the QMS;

- an analysis of organisational performance in relation to core business processes to identify potential areas for improvement;
- the production of a list of recommendations for QMS development.

General approach

To carry out the preliminary organisational review both quantitative and qualitative information will be gathered. A management review of current practices and a technical review of organisational processes should be undertaken. Information will be obtained from observing the company's organisational structure, business processes, methods of working and resources. Although the exact approach to such a review will differ among companies to reflect their particular organisational requirements, the generic aspects requiring consideration are as follows:

- organisational structure and the managerial hierarchy;
- management responsibilities, commitment and support mechanisms;
- activities within company business divisions, functional management discipline levels and support departments;
- compliance with the ISO 9001 standard requirements;
- core business processes including design, resources, production, delivery and distribution for the supply chain;
- technologies used in processes and alternatives available;
- management procedures and working instructions;
- operations, methods of working and job tasks;
- deployment and utilisation of resources;
- communication and dissemination of quality performance information to staff, the workforce and external parties;
- current audit procedures;
- existing management review;
- current continual improvement mechanisms.

Methods of data gathering and analysis

The company may gather information on its current position in relation to quality management through a number of well-recognised and applied methods. The most prominent of these are:

- questionnaire surveys;
- interviews with staff and workforce;
- workshops and group discussions;
- analysis methods, including SWOT (Strengths, Weaknesses, Opportunities, Threats), PESTE (Political, Economic, Social, Technological, Ecological) and PESTLE (as PESTE but including legal aspects);

- gap analysis – the identification of procedures in place or in need of modification or development;
- inspections – scheduled and random checks;
- audits.

The collection of information should be co-ordinated by the assigned company quality manager. Analysis may also be undertaken internally but is sometimes contracted to an external consultant. Three approaches are, therefore, usually employed:

1 *Self-analysis* – conducted by the quality manager, in-company departments or work sections.

2 *Organisational in-house analysis* – conducted by the quality management team.

3 *External analysis* – conducted by independent consultants or quality auditors.

Information from these sources will likely be collated into a number of sections and sub-sections as follows:

- *Marketplace and external influences:*
  - customer/client requirements;
  - service/product industry legislation;
  - local/international marketplace differentiation;
  - service, product, or project regulations of practice.
- *Quality management procedures:*
  - policy;
  - strategy;
  - responsibilities;
  - systems;
  - resources;
  - awareness, education and training;
  - record keeping, audits, reporting and review;
  - actions evaluation;
  - insurance and indemnities;
  - communications and public relations.

The development of systems frequently assumes that a company has existing organisational practices which can be modified and arranged into a standards-based QMS. Where there are no existing practices, a first step towards the preliminary organisational review should be to determine those areas of business activity which require review. This will involve a scoping exercise where an outline list would be compiled, and through detailed consideration the list would be refined until the key areas are identified. These would then form the basis of the review.

Excellence model

The application of an excellence model provides a company with a detailed self-assessment framework for measuring the strengths of organisational activities throughout the company, highlighting areas for improvement. Such metric-orientated models lend themselves well to preliminary organisational review as they can be configured to provide both qualitative information and quantitative assessment of performance. Moreover, they allow the examination of the functional management discipline central to quality while placing quality management within the context of the company's structure, organisation and resources and also the company's commercial operating environment. That is, an excellence model is able to assess the holistic scenario and perspective. Attention is drawn to the configuration and use of excellence models, which were described in a preceding section.

SWOT analysis

SWOT analysis is highlighted as an important method of reviewing and analysing organisational performance. The technique analyses the company's strengths, weaknesses, opportunities and threats, or SWOT. The technique is a basic management tool applicable to a wide variety of organisational analyses and is useful in the examination of performance and quality. Strengths focus on those aspects of the organisation and its management which are positive in nature. These are robust areas upon which the company can build its system and improve quality and performance. Weaknesses are areas of management which place organisational performance and quality of outputs at risk. Weaknesses will require improved management mechanisms to ensure compliance with company policies, procedures, system standards and customer expectations. Opportunities represent areas where new or improved practices could deliver business benefits, while threats are areas which could seriously impinge upon the standing of the company where quality and performance remain inadequate within the marketplace (see Table 2.2).

Preliminary organisational review report

Findings from the review are normally presented in the form of a report – the preliminary organisational review report. This covers the following aspects:

1 The nature and extent of quality and performance issues identified and the organisational priority and timeframe given to their remediation.
2 A clear and coherent programme of actions needed to address effectively the issues of (1) above and how the programme will be resourced and implemented.
3 Specifications for the QMS given (1) and (2) above to ensure continual improvement.

QMS

[Refer to ISO 9001 – Clause 4, sub-clauses 4.1 and 4.2]

Table 2.2 Simple application 'SWOT' analysis

'SWOT' analysis is a method by which a company can identify and analyse the **Strengths** and **Weaknesses** of the organisation, as well as the **Opportunities** and **Threats** from information gathered from the intra-organisation and external environment

The process consists of three steps as follows:

1 Internal analysis:
Examine the capabilities of the organisation by looking at the **strengths** and **weaknesses**

2 External analysis:
Examine the external environment by looking at **opportunities** and **threats**

Information from Steps 1 and 2 are collated in a simple table as shown below:

	Positive	**Negative**
Internal	*Strengths*	*Weaknesses*
External	*Opportunities*	*Threats*

3 Develop strategy:
Develop a strategy that *applies* the strengths and opportunities to *reduce* the weaknesses and threats and to achieve the business objectives of the organisation

SWOT analysis example:
A construction industry contractor considering the establishment of a quality management system (QMS)

	Positive	**Negative**
Internal	• Level of staff understanding • Good knowledge base • Willingness to adapt • Commitment to quality	• Image of bureaucracy • System complexity • Additional paperwork • Practical motivation
External	• Improved systems • Business advantage • Expanded markets • High uniform standards	• Stakeholder perception • Competitive disadvantage • Demanding customers • Requirements of standards

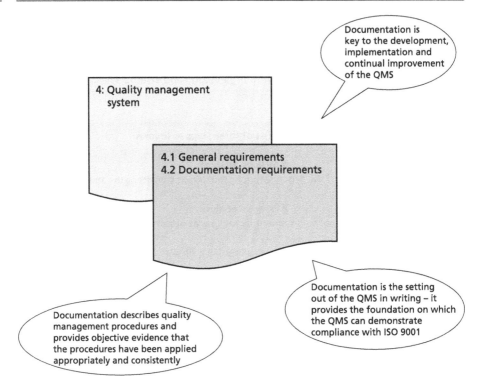

Figure 2.7 Key element – the quality management system (QMS)

Purpose of QMS documentation

Key to the development, implementation and continual improvement of a QMS is the system's set of 'documentation'. Documentation is the setting out of the system in writing within which the efficient and effective management practices which have been developed and are to be used in undertaking the business processes are explained. System documentation provides the foundation on which the QMS can demonstrate its compliance with ISO 9001. The standard requires that there is appropriate and applicable documentation to support the system in operation. The use of documentation needs to be managed to ensure correct application, and also so that the concept of 'controlled documentation' is established. This ensures the effective operation of all those procedures necessary to maintaining the desired level of quality performance in all aspects of the company's operation. Documentation describes quality management procedures and provides objective evidence that those procedures have been applied appropriately and consistently. This is fundamental to the long-term well-being of the QMS and a company's desire and requirement to ensure continual improvement to its development and application (see Figure 2.7).

The objectives of documentation

QMS documentation fulfils a number of fundamental and key objectives for the organisation. These are to:

- present a clear, written, quality management framework which arranges organisational resources effectively;

- formalise quality management through written documents with a uniform approach for consistency of operation;

- establish a written co-ordinating document – the quality manual – to frame, describe and explain all other QMS documentation;

- provide a written set of management procedures and working instructions that meet the requirements of ISO 9001 and the needs of the company's organisational practice and business outputs;

- provide documented evidence that the QMS in application has achieved what it has set out to achieve as stated in the company's quality policy and objectives;

- support continual improvement to quality through recording, auditing and reviewing organisational performance against stated goals and targets.

System documentation describes and explains how a company is structured, organised, resourced and operates. Therefore, QMS documents specifically detail how the quality-related dimension of a company's activities is configured and managed. This degree of organisational explanation is essential because, without explaining what is to be done and how things are to be done, the company would simply not function – it would become dysfunctional. Unfortunately, the need for and use of system documentation can become synonymous with bureaucracy, red tape and over-management. Therefore, QMS documentation needs to be written to an appropriate level of detail to allow the system to be implemented efficiently and effectively, but it should not become overloading and burdensome to the system users. Similarly, written management procedures and working instructions which direct system application must be simple to facilitate their use.

System documentation must be accessible to those who need to use it and it should be user-friendly. Although the documentation needed to describe and explain organisational processes and their management might be considerable, it should be broken down into sub-sections for ease of understanding and application. Documentation is an active constituent of the organisation and will change and evolve as the system matures through experience of use. Therefore, system documentation not only needs to carefully developed and applied, but must also be controlled and managed.

System documents

It was outlined earlier that the traditional and conventional approach to framing management system documentation is in a hierarchy, often termed the document pyramid. This is where one document at the top of the system is supported by an increasing number of documents throughout the lower levels, with implementation of the system at the base.

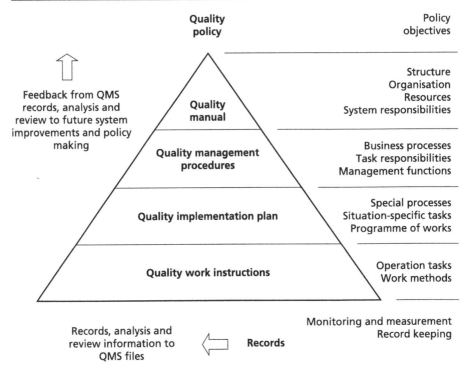

Figure 2.8 The QMS document pyramid

The framework for a QMS has four levels of documentation (see Figure 2.8):

1 Quality manual (generic) – level 1.

2 Quality management procedures (generic) – level 2.

3 Quality implementation plan (situation specific) – level 3.

4 Quality work instructions (generic and situation specific) – level 4.

The four levels of documentation provide the framework for the QMS in terms of detailed description and explanations of the system as it has been configured and how it should be applied. These are not the only documents which frame the system. There can be a host of accompanying and supporting documents describing other aspects of the total system. For example, at the top of the document pyramid there will be the company's quality policy document explaining policies on organisational performance and business output quality standards. At the base of the document pyramid are QMS records which gather documented information evidencing performance of the system in application. There may also be different and dedicated but interconnected documents such as the company's handbook of rules. These explain for example human relations management and general organisational dimensions of the company. These may need to be consulted in association with the QMS documents. System documentation can therefore be extensive, and for this reason it is best to think of QMS documentation as a 'set of documents' applied to a number of levels of management within the organisation.

Particular elements of the QMS will be implemented throughout the company and its activities continually and therefore much documentation can be generic in nature. Where the QMS has a particular application to cater for specific business outputs, for example services provided within the scope and contractual arrangement of a project, then aspects of the system will be situation specific and the documentation must reflect this. To accommodate such instances the manual (level 1) and management procedures (level 2) will be generic, the implementation plan (level 3), where required, will be situation specific, and working instructions (level 4) will be essentially generic but are influenced by situation specific characteristics.

The levels of documentation are also influenced by the management level within the organisation where they are to be implemented. The manual and management procedures are used predominantly to structure the corporate organisation of the company – they are organisation orientated. The implementation plan and work instructions are used to structure the operational organisation of the company's activities – they are output orientated.

Levels of document application

The quality manual (level 1)
The quality manual defines, describes and explains the:

- quality policy and objectives of the company;
- processes for quality management within the organisation;
- structure for the management of quality including commitments, responsibilities and authority;
- arrangement and deployment of resources to deliver quality;
- mechanisms for communication, awareness and training of system staff;
- product-realisation related quality objectives, where applicable;
- organisational mechanisms for monitoring, controlling and reviewing quality.

The quality manual therefore reflects the whole-organisation approach to quality management together with comprehensive details of how it will achieve this through the implementation of the QMS. The manual will embrace within its narrative those details necessary to describing the company's accommodation of the requirements specified by the clauses of ISO 9001. The manual will also include references to the management procedures (level 2) such that there is a logical link and trail of system documents from level 1 to level 2.

Because of the scale and complexity of many companies, including many within the construction industry, it is not uncommon to find that many use a framework manual which presents a listing of titles of those sub-documents which when viewed collectively form the system manual. The same approach is seen here where the quality manual provides a framework, albeit a detailed one, around which sub-documents such as the management procedures are developed and presented. This approach is useful, if undertaken carefully and correctly, as it allows flexibility to amend or add to system documents relatively easily without impinging upon the whole system manual.

Quality management procedures (level 2)

The quality management procedures describe:

- the processes to be managed;
- how the processes are to be managed;
- the responsibilities for managing the processes.

These are a set of procedures which will be implemented throughout all the company's organisational divisions, departments or operational sections and therefore embrace all of the company's organisational processes, business outputs and management mechanisms. Procedures describe in detail how things are to be done, how they will be done and who will do them. As the procedures apply throughout the whole organisation and all its activities, they are generic in nature, description and explanation. The management procedures will include references to the implementation plan (level 3) to establish the document trail from level 2 to level 3.

The set of procedures may be an included section in a singular quality manual or can be presented as a document in its own right, so forming a part of the document set. The latter approach is generally adopted as it allows the differentiation of organisation-orientated description from the output-orientated description. In application to construction-related organisations it is normal practice to have a dedicated document describing the management procedures. This is because there tend to be a great many procedures as a result of the technical requirements and complexity of construction works.

Quality implementation plan (level 3)

The quality implementation plan, sometimes termed a quality management plan, describes particular or special characteristics of any situation-specific application of the management procedures. This could occur in, for example, a service provided by the company within a project-based setting. This is a situation which happens routinely within the construction industry where services and products are provided to a construction project. The implementation plan contains:

- the situation-specific management plan;
- any special processes to be implemented;
- the programme for implementation;
- any special procedures needed to carry out the special processes.

The purpose of the implementation plan is to consider the requirements, if any, for specific management procedures. An example might be where a supplier of a product to a project must meet with specific quality requirements for offloading, handling, storage and protection of that product.

The plan will include references to work instructions (level 4) to ensure the documentation link and trail from level 3 to level 4. This reference is especially important at this level because work instructions will most certainly be influenced by the specifics of the situation such as the site location – the delivery point of service or product – and its physical conditions.

Quality work instructions (level 4)

The quality work, sometimes termed working, instructions:

* describe tasks;

* state how tasks are to be carried out.

Work instructions describe and explain quality performance criteria reflecting the requirements of the system in operation. They provide guidance to staff and operatives on how to carry out the tasks necessary to the undertaking of organisational processes in relation to expected performance. They also describe any training and instruction required for staff and operatives in order to carry out their tasks. To maintain the documentation trail, there should be references from level 4 to written record files. These records are an essential reference in the auditing and review of the QMS and in feedback to inform revised quality policies and objectives for the company in the future. In this way, during the implementation of each level of system documentation, information becomes available to add to the quality feedback loop and ensure that continual improvement is adopted as routine practice.

Work instructions are an essential part of a QMS in application to construction processes. They are fundamental to providing project-based supervisory, or first-line, managers and construction site operatives with detailed descriptions of how works should be undertaken with reference to the levels of quality and workmanship expected. Construction work on site is governed to a large extent by site rules and specified work practices, and the provision and use of quality-related work instructions are a composite part of such provision.

Method of documenting activities

There are a number of key steps to identifying and documenting activities within a QMS, as follows:

1 Identify which clauses of the standard are applicable to the organisation, operations and outputs of the company as an individual company will have many generic but some individual requirements.

2 Identify the management levels required to undertake the business processes throughout the whole company – corporate level through to service/product delivery point or project.

3 Consider the outline structure for the QMS – single manual or sets of documents – based on the scale and complexity of the company and its business.

4 Determine the requirement for generic and situation-specific documents based upon the nature of the company's business outputs and the way in which they are delivered.

5 Compile the quality manual and associated sub-documents, where required, to define, describe and explain the quality management approach both throughout and at each level within the QMS documentation.

Document format

The particular nature and presentation of a QMS will, obviously, vary from one company to another depending on their particular organisation and business requirements. Essentially, there are two approaches to formatting the system documents:

1 A 'single document' format – where the quality manual embraces all the documentation required for the QMS.

2 A 'multiple documents' format – where the quality manual is an overview document which co-ordinates a set of documents including: management procedures; implementation plans; and work instructions.

Whichever approach to presenting the quality manual and associated documents is adopted, it is important that the documents developed are presented in such a way that they are clear and unambiguous. The entire document, or set of documents, must link from the company's policy right through to post-implementation record keeping if the QMS is to operate completely and effectively. The documents can seek to achieve effective linking and tracking of system implementation by incorporating the following:

• a corporate flow diagram of organisational structure co-ordinating all of the QMS elements;

• flow diagrams to define management procedures and work tasks within the system;

• detailed descriptions of the management procedures and work instructions;

• specification of procedures and tasks to guide system users in a uniform and consistent way;

• job descriptions and performance specifications to define responsibilities and performance expected of managers and workers;

• monitoring mechanisms to ensure appropriate quality and performance;

• maintenance of operational records and filing for audit and review.

Control of documents

The control of documents, as specified in the standard, involves two elements of system management: (1) the organisation of documents; and (2) the control of documents.

In practice, the organisation and control of documents should reflect the following:

- The development and approval process.
- Version and revision changes.
- Accessibility to system users.
- Identification of documents to activities.
- External document identification.
- Prevention of unintended use or obsolete documents.

System documentation has to be developed and implemented sensibly. It must also be well organised and carefully controlled. Control of system documentation involves managers knowing what documents exist, why they were developed, where they are, who is using them, and the level of effectiveness in application. These expectations can be fulfilled only if system documentation is formally organised and managed. Each and every document used in the QMS must:

- have a purpose;
- have an owner for authorisation purposes;
- have currency and be the latest version;
- be readily available to the user;
- be traceable through the system;
- be archived for audit and review purposes.

A practical way of achieving the above is for the QMS to have a 'master list' of all documents used and for a 'master file' to be established within which every system document developed and used is catalogued and stored. In addition to the original documents, all subsequent versions including revisions and additions should be kept.

Management responsibility

[Refer to ISO 9001 – Clause 5, sub-clauses 5.1, 5.2, 5.3, 5.4, 5.5 and 5.6]

Management commitment

It is the responsibility of top, or executive, management to ensure that the appropriate emphasis, ethos and culture towards quality of infrastructure, organisation, processes and business outputs are developed and embedded within the company. Top management are charged with establishing the quality policy, which must permeate throughout the whole organisation and everything that it does. Appropriate policy is essential to the underpinning of the company's business objectives, goals and targets. Quality policy sets the tone and perceptions of how organisational activities and processes should be carried out, and serves in encouraging managerial and worker support and commitment to the delivery of quality.

Executive management must also give commitment to the appropriate and timely provision of all the resources necessary to undertake the business of the company. Although it is accepted that many managerial tasks will be delegated from

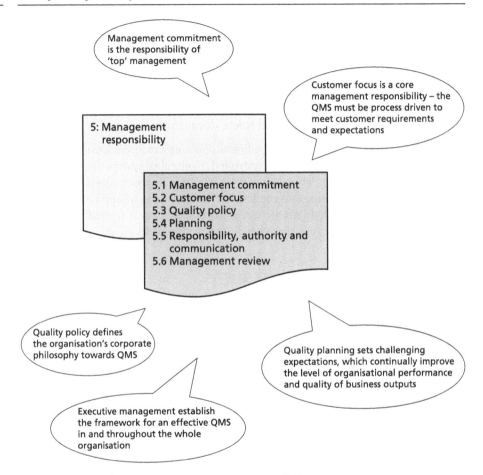

Figure 2.9 Key element – management responsibility

executive management to directive, operational and supervisory managers based throughout the company's organisation, it remains the duty and responsibility of executive management to ensure that the company has the wherewithal to commit to and support the delivery of quality in all its activities (see Figure 2.9).

Customer focus

A further dimension to embedding a culture of commitment and support for quality is to ensure that the whole-organisation philosophy embraces a customer focus. Because QMS is process driven towards meeting the requirements of clients and customers, the structure for delivering a high level of organisational performance and good-quality business outputs must have a committed focus to customer requirements and expectations. In addition, the system must incorporate as part of its cultural development the mechanisms to facilitate continual improvement. A focus on system improvement is a specified requirement of ISO 9001 and should be integral to the management arrangements contained within the company's structure. It is necessary therefore to include within the quality manual specific documents which carefully and clearly reflect these arrangements. Customer focus covers a range of organisational thinking from developing the company's policy towards delivering quality to a customer right through to continual improvement of the system with a view to enhancing customer satisfaction.

Quality policy

A quality policy is:

A published statement by a company of its intentions in relation to the quality and performance of its business activities and its outputs.

The purpose of a quality policy is to:

- define the company's corporate philosophy towards quality management in the context of its business activities;
- give authority to a written statement from executive management on the company's intentions for managing the quality aspects of its business.

To have relevance, currency and authority, a quality policy should meet a number of requirements as follows:

- Be relevant to the core business, support and assurance processes.
- Originate from top, or executive, management.
- Be supported by all management and workforce levels in the organisation.
- Encapsulate a culture of establishing, maintaining and improving the quality of all that the company undertakes.
- Be amenable to scrutiny by company stakeholders, systems certification bodies and the public.

The approach to developing a quality policy involves a number of stages. These are to determine the:

- current organisational performance and quality of outputs;
- envisaged future performance and quality of outputs;
- barriers to achieving the desired quality performance;
- organisational changes required to ensure the desired quality performance.

The content of the quality policy should incorporate the following aspects:

- *Corporate responsibility* – describes the range of responsibilities for the quality performance of the business activities.
- *Customer accountability* – describes the expectations of clients/customers associated with the outputs of the business activities.
- *Performance expectations* – describe the intended quality performance when carrying out business activities.
- *Communication requirements* – describe the need to facilitate clear communications and understanding within the organisation.

- *Improvement desires* – describe the intention to provide continual improvement to the QMS.

Quality statement

The published output from a quality policy is the *quality statement*. In many companies, the quality policy will seek to address a broad range of aspects, interests and intentions. These will depend upon the nature of a company and its business activities. Therefore, the policy will be as narrow or broad and as simple or detailed as the company needs to reflect its position and circumstances.

A quality statement should be:

- clear and understandable to a wide audience;
- presented in an unambiguous format;
- a true reflection of organisational intentions;
- linked to aims, objectives, goals and targets for the quality of operations and outputs;
- compliant with management system standards;
- commensurate with client/customer requirements and expectations;
- committed to continual improvement of quality management procedures;
- published with company identity;
- flexible for use in annual reports, publicity material and advertising media;
- authorised by signature of the chief executive officer (CEO).

The quality statement lays the foundation for development of a quality programme within a company and guides its performance setting. Policy should be configured with measurable criteria as an underpinning element. This is important in developing goals and targets which are likely to be quantitative and measurable. The policy sets the overall ethos and philosophy towards quality management, and as such the statement which encapsulates the policy should avoid becoming overdetailed – it should be clear and simple but not simplistic in tone and content. The management procedures underpinning and fulfilling the policy will be contained in the QMS manual and supporting sets of documentation.

Policy

The appropriate content and presentation of policy are fundamental to the effective establishment of a QMS as the policy defines and shapes and explains the company's overarching philosophy for organisational performance and business quality. It also forms the basis for a quality programme in which the deployment of management and workforce is configured. In line with establishing strong and clear leadership, the quality policy and statement must be developed at a strategic organisational level by executive management. Their perspective must then be cascaded throughout the whole organisation by each level of management to all employees. This is achieved through the implementation of QMS documentation – the system manual, the management procedures, the implementation plan and the work instructions, all as described earlier.

The policy should be sufficiently rigid to invoke the necessary change to working practices in meeting the changing demands placed on the company. Conversely, the policy should remain sufficiently flexible to accommodate fluctuations in organisational circumstances or marketplace demands of the customer.

The success of a QMS is heavily reliant on the acceptance, support and commitment of employees to its development and implementation. It is therefore important that policy is seen as clear and sensible by both the managers and the employees of the company. For policy to be embedded and become meaningful to employees, its intentions must be capable of translation into understandable management procedures and working instructions. Although policy is essentially a top-down management initiative, it must allow bottom-up involvement from the grassroots level. Methods will therefore be needed to cascade policy feedback, questions and ideas from lower management, and workforce levels can be incorporated which may improve the policy in the future.

Planning and objectives

QMS planning is concerned with the development of organisational goals, objectives and targets in relation to quality aspects of the business together with the formulation of a programme to ensure their planned delivery. The goals for quality are strategic in nature and form, and are designed to be fulfilled over extensive timeframes. Goals, therefore, form a part of five-year organisational business planning or even longer time periods. Some goals may be short term when directly associated with annual objectives and designed to be fulfilled within an annual operating plan. Quality objectives tend to be long term while targets are short term. Both should be quantified where possible to aid the measurement of achievement against planned performance. ISO 9001 specifies that quality objectives shall be measurable and also consistent with the policy for quality stated by the company.

Quality objectives need to be challenging to the organisation but must be realistic to become motivational for system managers and operatives. The overall aim of quality planning is to set performance expectations which continually improve the quality performance of the company. Objectives must be linked to measurable targets and with the time and resources needed for their fulfilment considered in a programme of delivery. Objectives and targets should be carefully and consistently monitored and reviewed to ensure that the company's goals are fulfilled over the long term. If quality objectives are not meeting performance requirements then the organisational goals may have been too ambitious.

Quality objectives can be broken down into a number of different types, each of which must be carefully created and with all being important to the fulfilment of the company's overall business and quality-related goals. These types are: *company and management objectives* – designed to meet long-term organisational goals through policies; *monitoring objectives* – designed to meet medium- and short-term objectives through control; and *improvement objectives* – designed to meet short-term objectives and targets through improvement actions.

A company's goals, objectives and targets should be formalised through documentation with realistic timeframes set for their accomplishment. Direct links between the quality policy and objectives need to be put in place with targets set for each. Supporting the commitment to quality within the organisation, objectives should originate from executive management and be seen clearly to underpin the policy and goals of the company. This aspect is essential in encouraging employees to see the relevance of their work in meeting the objectives. It is also a prerequisite for engendering support for the holistic business vision of the company.

Quality objectives form a crucial link between the strategic vision of the company and the operation of the company. The key links are as follows:

- the translation of the company's quality ethos and policy into principles of organisational practice;
- the configuration of principles of practice into organisational structure;
- the relationship of organisational structure to quality management and the QMS.

A company needs to find a way of planning how quality goals, objectives and targets will be developed, implemented, measured and improved. Many organisations seek to do this through the development and use of a *quality programme*, or *schedule*. This programme determines the following:

- the requirements for and development of quality objectives and targets in relation to quality policy and goals;
- the identification and designation of responsibilities for quality management within the organisation structure;
- the methods and timeframes by which quality objectives and targets will be met.

It is essential that the programme embracing the establishment of the goals, objectives and targets is supported by a realistic and workable scheduled programme. There must be a sensible timescale in which the actionable elements of the programme are to be carried out. Such programmes are often demanding but must be feasible. The programme should include all elements of organisational activity where objectives and targets have been set. For example, work to create an amended organisational process to deliver quality management to a set target will take time and resources, so this development must be included in the programme. Then time will be needed to pilot-test the process in operation followed by a shake-down review and modifications, and again these aspects must be accommodated. An attempt to incorporate all elements on a single programme may prove difficult, so where there are multiple objectives to be planned a master programme may be compiled, supported by sub-programmes to reflect individual objectives or groups of objectives.

Development planning programmes of this type can take many forms, but it is traditional to use a Gantt, or bar, chart to reflect activities against timescales in a simple and clear format. The bar chart also facilitates the monitoring and recording of progress where work on activities can be directly related to that which was planned. Other methods of development planning can be used to support bar charts such as network analysis. Networks examine the sequence, timing and interrelationship of scheduled elements in greater detail and with accurately quantifiable task durations present a useful numeric-based picture of the development programme.

Responsibility, authority and communication

It is the responsibility of top, or executive, management to establish the framework for effective quality management in and throughout the whole organisation. They must take a leading role in ensuring that the structure is clearly configured and communicated to senior, or directive, management. Senior management are then charged with cascading this structure to lower levels of management at operational and work task levels. It is absolutely essential that the structure which formalises responsibilities, authority and communication routes becomes, in time, part of organisational culture such that these aspects become a natural and intrinsic part of the company's day-to-day activities.

Establishing a workable and effective management structure that assigns appropriate responsibilities to designated personnel is a prerequisite for the successful introduction and embedding of a QMS. It is vital that everyone in the organisation knows what part they play in operating and supporting the system and their responsibilities towards fulfilling the objectives and targets set for quality and performance. The structure sets out the formal arrangement of roles, responsibilities, authority, communication and interrelationships.

Almost all companies reflect their QMS structure through the provision of one or more 'organisation charts'. The purpose of the organisation chart is to illustrate the company's organisation of human resources and the assignment of responsibilities for the management of those resources which are necessary for the implementation of the QMS.

In defining responsibilities and creating a QMS organisation chart, a number of fundamental principles are important:

- The executive directors of the company must be associated unequivocally with the QMS structure so that support and commitment are seen to come from the top.

- Specified management functions must be unambiguous to establish an organisation of champions for quality and performance within the whole-organisation structure.

- All personnel involved with the QMS should be included in the structure.

- The organisation of management should be reflected in the QMS documentation – in the quality manual and management procedures.

- The structure should show the relationship between managerial staff and operational personnel.

- Functional management responsibilities for the QMS together with linkages showing co-ordination functions and lines of communication should be included.

Management representative

It is conventional practice for almost all companies to appoint a dedicated quality manager within their organisational structure. This person is termed within ISO 9001 the *management representative*, and the appointment of this individual is a requirement of the standard. Although many functional managers will assume roles within the organisational structure for the management of quality, the management representative is the individual given the responsibility with assigned authority to oversee all aspects of quality management and the development and implementation of the QMS. The management representative is a senior staff member of the company, usually a director-level manager, but could be an executive-level manager. Their principal role and responsibility is to manage, on behalf of the company, all aspects involved with quality management as a functional management discipline within the organisation, and the development and implementation of the QMS.

The management representative's role is, ordinarily, extensive as it embraces the complete QMS including its application and evolution. Moreover, the role can be extraordinarily extensive when seen in perspective and application to a large and complex organisation. The role involves managing both the intra-organisational and inter-organisational dimensions of quality and its management. Within the company this involves an understanding of the concepts, principles and practice of quality management and systems applications as applied by the company. In almost all companies it generally requires a good working knowledge of the technical aspects and requirements of the business and its outputs, allied to a sound appreciation of the human resource dimension.

The role also entails good communication skills, as much of QMS application is concerned with the people who work within and around the company, processes and management system. Outside the company, the management representative will be responsible for dealing with all matters associated with quality standards, the adoption of ISO 9001, managing the external recognition of the system through certification, together with determining, interpreting and applying customer requirements, external regulation and auditing requirements. An important part of the role might therefore be said to border on public relations in so far as such communication is associated with the quality performance of the business.

In large companies, the remit of the management representative might be logistically overwhelming. It should not be inferred that the individual charged with the role of management representative is alone or works in isolation. Quality management as a functional discipline within an organisation and in relation to QMS may necessitate the deployment of a quality management team with delegated and devolved responsibilities. From a practical perspective, such an approach is desirable and encouraged, and will likely assist greatly in the day-to-day operation of the QMS. The point to remember is that, irrespective of the structure and practice of quality management within the organisation, the management representative retains oversight responsibility and accountability to the company. This rightly invokes single-point responsibility for one individual – the management representative.

Communication – internal and external

ISO 9001 specifies the requirements for internal communications. Effective processes of communication need to be established within the company with respect to the QMS. In practice, both internal and external communications are important to QMS implementation. Communication must be managed effectively to personnel within the company and to those outside who interact with or are influenced by the quality management of the company. Internal communication will focus on the implementation of a *communications plan* which, when implemented, continually informs and updates management and workers on the use of the QMS. In addition, it will assist and support training initiatives within the company which underpin the implementation and improvement of the system. External communication will focus on the provision of quality-related information to external parties including customers, suppliers, regulatory bodies and the public.

Effective internal communications rely upon management appreciating the following: the content and complexity of quality-related information; the required degree of interaction with receivers; and responses to such information. In relation to the developmental phase of the QMS, much initial communication will be related to generating awareness among managers, users and general company employees. In this situation, the degree of interaction between the sender and the recipient will be low and the content of the message will be simple to comprehend. As the QMS moves into the implementation phase the information to stakeholders becomes more detailed, whereupon the need for interaction rises and a greater response is expected. Because communication is dynamic and continuous, quality managers at all levels must consider carefully what communications are necessary, how they are to be made and their likely effect upon the receivers. All of these things have a bearing on the way that the QMS will be perceived and therefore could impinge upon its use and degree of success.

Quality managers will adopt a range of communication mechanisms, commensurate with the knowledge and skill needs of the workforce. These normally involve: general awareness campaigns; management team meetings; staff workshops; briefing papers; and newsletters. The purpose of these is to allow the company to develop and embed a culture of quality management supported by a good understanding of the organisational structure, the management function and implementation systems. Detailed knowledge and operational skills associated with QMS can be developed by appropriate training initiatives.

One of the key hurdles encountered by managers when implementing a new QMS is 'the fear factor' which can arise from misinformation, ill-timed information or poorly delivered information. Effective communication will gain the confidence of staff, increase the knowledge and skills of the workforce, and can engender support for the system. An important dimension to achieving successful communication is to link it with education and training, which is why training is a requirement of QMS within the standard.

For many companies, external communication means being open-handed with the provision of information concerning organisation performance and the quality of business outputs. Clients, customers and other business stakeholders are more demanding in their requirements for quality of services and products. Indeed, many require pre-qualification for quality standards prior to tendering or engagement of provision. Auditing of quality achieved is routine for many companies together with public-domain reporting of organisation and business performance. Much of this type of activity involves good communication with outside bodies and, moreover, the management of that communication. The key to effective external communication is to support a limited and defined route of information. In this way, communication and information can be structured and managed. This is important to ensure that correct information about quality is disseminated, but also to ensure that it is distributed in ways appropriate to the company's needs.

Management review

Within the requirements of ISO 9001, executive management have an obligation to review the QMS at planned intervals. Moreover, they have the obligation to seek improvement to the QMS. The system should be able to learn from the lessons of application and experience to inform policy making and objective setting in the future. Most companies adopt one of two approaches to the formal management review. They can undertake a *single annual review* – a comprehensive event undertaken once per year where the full QMS is reviewed. Alternatively, they may choose to carry out *multiple reviews* – where a series of short events is conducted throughout the year where key parts of the QMS are reviewed.

A small company will likely favour the single annual review. Such reviews take up a minimum of time and resources. Single review relies upon reviewing all aspects of the system throughout the organisation in one event, so they are highly cost effective but they can also be counterproductive in some respects. Where system managers or operators are unable to take part, or where many aspects are concentrated into a day or half-day meetings, then it is easy to gloss over or miss important details of system experiences. In such an event it can be easy for the company to fail in making vital or useful improvements to the system. Larger companies tend to use multiple reviews, which they conduct throughout the year. Such reviews can provide greater opportunities for learning from system experience as they can examine aspects in greater depth and across a wider section of organisational activity. It is likely that across a series of review events a higher proportion of staff will be able to attend, more issues will be identified and more of the QMS will be reviewed in greater detail.

It is important that quality performance data is reviewed in the context of the company's objectives and targets and traced back to quality policy and business goals. If any objective or target is not being met then the review must identify the reasons for this together with potential actions for improvement. If all objectives and targets are identified as being met and the policy is fulfilled, then the review should examine the potential for making whole-organisation improvements. Excellent performance will require the consideration of more stretching objectives and targets in subsequent years with a view to pushing the boundaries of goals. This may also involve a re-examination of management procedures and working practices to encourage incremental changes which seek greater effectiveness in day-to-day activities.

Key areas for quality management review are the following:

- quality policy;
- quality goals, objectives and targets;
- organisation structure for quality management;
- a quality planning programme;
- communications planning;
- awareness and training;
- management procedures and working instructions;
- monitoring methods, verification, testing and conformance/non-conformance;
- identification, application and monitoring or improvement actions;
- record keeping and reporting;
- audit results.

The overarching focus of the management review is to seek improvements to the QMS and, in so doing, improvements to the activities and processes of the organisation.

When carried out effectively a quality management review will ensure that:

- the system continues to meet the current standard;
- the QMS remains current, appropriate and effective for the company's needs;
- appropriate actions are developed and applied to address identified shortfalls in the QMS;
- improvements are made which enhance business processes and work practices;
- awareness, information, communication and training continue to meet the needs of managers and workforce;
- the QMS continues to provide the best approach to meeting the current and future challenges of quality requirements and expectations.

Resource management

[Refer to ISO 9001 – Clause 6, sub-clauses 6.1, 6.2, 6.3 and 6.4]

Provision of resources

A company must consider, determine and provide the resources needed to develop, implement and maintain the QMS; it should ensure that continual improvement is a tenet of its application, and that customer requirements lie at the centre of its focus. The identification and deployment of the system's management staffing will have been determined during QMS preparation, with designation of duties reflected in the organisation structure. As seen earlier, the structure adopted forms an essential consideration of Clause 5: Management Responsibility. The company must ensure that it gives full commitment to delivering these resources (see Figure 2.10).

In practice, a major element of the work of the *quality management representative* will be to secure appropriate resource commitment from executive management

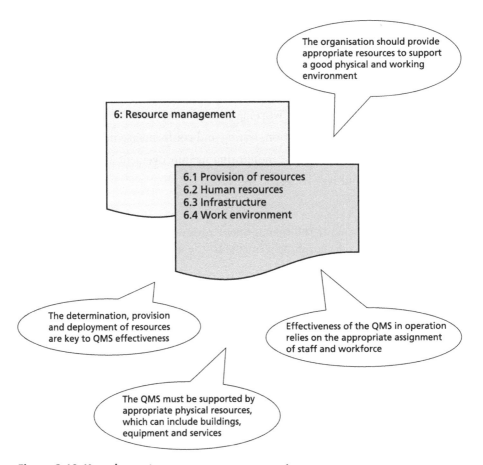

The organisation should provide appropriate resources to support a good physical and working environment

6: Resource management

6.1 Provision of resources
6.2 Human resources
6.3 Infrastructure
6.4 Work environment

The determination, provision and deployment of resources are key to QMS effectiveness

Effectiveness of the QMS in operation relies on the appropriate assignment of staff and workforce

The QMS must be supported by appropriate physical resources, which can include buildings, equipment and services

Figure 2.10 Key element – resource management

and translate this into operational resourcing and deployment to quality management of the business processes. This will require a good working sense of *corporate planning* and *human resource planning*, or what used to be termed, perhaps inappropriately, *manpower planning*.

Corporate-planning-related aspects for consideration include:

- government-related influences such as national, regional and local policies towards industry or sector employment;
- legislative and regulation-based influences on the business processes requiring specific employment requirements within the organisation;
- changes within the commercial operating environment such as whole-economy fluctuations and market sector variations;
- changes which influence marketplace position and competitiveness;
- variation in the availability and market cost of human and physical resources necessary to the business of the company;
- evolution of the business through technical advances which influence the knowledge and skill base required of the workforce;
- expansion or diversification of the core business base which influences the structure of the organisation, its activities and its workforce;
- intra-organisation changes, circumstances and situations which influence the smooth operation of the business.

The quality management representative will need to be aware of any and all of the potential corporate-planning-related aspects, as any change could be an influence upon the approach and operation of the company and thereby the QMS used in their connection. A good working knowledge of the possible implications of human resource planning is also needed, as such planning is a requirement of an associated sub-clause of QMS resource management.

Human resources

There is a direct and strong link between the organisational strategy, structure and organisation adopted by the company as determined by corporate planning and the concept of human resource planning. Human resource planning is concerned with identifying the staff and workforce requirements for the implementation and maintenance of the QMS. This must be commensurate with fulfilling the policy, goals, objectives and targets set by the corporate plan, with an effective approach to doing this governed by the configuration of structure and organisation.

At the human resource planning level, the quality management representative will be involved with ensuring the practical resourcing of the QMS in operation. This can involve consideration of:

- the knowledge, skills and competencies of those staff and members of the workforce with responsibilities for the implementation of the QMS;

- awareness and skills training to maintain, expand and diversify the knowledge and skill base associated with the operation of the QMS;
- succession issues associated with maintaining the knowledge and skill base for QMS application in the long term;
- monitoring of the labour-related turnover and maintenance of the ongoing skill base necessary to the continued application of the QMS.

In addition to the human resource planning dimension, consideration must also be given to *resource utilisation* and the methods by which the effective use of resources can be determined. Method study and work measurement are two traditional and well-recognised techniques that come within the collective concept termed work study. *Method study* involves: the analysis and simplification of work; developing the best ways to undertake works; and establishing the best physical layout of facilities which support business processes. *Work measurement* involves: determining how long works should take; the confirmation of optimum resource inputs to processes and tasks; and seeking improvements to the level of efficiency achieved when conducting work. The overall impact of these applications is to deliver higher productivity and improve organisational and work performance in business-critical areas. One such area is the delivery and assurance of good quality in all services, products and projects. The quality management representative should, therefore, take an active role in overseeing these important contributors to human resource planning.

Infrastructure

In addition to providing appropriate and adequate human resources to support the management of its activities through the QMS, the company must provide the necessary physical resources, or *infrastructure*. Infrastructure encompasses the provision of buildings, the equipment prerequisite to the execution of business processes and all the supporting services necessary to their underpinning.

The provision of infrastructure is really not determined by the quality management approach used or the QMS itself. It is influenced chiefly by the following:

- the core business of the company;
- processes which deliver the business outputs;
- requirements of the physical workplace and workspace;
- business support and assurance processes;
- welfare facilities for staff and workforce;
- supply chain input and output requirements, for example transport, handling and storage;
- communication and business information needs.

The focus on the appropriateness of infrastructure is directed to ensuring that there is, through the correct provision, uniformity and continuity to those processes which deliver the business outputs and conformity with the quality requirements of those outputs.

The aspects of infrastructure which can be said to impinge directly on the capability of the QMS to fulfil its function are: organisational communication; and business and process information systems. Good communication and clear information routes will be needed within and around the company for the effective management of quality. The physical size or location of workplaces and workspaces can be influential in this respect. Difficulties in the conveyance of information around the organisation, for example from one building to another, could be a hindrance to the smooth operation of the QMS. Where associated business processes are carried out at different venues, sometimes at geographically dispersed locations, information flow can be disrupted, with breakdowns in the effective implementation of the QMS.

Work environment

Leading on from the consideration of the infrastructure needed to support quality management, the company must ensure that those persons who work in and around the business processes and the QMS work within appropriate conditions, or what is termed the *work environment*. The work environment relates directly to those conditions under which company employees work on a day-to-day basis. The nature of the business outputs, whether they are services or products, will determine the processes with which employees engage. These, in turn, influence the general working environment and physical conditions which will be encountered.

The work environment includes any influence which impinges upon the physical being of the individual. Therefore, such influences as space, noise and lighting, to mention a few, must be appropriate to their interface with the worker. In addition to general, and what might be termed sensible, provision of the work environment, legislation and regulations will likely apply. Health, welfare and safety at work and environment-related requirements will be in force. So, many aspects of work environment provision are not left to the discretion of the employer; they are compulsory requirements.

Among an organisation's collective staff base, different employees will experience a variety of work environments and physical work conditions according to their particular role within the company. For the collective staff base the company needs to provide a good standard of work environment for all its human resources and one that facilitates effective operation of the QMS. For designated staff and workforce, who may undertake specific business processes or operate particular parts of the QMS, the company may need to provide special measures to meet a higher standard of work environment. For example, staff monitoring and measuring quality in a business process where there is a high level of physical human risk or danger may need additional safety and welfare measures to be taken.

Product realisation

[Refer to ISO 9001 – Clause 7, sub-clauses 7.1, 7.2, 7.3, 7.4, 7.5 and 7.6]

Special note

Product realisation is the one clause within the standard which is not all inclusive. Compliance with sections, or sub-clauses, is not compulsory where the requirements of the standard are inconsistent with the nature of a company's business and therefore the application of the QMS. It should be remembered that the underpinnings of a QMS should be commensurate with the organisation's own requirements. It is also worth reiterating that while the standard refers throughout to *product*, in the context of establishing the QMS a product is *any intended output* from the product realisation processes. Business outputs can be services, projects or both when delivered through a project delivery scenario, and therefore certain clauses from the standard will be relevant to QMS development while others are not, depending upon the type, nature and undertaking of the business processes (see Figure 2.11). It is emphasised that the individual characteristics of the construction process have a significant bearing on the practicality of product realisation. The construction industry has a diverse service and product range, and can have great variations in production methods, resource deployment and management structures from company to company and project to project. The diverse nature and transient of construction means that end outputs are realised in different ways, and this characteristic must be respected when devising and implementing this aspect of the QMS.

Planning of product realisation

Product realisation is concerned with the development of the processes involved in delivering the product to ensure that such planning is appropriate to the operation of the QMS. Many of the considerations which need to be made in respect of product realisation will have been determined during the preliminary organisational review stage. At that stage the company's consideration of quality policy, objectives, business processes and associated fundamental management aspects will have been determined. This stage transforms the intent of providing business outputs into plans for the actual delivery of those outputs.

Customer-related processes

The main underlying and underpinning tenet of a QMS is the management of processes which provide business outputs which meet the explicit needs, requirements and expectations of the customer. This means that the outputs of the business, the processes which deliver them and the management of those processes must be considered and arranged with a sharp appreciation of what the customer wants and how the customer interfaces with the provision. In their contractual arrangement with a service or product provider, a customer may have specific requirements which influence the business process and the quality management dimension of that process. For example, a customer might require a product to be packaged, handled and delivered in a particular way, and this would directly influence the production and distribution processes and their management. It might

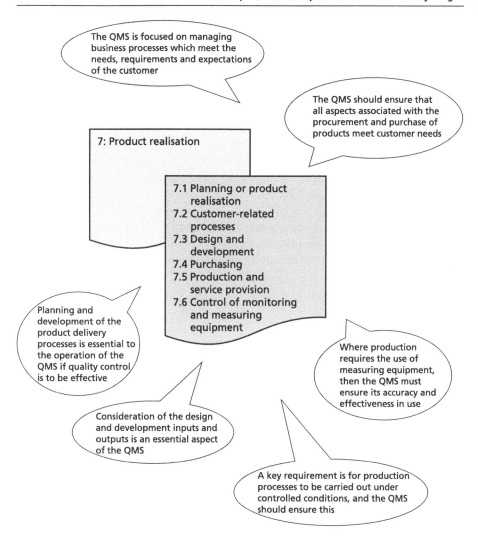

Figure 2.11 Key element – product realisation

also involve configuring processes and their quality management where there are customer specifications for intended use of the product. Here, environmental conditions, for example, may be stated, which again the QMS must accommodate by ensuring that receipt and start-up use are exercised in a particular way.

In association with customer requirements, the QMS should ensure that it meets current legislation and regulatory requirements insofar as they are influenced by the customer requirements. Provision may also have to meet customer-related contractual specifications, forms of contract and any additional requirements agreed between the customer and the provider. Such requirements are applicable to many goods and services where guarantees and warranties apply, or where there are periods of defect liability enacted, or where there are long-term commitments to post-delivery services, say for the maintenance and service of technical equipment.

Within the QMS there should be provision for the review of requirements of the actual products. This is normally undertaken before the commitment to provide the service or goods is entered into. Consideration, therefore, takes place during the procurement stage – during pre-tendering, or pre-selection, the submission of tender or at the time of contract agreement and documentation. Considerations must embrace the accurate definition of the provision and its specification together with a review to ensure that all commitments can be met. It is often the case that, over a period of time, the product requirements will change as customers evolve their business requirements. This means that the provider will need to adapt its outputs and perhaps its business processes and quality management approach accordingly.

Much of the above is dependent upon effective communication between the provider and customer. A company must ensure that its QMS provides for open, transparent and effective methods and lines of communication and information provision such that the company's processes and QMS support current requirements.

Design and development

Many companies will, chiefly, provide services to customers and as such there are no design and development dimensions to the organisation and its management. Their QMS will not be required to comply with this clause of the standard. Many companies will supply goods and products to customers, and there is a need to consider design and development. Much depends on the way the provider goes about its business. One company might act solely in the capacity of manufacturer while another might engage in design and manufacture. Others may undertake the manufacture of products but outsource the design element. So, arrangements are far from straightforward and the QMS must accommodate the needs of individual organisations as they are arranged to handle the design and development associated with their business outputs.

Where the design and development applies, a company's QMS must support its planning and control. Planning is seen as key to this, and therefore the company should clearly identify the various stages, put in place mechanisms to manage these stages and confirm the responsibilities associated with that management. Some large and complex companies may have multiple design and development facilities at dispersed locations. In such instances the company must ensure that there is effective interaction between these together with clearly defined communication mechanisms and designation of responsibilities.

Consideration must be given to both design and development inputs and outputs. Inputs should respect functional and performance requirements together with the requirements of legislation and regulation, while output requirements should respect the expectations associated with purchasing, provision and use. The QMS must allow for design and development review. This should evaluate the mechanisms used to manage design and development and ensure that, where problems or issues have arisen, actions are determined and applied. Review should also include mechanisms to facilitate the verification and validation of application to ensure that outputs are meeting the intended requirements. Where changes in design take place, which is a frequent occurrence in the design and development of many products, the QMS should have control methods in place to recognise, monitor, record and manage such changes.

Purchasing

There would be no point at all for a company to provide its business outputs in a form and to a specification which the customer does not want or has requested. The QMS needs to ensure that all aspects associated with the procurement and purchase of products meet customer needs. Purchasing covers: the purchasing process itself; the information describing the outputs to be purchased; and the verification measures to ensure that the product as purchased is as specified and expected by the customer when entering into the buying of it. Much of this aspect concerns the information describing goods and services which the customer receives during its discussions and negotiations to procure. It is absolutely essential that goods and service are as they are described and that, once supplied to a customer, they are fit for purpose. Anything short of this would, likely, lead to issues of misrepresentation. Consequently, a QMS must have measures which ensure that purchasing-related information is unambiguous and clearly and correctly conveyed to the customer.

Production and service provision

This particular clause is less related to product provision and associated more with the undertaking of production and the provision of services. As such, these aspects have much in common with the construction processes where services are provided and delivered within a project-based scenario. The main requirement is for the company to plan for and undertake production processes under controlled conditions, and this is exactly what the production phase of construction is challenged with accomplishing. The delivery of controlled conditions requires that the QMS incorporates the use of: management procedures; work instructions; appropriate equipment; monitoring and measurement mechanisms; and completion, handover and post-delivery procedures.

A QMS must be able to validate its activities, and this entails a mechanism to achieve this. Appropriate ways to support the delivery of services are to ensure that: the system has explicit criteria for review; the competence of staff and workforce is assured; equipment used is appropriate to the task and fit for purpose; sensible and effective work methods are adopted; and records are maintained to evidence all of these contributions. An important aspect of keeping records of the quality of provision is that there is traceability. In respect of products which are produced through business processes somewhat akin to assembly lines, then this may be quite straightforward. At any given point in the business process, monitoring can be undertaken systematically and continuously. For services, this is not so simple as services tend to be human resource orientated with a consequent greater variability in systematic and continuous delivery. Monitoring and record keeping which are inconsistent or untimely may inhibit traceability. So, methods within the QMS need to recognise the need for traceability of a parti-cular service and build in regular and practical means of monitoring and record keeping in the light of the inherent inconsistencies of human application.

Where the provision of a service could impinge upon the property of the customer, then the provider must ensure that all practicable means are taken to protect the interests of that customer. Again, this aspect is directly pertinent to project-based situations where the service is often delivered at the customer's site or facilities. Such sites tend to be business active with production or services conducted in and around the customer's ongoing activities. In addition, the provider will need to preserve the quality of products where goods form part of the pro-vision, reflected in packaging and handling procedures.

Control of monitoring and measuring equipment

Where the outputs of a company's business processes are amenable to, and there is a requirement for, the use of measurement equipment to determine quality standards, then the company must ensure that the measuring equipment and pro-cedures form part of the management of those business processes and outputs. Measuring equipment should be calibrated to appropriate standards and should be checked on an ongoing basis to assure its accuracy. Moreover, there should be records of such checks together with any required adjustment or recalibration. Within the scope of a QMS, all of this must underpin the capability of ensuring the required standard of performance in business processes and the quality of outputs.

Measurement, analysis and improvement

[Refer to ISO 9001 – Clause 8, sub-clause 8.2, 8.3, 8.4 and 8.5]

Monitoring and measurement

A company sets out to meet customer expectations and requirements through the performance of its organisation and the quality of its outputs. To achieve this it needs to know the perceptions of the customer, since only the customer is in a position to judge quality, as the direct recipient of the products or services. The essential question is: has the organisation, as provider, met all the requirements of the customer? This might not be just the provision to a minimum standard but can entail the sufficing of the wider package of provision – what customers often refer to as the added value. An efficient QMS will ensure that the standard of provision meets with the customers' requirements. A truly effective QMS will strive to deliver on expected requirements and then seek to provide an added-value element to its overall performance. This is what the aspiration – continual improvement – seeks to achieve. A QMS in the early period of operation will, if implemented correctly, deliver outputs to requirements. A QMS that has operated for an extensive period will gradually evolve through continual improvement to become more efficient and effective. System monitoring

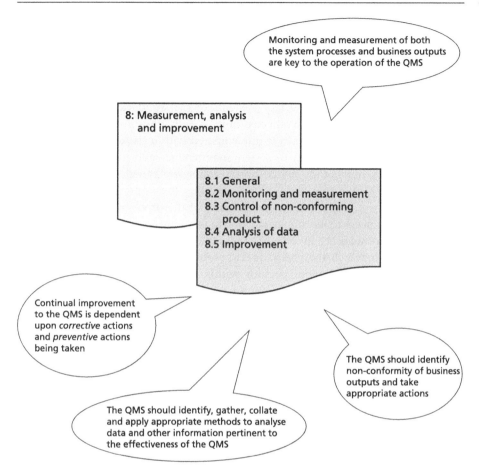

Figure 2.12 Key element – measurement, analysis and improvement

and measurement focuses on the performance of the QMS and its improvement, and for this reason the company needs mechanisms that gather objective evidence of its operation from the perspectives of customers. Common ways of gathering such information include: customer satisfaction surveys; user opinion surveys; guarantee claims; maintenance requests; and public network databases (see Figure 2.12).

In association with the above, the monitoring and measurement of both the system processes and business outputs should be a feature of the QMS. These focus on putting in place the mechanisms which record the performance of the QMS in application, and enable management to evaluate that performance and take actions to modify the system where problems are seen to exist. These attributes of an effective QMS need to be configured in the context of the level of conformity of the outputs to the requirements. This being so, the system must, obviously, monitor and measure the final quality of the product. Records of performance for both the quality of outputs and the QMS which manage the processes delivering them will need to be maintained for evidenced verification.

Auditing is an important management process for determining the performance of the QMS in application. The standard requires the company to undertake internal audits at periodic intervals for this purpose. Audit requirements vary from one QMS to another but annual audits are the norm with variations for more frequent inspections as part of a longer-term audit plan. An audit should ensure that the QMS conforms to the requirements of the standard and to those criteria set by the company itself, reflected in its system ethos and policy. An audit should also ensure that the QMS is implemented within the organisation on an effective basis and with respect to system maintenance and improvement.

The standard requires that an audit programme is planned, or scheduled, that the audits take place at specified intervals, and that this approach is commensurate with the nature, importance and criticality of the processes which are to be audited. Audits need to be impartial and highly objective in nature. For this reason, managers who are involved in the implementation of the QMS or parts of the system for which they have direct responsibility should not conduct the audit themselves. Independent persons within the company should undertake the audit. Companies should have employees with the necessary knowledge and capabilities to conduct audits, but where this is not the case then it is not uncommon for external consultants to be brought in.

The audit processes adopted and applied are a further responsibility of the management representative for quality, whose regular duties were introduced earlier in this section. It is important that the audit is structured and applied as a process of evaluation, review and reporting, and it usually involves the following:

- Confirming the scope – is it whole-system or part-system based and which organisational processes and products will be audited?
- Confirming the standards – does ISO 9001 and/or intra-organisational benchmarks apply against which compliance of the QMS will be evaluated?
- Confirming the management of the audit process – usually the QMS management representative will oversee the total process and practice of the audit.
- Specifying the lead auditor – a competent but impartial and objective member of the company should take the lead for the audit event.
- Identifying the audit brief – is the audit whole-process or part-process focused?
- Specifying the audit team – members selected according to audit workload.
- Planning the audit – determining the schedule of activities and duties within the team.
- Reviewing the QMS documentation – involves scrutiny of the current system manual, management procedures, implementation plans and work instructions.
- Conducting a preparatory meeting – where auditors meet the QMS managers and operators to plan the audit, its activities and the schedule.
- Collecting information – the collection of QMS records and observation of practices to determine performance of the system, both good practice and non-conformance.

- Reviewing the evidence – to confirm practice performance.

- Concluding the audit – through verbal feedback given at an audit closing meeting.

- Reporting – via written reports, first a draft provided to the QMS staff for verification of the accuracy of information and then by a final report confirming the outcomes of the audit and actions required.

- Implementing subsequent actions – to accommodate matters identified by the audit undertaken.

The written outcomes from internal audits should be catalogued and filed. The information gathered from such audits will, likely, form an important element of any external audits that are undertaken. In addition, internal audit records will be scrutinised as part of any QMS certification process.

Control of non-conforming product

The internal audit process frequently identifies processes within the QMS where there is non-conformance with the standards. Such instances are not unexpected as occasionally all management systems will experience shortfalls in their per-formance. As a result, the output quality standard of the product will not always meet the product requirements. Notwithstanding, the QMS must employ proce-dures to rectify the manifestation and effects of non-conformance. There must be mechanisms to: take action to address the non-conformance; take action to discontinue product use until matters have been addressed; take action to authorise continued use subject to concessions with the agreement of users and other stakeholders; and take action to rectify matters where the product has already been supplied or been used. Again, records of the actions taken in respect of the system's handling of non-conformities must be maintained and stored for future reference.

Analysis of data

The company is required to identify, gather, collate and apply appropriate methods to analyse data and other information pertinent to the effectiveness of the QMS. The focus should be on analysing and evaluating the performance of the QMS with a view to perpetuating continual improvement. Much of the data will have been acquired from system audits and system records in addition to third-party-based information such as customer survey feedback. The main task within this requirement of the standard is to ensure that all such data is avail-able, collected and in a form that aids meaningful analysis.

Improvement

With the principal goal of achieving continual improvement of the QMS, considerations focus upon those actions needed to fulfil this. Two sets of actions influence this capability greatly: the application of *corrective* actions; and the use of *preventive* actions. Preventive actions entail the implementation of preventive controls and seek to prevent non-conformances occurring within the quality processes and procedures used. Corrective actions involve those corrective controls used to rectify non-conformance following their occurrence. The consideration of both types of actions is essential. It would be foolish to assume that the use of preventive measures was enough to eliminate all non-conformances, since they will at some time occur and likely reoccur during the long-term implementation of the QMS. Rather, it is best to have preventive controls in place with a view to minimising non-conformance and also to have corrective controls available to mitigate non-conformances quickly and effectively when they occur.

An efficient and effective QMS will consider and implement actions to ensure that all of the main clauses and sub-clauses, discussed in this section, are accommodated and that the requirements of the standard are met. It should be remembered that a company will need to consider and configure its QMS around the nature of its own business within the context of its particular business environment. Therefore, the standard represents a management tool for appropriate and sensible application. A later section in this chapter describes and explains how a construction-industry-related company can develop and implement a QMS appropriate to its own needs within the context of the construction industry and its processes.

QMS: independent assessment and recognition

This section examines the need for, benefits of and approach to procuring independent assessment and recognition, or certification, of a QMS by the organisation.

Purpose of independent, or external, assessment for QMSs

Purpose

The quality of a service or product is not judged by the provider – it is judged by the customer. Although it may not always appear fair to say, it is pertinent and correct to say that one's performance is only as good as one's last job. Clients and customers set great store by making judgements on competence, performance and quality based upon what they perceive, have seen or been told. A company which says it can deliver good quality, does deliver good quality and develops a reputation for delivering good quality will gain a premium and competitive profile in its industrial or commercial marketplace. For many companies, being able to say that their quality of service or project is excellent is simply not enough. The customer or client is on the receiving end of the provision and is in the best and only appropriate position commercially to judge on quality. Although it will be seen subsequently that there are different ways of assessing quality, the only true measure of a company's quality of performance and outputs is when they are independently, or externally, assessed.

Not all companies either want or need to have their performance in quality management competency assessed. Some companies may be delivering their services and products to customers who are perfectly satisfied with the provision, even though the organisational systems managing the provision are not independently recognised or have been formally assessed for standard of management. The want and need for companies to acquire independent recognition of the QMS that they use is dependent upon a host of business and organisational considerations. Not least of these is the effect that recognition of their management system will have upon the operation, well-being and future business of the company. A small company, such as a self-managed or family business, may secure its work from clients and customers who have no want or need themselves to demand formal quality assessment from their service or product providers. In this case, there will be no commercial market pressure for a formal and structured QMS to be used. Conversely, a larger company which serves major private-sector and public-sector clients may well find that the demonstration of a QMS is a requirement set by those clients.

So, for first-order providers of services and products, there may be an inherent requirement within their business relationship with their clients to provide and demonstrate the application of a standards-based and independently assessed QMS. For second-order providers there may be some requirements for quality management but not to the level of a full-blown QMS. For third-order providers there may be no requirements at all. ISO 9001 places no demands upon companies to establish a QMS to a level of development where it is amenable to independent assessment and recognition. In fact, the standard propounds that a company must determine for itself the commercial benefits or otherwise of such approach. What the standard does do is provide a specification for a QMS which if comprehensively considered can place a company in an appropriate position to have its QMS independently assessed and recognised should it so wish.

The assessment of a QMS for companies operating in some sectors of industry, business and commerce is exceptionally important and, in fact, is virtually a prerequisite to their existence and livelihood in their business marketplace. Within the public sector, government authorities at national, regional and local levels consider quality management and the use of QMS as one of a range of criteria when selecting or engaging the providers of goods and services to publicly funded developments and projects. In the private sector, large client organisations use pre-qualification criteria similar to those used in the public sector. As quality is seen as one of the trinity of value for money – time, cost and quality – providers need to be able to demonstrate their capability and commitment to delivering on quality expectations.

For those companies where formal quality management is not as important an organisational performance indicator, it is not true to say that QMS is an irrelevance. It is the case that QMS is not a compulsory requirement of marketplace positioning but may be a voluntary commitment to seeking to deliver their provisions to a better than baseline standard. Ignoring the external dimension to QMS recognition, there may be good and valid reasons for QMS to support intra-organisational improvements to business processes, management procedures and working practices. Irrespective of its orientation, a QMS can have a positive part to play in the organisation and evolution of any company.

Types of assessment

There are three types of assessment associated with the recognition and assessment of a QMS. These range from what is essentially self-assessment through to independent assessment by an external party. Moreover, independent assessment is associated with the entry of the QMS in a register of companies of assessed capability – the concept of *certification*. The three types of assessment for a QMS are as follows:

1 *First-party or self assessment* – where a company has an established internal system and notifies interested parties when seeking to work with them.
2 *Second-party or collaborative assessment* – where a company has an established internal system which is open to the scrutiny of interested parties when seeking to work with them. In some business sectors companies may develop a system in collaboration with a key client or customer.
3 *Third-party or independent assessment* – where a company has a QMS which has been awarded a certificate from an independent, or third-party, body.

With first-party and second-party assessment it is not always the case that the provider company will have a formalised and comprehensive QMS in operation. It could be that it has a rudimentary quality system in place or a system which is applied only to discrete parts of the company's operations. It may also be the case that its procedures cannot be termed a QMS at all, but may have undertones of a sensible general management approach to quality. What is important with both these types of assessment is that when a client or customer reviews the provider's quality management approach, it is satisfied that the approach will give it what it seeks within the business relationship that will be established. With third-party assessment the judgement of the QMS is based on a uniform standard for quality, is determined independently of the provider's or customer's perceptions of quality, and is recognised by listing in a register of capability for quality.

Certification

The term certification is used in some countries instead of registration. The certificate awarded to the QMS following assessment confirms that the QMS has been awarded a certificate of registration and that the QMS meets with the requirements of the standard used in association with its development. The holding of a certificate does not and cannot guarantee that the company can and will deliver 100% quality in all its business activities. The certificate confirms that the company's QMS meets with a standard, and that the system places the company in a position of being able to provide a standard of quality commensurate with its business and customer requirements.

Benefits of QMS certification

The benefits of achieving certification of an ISO 9001 standards-based QMS were identified earlier in this chapter. There can be little doubt that a company which submits a QMS for independent assessment and achieves the recognition of certification may accrue a wide range of intra-organisational and external benefits and rewards.

The importance of independent recognition of a QMS and indeed any other management system is propounded by Davis Langdon Certification Services (DLCS):

> Certification makes for a more competent business, improved corporate governance and knowledge through transparent assessment reports and risk management principles, reduces staff turnover and assures outcomes for you, your stakeholders and clients.
>
> (DLCS, 2010)

While third-party assessment may demonstrate the optimum level of organisational management system performance, second-party assessment is often as important to customers and providers. Davis Langdon Certification Services (DLCS) notes:

> Second party audits are becoming increasingly prevalent in the property and construction industry. We undertake these on behalf of our client's suppliers/service provider's management systems to determine the level of conformance to the specified requirements as part of a supply contract and/or major project.
>
> (DLCS, 2010)

Framework for certification

There is a formal and structured framework for the certification of a QMS. 'Certification' bodies, which assess the QMS of companies, have their activities and conduct controlled by 'accreditation' bodies. In turn, the activities and conduct of accreditation bodies are monitored and controlled by government. In this way there is a hierarchical framework of accountability. This framework exists to ensure that all certification bodies – and there are many covering different sectors of industry and commerce – act professionally. They must meet strict codes of conduct to act impartially in all their activities and exert an appropriate level of competence to enable effective management systems assessment.

In the UK, the United Kingdom Accreditation Service (UKAS) is the regulatory body governing the activities of certification bodies which provide accredited certification services to companies seeking registration of their QMS. UKAS reports to and is accountable to the government and is responsible for overseeing the activities of certification bodies. Certification bodies are themselves represented as a service group by the Association of British Certification Bodies (ABCB). Certification bodies may cover a wide range of industry sectors or be specialist industry sector focused. Certification procedures are conducted by approved auditors working on behalf of the certification body and operating with the client organisation where the QMS is being assessed. The most prominent of all certification bodies within the UK is the British Standards Institution (BSI). There are also industry-sector-specific certification bodies, the most prominent within construction being: the Quality Scheme for Ready Mixed Concrete (QSRMC); the Timber Research and Development Association (TRADA); and the UK Certification Authority for Reinforcing Steels (CARES).

Internationally, the specifics of accreditation and certification frameworks have differences in other countries and continents to reflect local operating requirements and practice. Nevertheless, developing global harmonisation tries to ensure that the implementation of ISO 9001 is uniform throughout. Bureau Veritas (2010) is a name synonymous with management systems worldwide. The Bureau Veritas Group operates in 150 countries through 900 offices and, as such, it is one of the world's leading certification bodies, providing systems assessment and certification across a wide range of industry sectors including the construction sector. Bureau Veritas provides for the recognition of quality systems to ISO 9001 and also handles ISO 14001 and OHSAS 18001 systems in dual and triple applications with QMSs.

The certification process

A company which seeks to obtain the award of a certificate for a QMS which meets ISO 9001 must apply for assessment to be undertaken by an accredited certification body. The QMS will be assessed within a prescribed process of certification. The certification process involves the following activities:

1 *Pre-audit* – initial assessment of the QMS.

2 *Desk-top study* – assessment of the QMS documentation.

3 *Certification audit* – assessment of the QMS in operation.

4 *Certification outcome* – results of the QMS assessment leading to the award of the certificate or alternative outcomes.

5 *Surveillance* – six-monthly checks on the operation of the QMS by sampling parts of the system.

6 *Certificate renewal* – three-yearly detailed scrutiny of the QMS by extended surveillance or full certification audit.

Note that Stage 2: Desk-top study is sometimes carried out as part of Stage 1: Pre-audit.

Stage 1: Pre-audit

The purpose of the pre-audit stage is to ensure that:

• business processes and the QMS are at an appropriate state of implementation to engage in assessment for certification status;

• documentation of the system has been established to the appropriate level for assessment;

• the QMS is operating at a level of fitness for intended purpose;

- internal audit procedures are appropriate and applied correctly within the system;
- the company is prepared appropriately for subsequent stages of assessment;
- arrangements and resources for subsequent stages of assessment are considered and agreed.

When a company applies for QMS assessment, the certification body will appoint a lead auditor to begin the process of assessment with the company. Subsequently, a team of auditors may be involved depending upon the magnitude and complexity of the certification audit. At the pre-audit stage the auditor will engage with the company in discussions of the background to the establishment of the QMS. This will involve the concept, context and methodology of system development, documentation and implementation. The focus is on determining if the QMS has been established in an appropriate way and for the right reasons rather than examining explicit intricacies of the system. Detailed evaluation of system documentation and application will follow during subsequent stages of assessment. It is essential that the QMS focuses on identifying the environmental effects of the business, determines the impact of these effects and has applied the approach and resources necessary to ensure their effective management. The question which needs to be asked is: does the QMS meet the intended fitness for purpose?

The pre-audit will examine the company's internal audit procedures. It is paramount that the system identifies and handles appropriately any non-conformances within its business processes. In addition to examining audit procedures, audit plans and programmes will be reviewed to ensure that audits are frequent, timely, and facilitate appropriate actions where needed. Internal audits are not simply a matter of checking organisational activities against clauses in the standard nor routinely checking processes in turn to see how they are operating. Internal audits are concerned with checking that the QMS actually works. It is highly likely that improvements could be made, and the auditor will highlight these where identified. Much of the pre-audit attention to internal audit procedures will be given to assessing the capabilities of the company to undertake the audit effectively.

The pre-audit will also determine the company's state of preparedness for the subsequent stages of the assessment. There is little point in the company engaging the next stage if the QMS cannot satisfy the basics of examination by pre-audit. The auditor must determine if the system has a considered chance of success under the more searching aspects of assessment which will follow. The auditor will evaluate evidence to determine if the QMS does comply with the standard and also that it fulfils the company's quality policy, goals, objectives and targets.

It is essential that the auditor sees the commitment to continual improvement of processes and system procedures. The auditor will provide feedback to the company on observations made and make suggestions for further developments if these are necessary prior to progressing to the next stage of assessment. It may be that fundamental deficiencies exist with the QMS such that progress to

Stage 2 is not recommended at that time. Alternatively, minor amendments to the system may be recommended, which after rectification allow the assessment to proceed.

If the auditor is satisfied that the fundamental elements of the QMS are in place a recommendation to proceed to Stage 2, the desk-top study of system documentation, will follow. In concluding the pre-audit assessment, the auditor will discuss and agree the arrangements and resources for the desk-top assessment together with dates and schedules of activities to be conducted. This discussion will be confirmed in writing following completion of the pre-audit. Where a recommendation is made to modify and improve the system prior to progression to the next stage, then this will be reflected in the verbal feedback given and confirmed in writing thereafter.

Stage 2: Desk-top study

The purpose of the desk-top study stage is to ensure that:

- there is a detailed review of the company's QMS documents against the standard;
- the QMS documentation has addressed all the requirements of the standard;
- any interpretations of the standard in relation to particular organisational implementation are grounded in the generic standard;
- documentation is structured in such a way that procedures are traceable against the standard.

Desk-top study of the QMS documentation can be extensive and involved as it must ensure that comprehensive examination and evaluation of all aspects of the system against the standard are achieved. To simplify the overall task, the documentation is normally divided into sections and sub-sections as follows:

- *Documents*:
  - Quality policy (statements of goals, objectives, targets)
  - QMS manual
  - Quality management procedures
  - Quality implementation plan (where applicable)
  - Quality work, or working, instructions
  - Process, operation and task guides for outputs (products and services).
- *Legislative and contractual framework documents*:
  - Legislation governing business processes and procedures
  - Regulation of processes and procedures
  - Contractual requirements specified by clients and customers.
- *Background documents*:
  - Business processes and how they are undertaken
  - Customer requirements and their influence on processes and procedures

- Special requirements influencing processes (e.g. packaging and supply of products or components)
- Special requirements on services (e.g. project site working practices).

The auditor's focus for the desk-top study is to undertake a comparison of the company's documented QMS against the requirements of the standard. The auditor may carry out the assessment at the auditor's office or in the workplace of the client. The advantage of conducting the study in the workplace is that any matters requiring clarification can be raised with the client almost at once. Such convenience must be balanced with the auditor's need to conduct the study without distraction and in comfort, whereupon the auditor's office might be preferred. The location for the desk-top study is therefore at the discretion of the auditor.

The outcome of the desk-top study is normally verbal feedback followed by a written report. Within this, points of clarification will be identified, questions may be asked, omissions will be highlighted and any matters to be followed up during the certification audit will be confirmed. The company has the opportunity to reflect on the auditor's desk-top study and take corrective action to remedy any system deficiencies reported. The final aspect of the desk-top study stage is for the company to consider and agree the required time interval between completion of Stage 2 and the commencement of Stage 3, the certification audit. This can be to a timeframe of weeks or months as deemed appropriate.

Stage 3: Certification audit

The purpose of the certification audit is to ensure that:

- the QMS complies with the requirements of the standard in real application;
- the QMS meets legislative, regulatory and customer requirements and supports continual improvement to the system.

It was mentioned previously that while the pre-audit and desk-top study stages may be conducted by a single auditor, the certification audit may be carried out by an audit team. This will be determined by the extent and characteristics of the work involved. The certification body will determine the resources it devotes to this stage of assessment, consulting with the client where necessary. The certification audit will focus on the following broad areas of examination and assessment:

- management responsibility in fulfilling the company's quality policy;
- determination of quality goals, objectives and targets in relation to quality policy;
- monitoring of quality against objectives and targets including measurement, reporting, record keeping and review;
- operational control measures implemented to maintain quality;
- internal audit plans, procedures and schedules to maintain the effectiveness of the QMS and perpetuate continual improvements to the system.

The certification audit is probably the most involved stage of the assessment process. Whereas the pre-audit and desk-top study stages are carried out in what might be described as a semi-advisory and developmental atmosphere, the certification audit is formal, highly structured and clearly focused on determining whether or not the QMS is compliant with the standard. During the certification audit the auditors cannot offer advice or suggestions; rather they inspect and test the system.

The process of the certification audit normally involves the following:

- *The certification audit opening meeting* – where the lead auditor will: outline the assessment process; recap on the pre-audit and desk-top study stages; explain how the certification audit is to be conducted; agree an itinerary; assign audit activities; and formalise communication mechanisms between the audit team and the client.
- *The certification audit* – where the audit team undertakes the detailed assessment of the QMS in operation against the requirements of the standard.
- *The certification audit closing meeting* – where the lead auditor will report verbally on the findings of the audit team and give judgements on the efficacy of the QMS.

This is a structured and formal activity but best undertaken in a collegiate spirit with information and discussion facilitated by both parties throughout the assessment. All aspects of the QMS will be examined in some way and to some level of detail as deemed necessary by the audit team. The focus is on determining whether system aspects comply or otherwise with the standard.

Where an apparent deficiency results from the investigation of a particular aspect of the system, the auditors can request the client to check the nature and extent of the problem. The problem might just be a rogue activity among many compliant activities or a fundamental problem that inhibits system operation and effectiveness. Such checks are termed corrective action requests and can be minor or major depending upon the severity of deficiency identified. A minor corrective action would not prevent the system being certified, unless there were many such lapses in the system. Conversely, a major corrective action would prevent the system from being certified. A major non-compliance would likely be a documented procedure which is deficient to an extent that there is a high risk that quality will suffer.

It is correct and appropriate to say that considerable ambiguity can surround the determination of severity for any identified system deficiency. The decisions that auditors reach in their assessments must be treated by the client as objective and balanced judgements based upon the information provided and the information seen. The benefit of a team approach to assessment does mean that there is a knowledgeable, transparent and collective judgement made on all issues. In addition, the judgements reached are normally arrived at during or following extensive discussions with the client rather than in isolation. In addition to corrective action requests the audit team can make observations. These focus on making sensible improvements to the QMS. While corrective action requests must be responded to by the client, observations do not. In practice, observations are presented as opportunities for system improvement, and any actions taken can be reviewed during surveillance visits. The auditors will provide a verbal report of the proceedings at the certification audit closing meeting and later confirm their observations and recommendations in writing.

Stage 4: Certification outcome

The verbal feedback from the auditors on the certification audit will be confirmed in writing to the client following the event. A detailed report is also issued to the client from the lead auditor. The report covers all aspects of the certification audit including the specifics of corrective actions and observations. The recommendation made by the lead auditor to the certification body results in one of the following outcomes:

1 Award certification.

2 Award certification with minor corrective action(s).

3 Delay certification until improvements are made to the QMS in response to major corrective action(s).

The auditors are not empowered to award a certificate. They make a recommendation based on the judgements from the certification audit. The certificate is awarded by the accredited certification body.

Stage 5: Surveillance

The purpose of surveillance is to ensure that:

- the certification body maintains over the three-year period of the certificate a series of visits, which can be termed interim audits, to confirm that the QMS continues to operate effectively.

Surveillance visits are normally carried out at six-month intervals, with at least one visit taking place within any twelve-month period. Ideally, the lead auditor or member of the audit team which conducted the certification audit will carry out the surveillance visit. This is helpful to both the certification body and the company as there is existing knowledge and familiarity with the company and its QMS. The breadth and depth of examination carried out depend very much on the perceptions of the surveillance visit team, but the focus is always on the continuing need to ensure that the QMS complies with the standard and quality policy of the company. Surveillance will likely involve sampling selected parts of the QMS rather than the whole. Nonetheless, a number of key aspects will be looked at, as follows:

- the effectiveness of the QMS;
- the efficacy of quality management and operations;
- activities directed to enhancing system performance through continual improvement to the QMS;

- response to corrective actions and observations made during the certification audit or previous surveillance visits;
- continuing procedures used for internal audit activities.

Where the surveillance visit raises no aspects of concern, the auditor will advise the client verbally and follow up with written confirmation accompanied by a surveillance visit report. The report will: clear non-conformances from previous visits, if applicable; make any observations of system operation to promote improvement; and outline the audit activities undertaken during the visit. Where the visit raises aspects of concern, the auditor will advise the client, follow up in writing and provide a report which details: any non-conformances; observations; corrective actions; and matters to be examined during the next surveillance visit.

Stage 6: Certificate renewal
The purpose of certificate renewal is to ensure that:

- a comprehensive and detailed reassessment of the QMS is undertaken every three years.

The reassessment visit can take the form of an extended surveillance visit or a certification audit. Much depends upon the status of the QMS as seen by the auditors during the three-year period of surveillance visits. The activities involved in a reassessment visit follow those previously outlined for certification audits and surveillance visits.

The outcome from a reassessment, or certificate renewal, visit can be one of the following:

- renewal of the certificate and continued registration of the QMS;
- qualified renewal with a prescribed timescale for the implementation of specified corrective actions;
- recall of the certificate and discontinuation of registration.

Minor corrective actions resulting from a reassessment visit are commonplace, and most companies will respond positively and rectify matters such that certification renewal is achieved. Only in a very small number of cases does a reassessment result in the recall of a certificate. While the process involved in obtaining and maintaining a certificated QMS may appear involved and perhaps burdensome, the true value of certification is founded in the structure and integrity of the process. Moreover, the regime imposed upon companies to manage quality through a unified and recognised management system is based in the value of holding a valid certificate.

QMS: application to construction

This section examines how an effective framework for a QMS may be configured by a principal contracting organisation operating within the construction industry. It is based on case study material.

QMS case study example: Developing and applying a QMS

Introduction

This section presents a framework for the main elements involved in developing and implementing a QMS meeting the requirements of ISO 9001. As the full set of documents forming a QMS can be voluminous, it is not intended to provide a full set but rather present simple examples for each main element. Readers who require complete sets of documents and template pro formas for system development are directed to the many private-sector organisations and consultancy practices which provide system development services on a commercial basis. The example presented should be read in conjunction with Chapter 5 of this book, specifically the second and third sections, and the quality systems standard.

QMS focus of application

The example presented is a QMS for application by the principal contracting organisation. The system is a composite of applications of QMSs obtained from six principal contractors. As such, the system reflects current industry practices. The approach, structure and organisation depicted form a useful framework for QMS development and implementation by any principal contractor. They may also provide guidance to contractors/sub-contractors.

QMS conceptual approach

The example reflects a company-wide 'umbrella' QMS for application throughout the corporate organisation and with application to the many 'project-based' construction projects that might be undertaken in the normal course of the business activities of the principal contractor.

Core business and processes

The core business of the principal contractors in the case study is, ostensibly, building work in the new-build construction sector, augmented by some repair, refurbishment and maintenance contracts. The business processes in all cases are delivered by on-site production through conventional construction projects using broadly similar and traditional construction techniques and methods of management.

Standards

The companies all implement ISO 9001 compliant QMSs.

Relationship with other organisational systems

Five of the six companies use ISO 14001 compliant EMSs and all have OHSAS 18001 compatible H&SMSs.

System structure and main elements

System structure

The QMS structure shown in Figure 2.13 is a corporate-based umbrella system embracing all business activities and focusing down to outputs, or services,

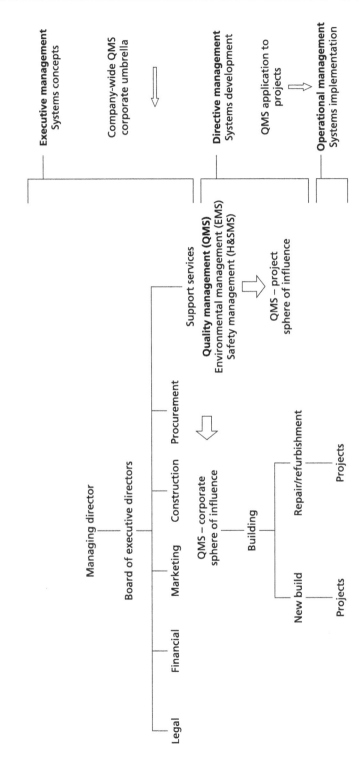

Figure 2.13 Company-wide corporate umbrella-based QMS applied to construction projects

provided at the construction project production site. This system has the advantages of: (1) application to the whole company; (2) setting a quality standard of performance throughout the organisation; and (3) embedding a set of uniform management procedures and working instructions for use on all the company's construction projects.

Organisational structure

Central to the formation of the organisational structure for QMS establishment is the appointment and designation of the *quality manager* – the management representative specified by ISO 9001. The quality manager forms the focus of management for both the corporate perspective and project applications. The essence of the QMS is to apply the management procedures and work instructions developed in the system documentation to the organisational processes. The role of the quality manager is crucial to the successful implementation of the system as the role entails the assurance of open, timely and appropriate communications between senior management and project-based and first-line supervisors servicing the construction projects. Figure 2.14 illustrates the organisational structure adopted by the case study organisations, with the important positioning of the quality manager within that structure.

The role of the quality manager is an oversight management one and is supported at various project levels. For small construction projects the site manager will assume responsibility for quality, while on larger projects the site manager will, likely, be assisted by a site-based quality manager. This person may be based on the site or simultaneously service a number of construction projects. In line management terms the site-based quality manager is accountable to the site manager, who has single-point responsibility for the project. In portfolio management terms the site-based quality manager will link closely to the quality manager. The quality manager must ensure that the QMS is established and maintained to the requirements of ISO 9001 and that quality management procedures are applied to all construction processes. The site manager and site-based quality manager must ensure that the system's corporate quality procedures are effectively translated into work instructions for implementation during the construction works on site.

Business positioning

The principal contractors involved in the case study used a variety of mechanisms to analyse organisational performance in readiness for establishing their QMSs, with SWOT analysis and excellence modelling used. Customer focus is central to organisational thinking and business positioning. This is particularly true within construction, where contractors have to meet increasingly demanding client needs, requirements, technical specifications and expectations.

Business factors

The dominating business factors recognised by the case study organisations form the key aspects to be considered when developing an effective QMS. These are as follows:

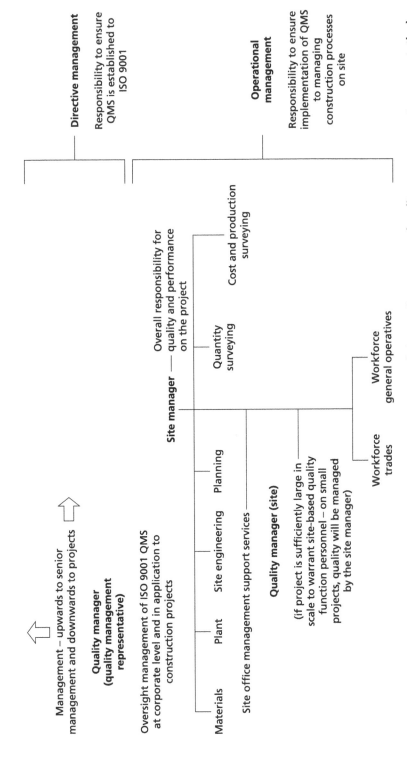

Figure 2.14 Project quality management illustrating the position of the quality manager (quality management representative) and site-based quality management

- The core business activities, including their management, resourcing and production, including supply chain implications.

- The processes involved in delivering the business activities, the technologies and management methods used, including potential alternatives.

- The specific requirements of the QMS relating to the standard – ISO 9001.

- The generic and specific quality requirements of the organisation's clients and customer base.

- The requirements of industry-specific legislation which impinge upon business activities.

- The adherence to general and industry-specific regulation of business practice – national and local.

- The activities of competitor organisations within the industry marketplace in terms of their organisational performance and quality of outputs.

QMS development – key elements

Six key elements of QMS development are identified and considered by the case study organisations, as shown in Figure 2.15. They are as follows:

1 quality policy

2 organisation for quality management

3 risk assessment of performance and quality

4 planning with a focus on the project quality plan (PQP), sometimes termed a quality site plan or quality implementation plan

5 implementation of management procedures and work instructions

6 auditing and review

These elements are consistent with the requirements of the ISO 9001 clauses described earlier. Each element is sequential in terms of organisational thinking, consideration and development and all are interdependent.

System documentation

The structure for QMS documentation is illustrated in Figure 2.16. The sequence of document development follows that described earlier in terms of the conventional hierarchy, or document pyramid. The structure outlined, used in principle by all the case study systems, highlights four levels of documentation:

1 Quality manual

2 Quality management procedures

3 Quality plan

4 Quality work instructions

For the case study organisations a range of terminology was used to define the procedures, although in operation all derivations performed the same function

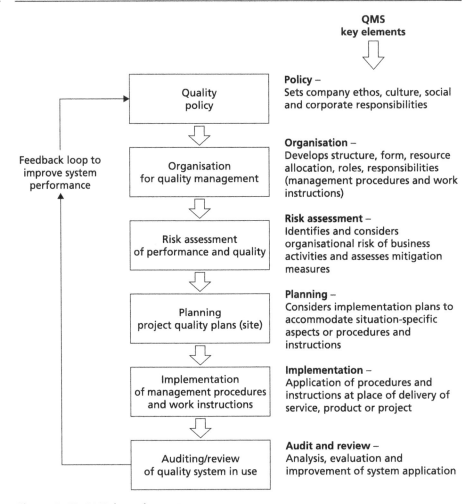

**QMS
key elements**

Policy –
Sets company ethos, culture, social
and corporate responsibilities

Organisation –
Develops structure, form, resource
allocation, roles, responsibilities
(management procedures and work
instructions)

Risk assessment –
Identifies and considers
organisational risk of business
activities and assesses mitigation
measures

Planning –
Considers implementation plans to
accommodate situation-specific
aspects or procedures and
instructions

Implementation –
Application of procedures and
instructions at place of delivery of
service, product or project

Audit and review –
Analysis, evaluation and
improvement of system application

Quality
policy

Feedback loop to
improve system
performance

Organisation
for quality management

Risk assessment
of performance and quality

Planning
project quality plans (site)

Implementation
of management procedures
and work instructions

Auditing/review
of quality system in use

Figure 2.15 QMS key elements

– to encapsulate what were, conventionally, standing procedures used throughout the companies and applied to their output services on construction projects.
It should be noted that the quality plan relates in systems terms to an 'implementation plan' and sometimes is referred to as a project quality plan. The core
value of the quality plan is to translate company-wide system aspects into
situation- or project-specific aspects, leading the way for management procedures
and work instructions to be applied on construction projects. This quality plan
will have some generic and specific characteristics to it, and there will therefore
be a particular plan for each construction project.

The precise nature and content of the QMS documentation will, obviously, be
influenced by the orientation of the business activities and processes used by the
organisation. In all cases, the documents will be informed by the quality policy,
which establishes the ethos of the company and sets the foundations for quality
goals, objectives and targets. System feedback will be common to any approach

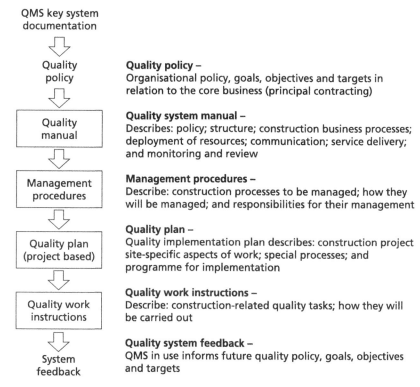

Figure 2.16 QMS documentation structure

such that continual improvement – a tenet of ISO 9001 QMS – is positively encouraged.

Policy, goals, objectives and targets

These aspects of a QMS will vary considerably among organisations using QMSs. Notwithstanding, any company should set out its quality stall with a sound, robust and transparent *quality policy statement* of its intentions towards quality in the context of its business. See Figure 2.17, which is annotated to provide greater explanation.

System implementation

Implementation of the QMS confirmed two distinct requirements:

1 A method of defining, describing and explaining the approach to delivering the system elements – *documents*: management procedures, implementation plan and work instructions.

2 A method of depicting procedures, plans and instructions in a written, practical and usable form – *implementation pro forma*, sometimes called *administrative forms*.

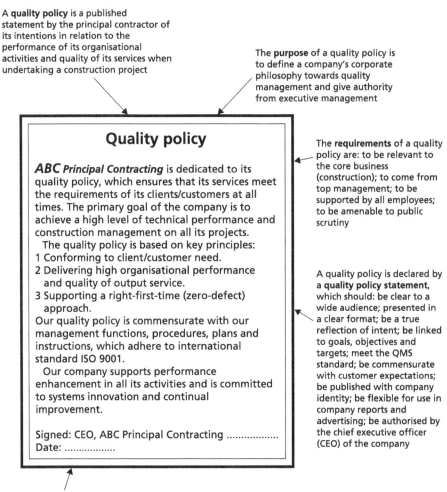

A **quality policy** is a published statement by the principal contractor of its intentions in relation to the performance of its organisational activities and quality of its services when undertaking a construction project

The **purpose** of a quality policy is to define a company's corporate philosophy towards quality management and give authority from executive management

Quality policy

***ABC** Principal Contracting* is dedicated to its quality policy, which ensures that its services meet the requirements of its clients/customers at all times. The primary goal of the company is to achieve a high level of technical performance and construction management on all its projects.
   The quality policy is based on key principles:
1 Conforming to client/customer need.
2 Delivering high organisational performance and quality of output service.
3 Supporting a right-first-time (zero-defect) approach.
Our quality policy is commensurate with our management functions, procedures, plans and instructions, which adhere to international standard ISO 9001.
   Our company supports performance enhancement in all its activities and is committed to systems innovation and continual improvement.

Signed: CEO, ABC Principal Contracting ....
Date: ....

The **requirements** of a quality policy are: to be relevant to the core business (construction); to come from top management; to be supported by all employees; to be amenable to public scrutiny

A quality policy is declared by a **quality policy statement**, which should: be clear to a wide audience; presented in a clear format; be a true reflection of intent; be linked to goals, objectives and targets; meet the QMS standard; be commensurate with customer expectations; be published with company identity; be flexible for use in company reports and advertising; be authorised by the chief executive officer (CEO) of the company

The **content** of a quality policy should include: corporate responsibility; customer accountability; performance expectations; communication requirements; and improvement desires

Figure 2.17 Typical policy statement – annotated for additional explanation

As discussed previously, the documents can be of singular form – a quality manual – or can be a set of documents – typically a set containing: a quality manual; quality management procedures; quality plan; and quality work instructions. In the example, all of the contractors used multiple documents which were broadly similar to reflect their QMSs.

Requirement 1: System documents for construction organisation
The following list illustrates the groups of elements which can be used to convert QMS documents into construction-related documents:

1.0   Quality management system control
2.0   Corporate quality policy

3.0   Company quality organisation
4.0   Company standing quality management procedures
5.0   Project quality plans
6.0   Quality work instructions

These are incorporated into the QMS documentation as follows:

1.0   Quality management system control (in the manual)
2.0   Corporate quality policy (in the manual)
3.0   Company quality organisation (in the manual)
4.0   Company standing quality management procedures (in the manual *or* in a separate document)
5.0   Project quality plans (in a separate document)
6.0   Quality work instructions (in the manual or in a separate document)

These elements are presented to a standard and uniform method of presentation within the quality manual and/or other system documentation as shown in the sequence of Figures 2.18 to 2.25 inclusive. It is seen in each case that a standard format is used throughout, with each page of any document containing: logo, name and details of the company; document numbers and dates of application; issue and revision details; links to other system documents; and authorised signatories. The same or a similar format may be used throughout a complete set of QMS documentation.

Each clause of ISO 9001 must be embraced within the set of documents, so each of the six sections shown in the system contents (Figure 2.19) would need to be greatly expanded upon. The vast majority of the document content will be contained in section 4.0 – the company standing quality management procedures – where all of the ISO 9001 clauses will be considered and configured into practical procedures for implementation. The precise detail of any management procedure does not have to be described with the main documents but can be referenced to associated documents. For example, Figure 2.23 shows how standing procedures for concrete works can be related to guidance documents (GDs) such as GD/010/00: a specification for testing concrete samples. Likewise, special site-based procedures given in the project quality plan (Figure 2.24) or work instructions (Figure 2.25) can relate to separate documents as shown.

Linking construction technology and management to the system documents – the method statement
System documents must be based upon and configured around the business processes of the company. Within construction the business is based upon a great many technological processes and their associated tasks, or operations, all of which need to be determined, described and scheduled in association with the resources deployed on a project. An approach needs to be adopted which reflects the construction process inputs and outputs in an uncomplicated way. The traditional and most commonly used and understood mechanism for this is the *construction method statement.*

ABC	ABC Principal Contracting (International) Ltd	Document no.:
		Page no.:
	QUALITY MANUAL	Date:

Quality Manual Cover Sheet

Company Quality Manual

This Company Quality Manual is the property of ABC Principal Contracting (International) Ltd.

The manual does not form part of any contract.

The company reserves the right to amend its standing instructions, management procedures, plans and work instructions in order to comply with individual contract requirements, management system standards, legislation and regulation.

Issue no.:	Links to other system documents:	Authorised by Quality Manager:	Date:
Issue date:			
Revision no.: Date:		Authorised by Chief Executive Officer:	Date:

Figure 2.18 Quality manual – example cover sheet

The construction method statement is a two-dimensional chart which, in its simplest format, lists construction operations on the Y-axis and outlines on the X-axis, in a series of boxes, the construction methods which will be used to complete the operations, together with an outline of the resources which will be deployed. The method statement highlights the main operations within the total construction process, which can then be broken down into smaller components of the process, or tasks, for the purpose of detailed planning.

It is essential to remember that the management procedures and work instructions define and describe the 'quality management' aspects associated with undertaking the construction processes, operations and tasks as identified in the construction method statement. They do not define and describe the 'building

ABC	ABC Principal Contracting (International) Ltd	Document no.:
		Page no.:
	QUALITY MANUAL	Date:

Quality Manual Contents

Contents

Section:

1 Quality Management System Control

2 Corporate Quality Policy

3 Company Quality Organisation

4 Company Standing Quality Management Procedures

5 Project Quality Plans

6 Quality Work Instructions

Issue no.:	Links to other system documents:	Authorised by Quality Manager:	Date:
Issue date:			
Revision no.: Date:		Authorised by Chief Executive Officer:	Date:

Figure 2.19 Quality manual – example contents

technology and construction methods', which are detailed within construction/production manuals if so needed.

Each operation or group of operations contained within the method statement will need to be considered in relation to the QMS requirements specified by ISO 9001 in its set of clauses as presented earlier and noted again below. All sections of the standard apply with Clause 7: Product realisation, pertinent where required in the context of a 'service' related output:

1   Scope
1.1   General
1.2   Application

ABC	**ABC Principal Contracting (International) Ltd**	Document no.:
		Page no.:
	QUALITY MANUAL	Date:

Quality Manual System Control

Section 1: Quality Management System Control

Sub-section 1.2: Change Control

- This manual is a controlled document.

- This manual is revised periodically to maintain currency of application.

- The issue number and date and revision number and date are specified.

- Obsolete and uncontrolled copies will be removed from circulation, with base copies, duly marked, forwarded to system files.

Issue no.:	Links to other system documents:	Authorised by Quality Manager:	Date:
Issue date:			
Revision no.: Date:		Authorised by Chief Executive Officer:	Date:

Figure 2.20 Quality manual – example QMS control

ABC	**ABC Principal Contracting (International) Ltd**	Document no.:
		Page no.:
	QUALITY MANUAL	Date:

Quality Manual System Policy

Section 2: Quality Policy

Quality Policy

ABC Principal Contracting is dedicated to its quality policy, which ensures that its services meet the requirements of its clients/customers at all times. The primary goal of the company is to achieve a high level of technical performance and construction management on all its projects.

The quality policy is based on key principles:
1 Conforming to client/customer need.
2 Delivering high organisational performance and quality of output service.
3 Supporting a right-first-time (zero-defect) approach.

Our quality policy is commensurate with our management functions, procedures, plans and instructions, which adhere to international standard ISO 9001.

Our company supports performance enhancement in all its activities and is committed to systems innovation and continual improvement.

Signed: CEO, ABC Principal Contracting ....
Date: ....

System version no.: ....

Issue no.:	Links to other system documents:	Authorised by Quality Manager:	Date:
Issue date:			
Revision no.: Date:		Authorised by Chief Executive Officer:	Date:

Figure 2.21 Quality manual – example quality policy statement

6   Resource management
6.1   Provision of resources
6.2   Human resources
6.3   Infrastructure
6.4   Work environment

	ABC Principal Contracting (International) Ltd	Document no.:
ABC		Page no.:
	QUALITY MANUAL	Date:

Quality Manual Company Organisation

Section 3: Company Quality Organisation

Sub-section 3.1: Quality Management Structure (Corporate)

- The company-wide corporate umbrella QMS is shown in system Figure 2.13. Quality management is the oversight responsibility of the quality manager at the executive organisational level.

Sub-section 3.2: Quality Management Structure (Project)

- The structure for project quality management is shown in system Figure 2.14. The quality manager (executive) has oversight responsibility for implementation, maintenance and improvement of the QMS as applied to all projects.
- The quality manager (site) has responsibility (under line management by the site manager) for managing all aspects of quality on project sites.
- The site manager assumes overall responsibility for quality matters on and throughout the project.

Sub-section 3.2: Quality Organisation Charts

- Refer to Figures 2.13 and 2.14 for QMS organisation charts.

Issue no.:	Links to other system documents:	Authorised by Quality Manager:	Date:
Issue date:		Authorised by Chief Executive Officer:	Date:
Revision no.: Date:			

Figure 2.22 Quality manual – example company quality organisation

7   Product realisation
7.1   Planning of product realisation
7.2   Customer-related processes
7.3   Design and development
7.4   Purchasing
7.5   Production and service provision
7.6   Control of monitoring and measuring equipment

ABC	ABC Principal Contracting (International) Ltd	Document no.:
		Page no.:
	Quality Management Procedures	Date:

Quality Procedures
Company Standing QM Procedures

Section 4: Company Standing Quality Management Procedures

Sub-section 4.8: Concrete Works
            4.8.1: Concrete in Floor Slabs

Where concrete works take place the following guidance documents must be consulted and standing procedures applied:

GD/001/00   Approval of concrete supplier
GD/002/00   Approval of reinforcement supplier
GD/003/00   Approval of transportation and delivery
GD/004/00   Specification for setting out
GD/005/00   Specification for reinforcement installation
GD/006a/00   Specification for timber formwork
GD/006b/00   Specification for metal formwork
GD/007/00   Specification for protection and curing
GD/008/00   Specification for cleaning works
GD/009/00   Specification for inspection of works
GD/010/00   Specification for testing concrete samples

Issue no.:	Links to other system documents:	Authorised by Quality Manager:	Date:
Issue date:			
Revision no.: Date:		Authorised by Chief Executive Officer:	Date:

Figure 2.23 Quality procedures – example company standing quality management procedures

8   Measurement, analysis and improvement
8.1   General
8.2   Monitoring and measurement
8.3   Control of non-conforming product
8.4   Analysis of data
8.5   Improvement

ABC	ABC Principal Contracting (International) Ltd	Document no.:
		Page no.:
	Project Quality Plan	Date:

Project Quality Plan
Special Site-Based Procedures

Section 5: Project Quality Plan

Sub-section 5.3: Special Site-Based Procedures
         5.3.8: Special Procedures – Concrete Works

Where concrete works take place *on this project* the following special procedures documents must be consulted and special procedures applied:

SP/001/00   Works in confined spaces

The following applies (examples):

Task requirements (SP/002/00)
• Risk assessment
• Buddy-system working
• Material handling

PPE requirements (SP/003/00)
• Hearing protection
• Respiratory protective equipment
• Eye/face protection
• Protective clothing
• Skin protection
• Safety harnesses

Issue no.:	Links to other system documents:	Authorised by Quality Manager:	Date:
Issue date:			
Revision no.: Date:		Authorised by Chief Executive Officer:	Date:

Figure 2.24 Project quality plan – example special site-based procedures

Developing management procedures and work instructions
For any operation the QMS should contain an appropriate quality management procedure and a quality work instruction. The quality management procedure will be a standing generic procedure reflecting the same or similar applications throughout the organisation and all the projects the system is applied to. The quality work instruction will be project specific, reflecting the particular requirements of

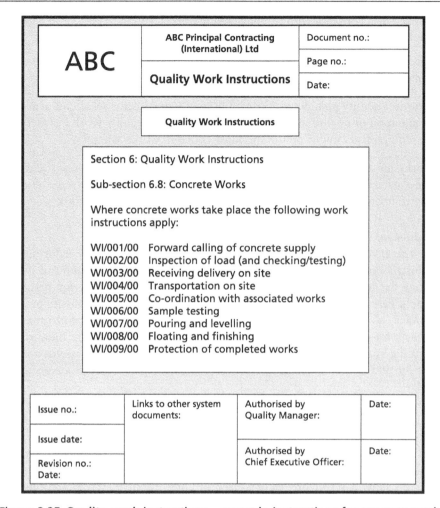

Figure 2.25 Quality work instructions – example instructions for concrete work

the project. The particularities of the project will be reflected in the project quality plan. So, for example, a quality management procedure for concrete work will specify that all concrete deliveries will be:

- checked for quality prior to acceptance on site;
- inspected and tested to specified technical and contractual criteria;
- received on site to process the specified procurement order/delivery/receipt/ invoice documentation.

The quality work instruction will specify that all tasks associated with concrete work will be:

- carried out in accordance with current codes of practice for concrete work;

- fulfilled in accordance with technical requirements, specifications, drawings and applicable contract documents;

- checked, inspected and tested to specified measuring and testing criteria and using appropriately calibrated measuring equipment.

The project quality plan may specify additional requirements which influence the determination of quality management procedures and quality work instructions. Such requirements might be for specialist concrete works where:

- works must be carried out in conjunction with the client's special schedule for concreting activities to accommodate working within an active production environment (contractor to refer to client specification CS/001/00).

Requirement 2: Implementation pro forma (administrative forms)
Once the QMS has been developed in the form of a manual, procedures, plan and instructions, it needs to be implemented. Ways need to be found of implementing the documentation in association with site-based construction practices. It would be inconvenient to use the manual, for example, as an on-site practice document, so an *implementation pro forma* is often used. Such pro formas are the basis for the undertaking of construction operations, for the monitoring, measuring and recording of performance in relation to those operations, and for reporting and initiating action in relation to the management of processes to the standard.

The essential characteristic of implementation pro formas is that they are kept simple for actual use on site. Often they take the format of simple charts, spreadsheets or checklists which quality managers can take on site and complete at the workplace. Obviously, there is a direct link between the requirements of the QMS standard, the QMS documentation set and the pro forma. Figures 2.26, 2.27 and 2.28 show a set of pro formas used in application to system auditing – a requirement of Clause 8: Measurement, analysis and improvement, sub-clause 8.2.2: Internal audit. These illustrate the requirements of audit planning, audit reporting and corrective action following an audit.

Again, for any aspect of construction work undertaken within the total process and in association with the requirements of the standard, an implementation pro forma should be established. A common way of achieving this is to use a uniform template, and such an approach is reflected in the examples presented. Using an individual pro forma for each and every construction operation and separate task would be unwieldy and unworkable. So a pro forma is applied to key construction processes and, where necessary, identified key groups of operations. The important point with the use of pro formas is that they should be kept as simple as possible for effective use, be fully amenable to routine and consistent application, and, as a single pro forma or one of a set of pro formas, follow a process or operation through from initiation to completion. Such pro formas can be configured not only to reflect monitoring and reporting mechanisms, but also as the basis of file copies for the QMS project files.

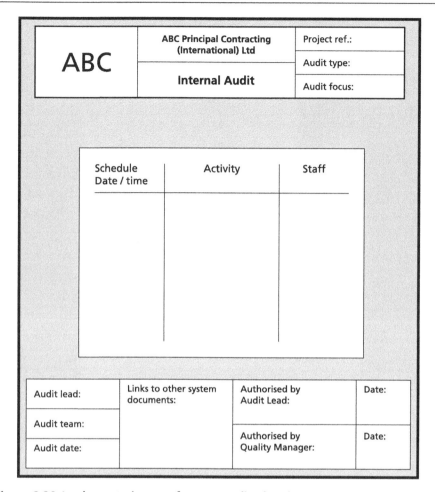

Figure 2.26 Implementation pro forma – audit planning

Auditing and review

All of the case study contractors implemented their QMS to the ISO 9001 standard and therefore all pursued a QMS continual improvement philosophy. In all cases the QMS implemented supported the Deming methodology (PDCA). As all the organisations had been using QMSs over a long period, the continual improvement philosophy was embedded, and there had been evidenced results from clients of improved organisational performance and quality of service.

A final thought

ISO 9001 is the most widely accepted and sought-after European quality management system. Compliance sends a very strong message: Your company takes quality and customer satisfaction seriously, which will influence potential customers to choose your company over competitors.

(ISO, 2008)

ABC	**ABC Principal Contracting (International) Ltd**	Project ref.:
		Audit type:
	Internal Audit	Audit focus:

Scope of audit:
(output/service)

Generic aspects of audit:

Specific aspects of audit:

Audit summary:

Distribution list:

Audit lead:	Links to other system documents:	Authorised by Audit Lead:	Date:
Audit team:		Authorised by Quality Manager:	Date:
Audit date:			

Figure 2.27 Implementation pro forma – audit report

Quality management: key points, overview and references

This section presents: a summary of the key points from the collective sections of Chapter 2; an overview of Chapter 2; and a list of references used in the compilation of Chapter 2.

Key points

- To survive and prosper in the business marketplace, any provider of products or services must, first and foremost, ensure that it meets the needs and expectations of its clients and customers.

ABC	ABC Principal Contracting (International) Ltd	Project ref.:
		Audit type:
	Internal Audit	Audit focus:

Audit ref.:
_____

Audit report ref.:
_____

Audit findings:

_____

Quality management response:

_____

Agreed corrective actions:
(including completion dates)

_____

Follow-up
(including sign-off)

Audit lead:	Links to other system documents:	Authorised by Audit Lead:	Date:
Audit team:			
Audit date:		Authorised by Quality Manager:	Date:

Figure 2.28 Implementation pro forma – corrective action

- Within the construction industry, the organisational performance of the provider (designer, contractor, supplier, or other) and the quality of its business outputs (products, services, projects) are perhaps the most prominent differentiating characteristics of the standard of performance within the commercial marketplace.

- Construction clients, customers and stakeholders expect the provision of a high standard of quality in the finished product in addition to delivering the product on time and to budget – they expect 'value for money'.

- A carefully considered, structured and systematic approach to the management of quality is prerequisite to addressing the demands of clients and customer for better organisational performance and quality of business outputs.

- A quality management system (QMS) can help those companies which provide goods and services to improve the quality of their provision and enhance customer satisfaction.

- An effective QMS focuses on delivering better managerial, supervisory and operational performance of its organisational resources, more efficient execution of its business processes and the highest quality for its business outputs.

- A QMS conveys to the business marketplace that a company has a genuine commitment to customer-focused quality of outputs.

- Eight principles of quality management can assist a company to achieve better organisational performance: (1) customer focus; (2) leadership; (3) involvement of people; (4) process approach; (5) system approach to management; (6) continual improvement; (7) factual approach to decision making; and (8) mutually beneficial supplier relationships.

- The process model can be applied effectively to configure a QMS as it intrinsically links organisational processes with a view to continual management system improvement while meeting customer requirements.

- A company must develop, implement and maintain a QMS based upon its particular needs and requirements, business situation and circumstances.

- ISO 9001 Quality Management Systems – Requirements, one of the ISO family of standards relating to quality management, provides guidance to companies seeking to apply a QMS to their organisational activities and business outputs.

- There are many business advantages for companies establishing an ISO 9001 compatible QMS through a range of external and intra-organisational benefits.

- Complementary business approaches to ISO 9001 related QMSs exist, with the most prominent of these being total quality management (TQM) and the European Foundation for Quality Management (EFQM) Excellence Model – both have attributes which can be considered as part of QMS development and implementation.

- ISO 9001 QMS – Requirements contains eight main section clauses which provide guidance on: scope; references; terms and definitions; quality management system; management responsibility; resource management; product realisation; and measurement, analysis and improvement.

- In applying the standard to a QMS there are six key elements for an organisation to consider: (1) quality policy; (2) organisation; (3) risk assessment; (4) planning; (5) implementation; and (6) auditing and review.

- A QMS is, normally, configured as a 'soft', or paper-based, management system founded on a set of system documentation.

- Documentation for a QMS takes the form of the 'document pyramid' embracing: a quality system manual; management procedures; situation-specific implementation plans; and work instructions.

- A QMS can be first-party, second-party or third-party (independently) assessed and in the case of third-party assessment can be subject to the process of certification and registration of the QMS with a government-accredited management system certification body.

- QMS certification can bring very considerable benefits in the business marketplace as customers recognise the importance of a provider of products and services being able to demonstrate effective organisational performance and better quality of business output.

- QMS certification is virtually a prerequisite in most sectors of business and commerce where clients and customers place very considerable importance on the pre-qualification of companies when procuring goods and services.

- The use of a QMS is firmly embedded within the construction industry with a great many service providers (principal contractors/sub-contractors) and suppliers of goods and materials operating certificated QMSs within their organisations and applied to their projects.

- Almost all public-sector clients and a great many private-sector customers demand or expect procured inputs to the construction processes to maintain a certificated QMS and apply it to their contracts and projects

Overview

Quality is a characteristic of products and services which customers today simply expect – nothing less than good quality is a basic business requirement. Quality is more than delivering an inherent characteristic of a product or service. It is about demonstrating high organisational performance of the core business, focusing on the needs, requirements and expectations of customers and adding value to the business relationship between the provider and customer. Such a standard of provision does not simply happen or happen by chance. Quality must be an organisational desire, challenge and commitment. Quality must be managed. Quality management is one of the key functional management disciplines in any industry, business or commercial sector. It is central to delivering high standards of performance within the construction industry, where the quality of the finished product is a key determinant in judging the overall success and value of the project. The establishment of a formalised approach to quality management and in particular the use of a quality-focused management system can place an organisation in an advantageous position to respond best to the demands of its customers. A carefully conceived, well-structured and effectively implemented QMS can focus the activities of a company towards delivering better managerial and operational performance, enhance business processes and promote the highest quality of business outputs. Moreover, accomplishments and success in the management of quality can positively influence management performance in other areas of the business, so promoting whole-organisation continual improvement. The bottom line is that better quality business processes leads to better-quality outputs, and better-quality outputs leads to better-satisfied customers – and customers are the future business.

References

BQF (2009). *The EFQM Excellence Model*. British Quality Foundation (BQF). www.bqf. org.uk

Bureau Veritas (2010). *Bureau Veritas Services*. www.bureauveritas.com

Crosby, P. (1979). *Quality is Free*. McGraw-Hill, New York.

Davis Langdon Certification Services (DLCS) (2010). *Certification Services*. www.davis-langdon.com

Deming, E. (1950). *Some Theory of Sampling*. Wiley, New York.

DTI (2009). *TQM*. Department of Trade and Industry (DTI). www.dti.gov.uk

Feigenbaum, A.V. (1951). *Quality Control: Principles and Practice*. McGraw-Hill, New York.

IiP (2009). *Investors in People Standard*. www.investorsinpeople.co.uk

Imai, M. (1997). *Genba Kaizen: A Commonsense, Low-Cost Approach to Management*. McGraw-Hill, New York.

Ishikawa, K. (1984). *Guide to Quality Control*. Asian Productivity Organisation, Japan.

ISO (2008). *BS EN ISO 9001:2008 Quality Management Systems – Requirements*. International Organization for Standardization, Geneva.

Juran, J. (1964). *Managerial Breakthrough*. McGraw-Hill, New York.

Kaplan, R.S. and Norton, D.P. (1996). Using the balanced scorecard as a strategic management system. *Harvard Business Review*, Jan/Feb, 75–85.

Smith, G. (1993). The meaning of quality. *Total Quality Management*, 4, (3), 235–245.

UK Government Cabinet Office (2009). *Charter Mark*. www.cse.cabinetoffice.gov.org

CHAPTER 3

Environmental management systems

Introduction

Chapter 3 focuses on *environmental management systems*, or EMSs. Companies throughout the industrial, manufacturing and commercial sectors, including the construction sector, are under increasing pressure to make their business activities and outputs more environmentally acceptable. Chapter 3 examines organisational awareness of environmental matters in relation to business and organisational need and how an effective EMS can assist the organisation to meet its environmental responsibilities. With detailed reference to international system standards, in particular ISO 14001 (ISO, 1996), Chapter 3 examines system development and implementation, third-party system assessment and certification, environmental supply chains, and systems application within construction.

Environmental management and organisational awareness

This section examines: business and organisational needs for environmental management; awareness and recognition of environmental matters and sustainability; the corporate social responsibilities and obligations of companies and organisations for ensuring environmental safeguards; the benefits of effective environmental management; and the need for a standards-based systems approach to environmental management.

Business and organisational need

Addressing the environmental requirements of a company's activities, whether their provisions are services or products, is absolutely crucial to maintaining a competitive advantage within the business marketplace. In addition to satisfying the needs and wants of clients and customers for better quality and more timely delivered and more cost-effective services and products, a company must deliver much better environmental performance. Environmental standards for business operations, increasingly stringent legislation, closer industry regulation and wider public expectations have all converged to put unquestionable pressure on the ways in which an organisation perceives and manages its business activities. All sectors of industry, commerce and business have become seriously challenged to be more environmentally aware, to recognise the potential effects that their activities can have on the environment, to conduct business in an environmentally friendly way and to promote sustainability. This is particularly true within construction, which has an inherent and irrefutable impact on the environment. In the past, a company might have been oblivious to the physical and natural environment. Conversely, it might have been aware but have simply turned a blind eye. Today, it simply cannot afford to do so.

Environmental impact of business, responsibilities and response

The paradigm shift in awareness of environmental matters within business, industry and commerce is reflected in the 'Green List' (*Sunday Times*, 2009). The list considers the most proactive companies and organisations striving to improve their environmental performance: 'The 60 companies listed are all at the forefront of efforts to produce real environmental change – enterprising, enlightened and fizzing with new ideas.' Participants vary from small supply companies employing a few people to supermarket giants employing several hundred thousand staff. The identifying characteristics common to all those listed are their awareness of the environmental dimensions of their business, their recognition of the potential impacts of their organisations upon the environment, and the proactive measures they take to obviate and mitigate those impacts.

It is perhaps not surprising to see that construction companies feature prominently in the Green List. These include national–international contracting organisations, building materials suppliers, component supply companies, architecture design practices and property development organisations. *'Bigger companies with high environmental impact'*, as categorised in the Green List, include the construction companies Willmott Dixon Group, Skanska UK, Carillion and the Wates Groups together with Kingspan and Johnson Tiles, both leading UK building materials suppliers. Recognising the environmental significance of their businesses, all of the above feature in the top 30 of the Green List, with Willmott Dixon and Skanska ranked in the top 10.

From an environmental management perspective, forty-four of the sixty organisations in the Green List, including all the construction-related companies previously identified, have established an EMS that meets the ISO 14001 international standard. It is clear that these highly progressive construction industry organisations are serious about their green commitments and place environmental matters at the heart of what their companies do. Moreover, these, and indeed other, leading organisations within construction are committed to sustainable building and sustainable practice throughout the total building process and construction supply chains.

Sustainable development and sustainable construction

Through a succession of United Nations (UN) 'Earth Summits', the UN Framework Convention on Climate Change (UNFCCC) and 'Agenda 21' programmes, 'sustainable development' has become a prominent worldwide theme. Sustainable construction is the application of sustainable development to the construction industry:

> There is an increasing demand, in both the private and public sectors, to understand sustainable construction practices. This demand is driven by a realisation that sustainable practices make sense to both owners and operators. The practices not only help the environment but can also improve economic profitability and improve relationships with stakeholder groups.
>
> (WRAP (Waste and Resources Action Programme), 2010)

The term sustainable development is often associated with environmental issues, especially within construction. However, its scope is much wider than that. Sustainable development is concerned with ensuring a better quality of life for all within society and for generations to come. The main driver of sustainable development emanates from the concern for the negative environmental effects that have blighted manufacturing, industry and commerce over the last two centuries (WRAP, 2010). Legislation, regulation and change in perception have perpetuated greater regard for the environment and how business responds to upholding its safeguard. This is no more so than within the construction industry, where it is well accepted that environmental effects and impacts are an inevitable and significant consequence. Organisations within the construction industry are, more than ever before, cognisant of the carbon footprint, electricity, gas and water consumptions, and the generation of waste products from their business activities.

Governments take a long-term view of how the construction industry can become more sustainable, and have adopted a variety of initiatives to perpetuate sustainability. In the UK a prominent initiative is 'Constructing Excellence' – a member-led stakeholder organisation charged with driving the change agenda within construction. Its activities have led to reviews of procurement methods, project practices, managerial approaches and a greater appreciation of corporate social responsibilities (CSRs) and triple-bottom line (TBL) – economic/social/environmental – analyses as important measures of business and organisational success. The contributions of CSR and TBL are pertinent to environmental management and, in particular, the establishment of an EMS. They provide the all-important business focus to developing company environmental ethos, culture, policy, goals and principles of structure and operation while key elements of their concepts help shape environmental management practices and organisational procedures. As such, these initiatives can be a key driver for companies seeking to accommodate sustainability in the management system approach to delivering their business.

CSR

The Chartered Institute of Building, or CIOB (WRAP, 2010), comments that:

> recent demand from shareholders, the government, consumer groups and the public for products and services to be 'socially responsible' extends and develops these issues to

an extent that can no longer be assumed to be covered by standard construction procedures.

CSR is a key driver for any company seeking to consider and embed sustainability within its business activities. There are benefits to incorporating socially responsible behaviour within the structure and operation of an organisation, including: industry recognition; client appreciation; marketplace advantage; and public acceptance. By seeking to uphold business integrity, transparency, fairness and diversity, the organisation which exhibits good CSR has the capability to exceed the minimum standards of performance across a range of expected business performance indicators. One of these indicators is indeed the standard of an organisation's environmental performance.

TBL

The concept of TBL refers to the assessment and reporting of environmental performance and social performance in addition to the traditional economic performance of a company. While social performance will be assessed against CSRs and economic performance against business targets and budgets, environmental performance will be judged against environmental policy, goals and objectives. The effective implementation of the company's EMS is central to environmental success.

Environmental awareness and recognition

The environment is an eclectic phenomenon, wide ranging and complex in the extreme. The influence of the environment upon the activities of many companies will have an obvious and clear manifestation while for others it will present in ways that are vague and open to misinterpretation. It is not that organisations are unaware of their environment or that they consciously seek to disregard its interface, although it is sometimes argued that industry adopts the principle of 'out of sight, out of mind'. It is more likely that organisations misinterpret the environment within which they exist and operate. It is easier to disregard or ignore a phenomenon that one does not fully understand than to embrace it. It is easier to see the environment as a threat rather than an opportunity.

What is clear, if not fully understood and embraced by some companies, is that the provision of almost all services and products has an effect on the environment. Environmental effects can range from indirect and minor to direct and highly manifest, but, notwithstanding, environmental effects are always evident. With each and every environmental effect there is some degree of environmental impact. Moreover, these impacts can exist long after the delivery of the service or product. The construction industry, by its very nature, is synonymous with short-term and long-term environmental effects and their impacts. Perhaps the most extreme example of this can be seen in the provision of infrastructure development.

An overt and clear example is the construction of any larger retail outlet such as a supermarket. The completed building physically occupies land, the construction will entail some degree of disruption to the locality and its community, the structure requires physical and natural resources, and when completed it will through occupation and use generate waste products and emissions. The effects do not stop, however, upon completion of the building. Shoppers arrive in cars which create pollution, they purchase products which require replenishment, so a cycle of manufacture and distribution is set up, and the products themselves produce unwanted packaging and waste. The waste goes to landfill or to incineration, which produces more environmental effects and potential impacts. Therefore, the building of the supermarket, which may take little longer than nine months to construct and may be seen at the time of construction to have short-term environmental effects, will likely have decades of subsequent environmental impacts. The key point is that almost all business activity gives rise to environmental effects and impacts, and these must be recognised and managed.

Environmental obligations

It would be wholly wrong to suggest that companies are oblivious to the environment and the effects and impacts that their businesses can bring. In fact, great efforts have been made and are continuing to be made within many business sectors across industry in general to address environmental issues and make environmental improvements to practice. This is being pursued avidly because:

- customers are expecting improved environmental performance of the products they purchase;
- almost all public-sector clients and major private-sector clients request evidence of environmental capability before engaging with companies for services and products;
- company executives are demonstrating their environmental vision and commitment by publishing environmental policies, objectives and targets;
- organisations are formalising their approach to environmental matters through the development and implementation of management systems.

These perceptions are influenced by two key factors: first, there are very costly implications through legal and financial penalties for companies which do not respond to their environmental obligations; and, second, there are major benefits to be gained through improved business competitiveness by those companies which promote environmental safeguards.

Environmental, economic and organisational benefits

There are a number of discernible benefits from openly recognising that the activities of a company impinge upon the environment, and that a positive and active response should be made to alleviate or manage those effects. Benefits are achieved directly, through a reduction in environmental impacts and their associated costs, and indirectly, through efficiency gains and improvements made within the company. Benefits fall into three groups: (1) environmental; (2) economic; and (3) organisational.

1 Environmental:

- *Reduced impact on the natural environment* – through reductions in environmental effects such as: inappropriate land use; air, land and water pollution; water consumption; disruption caused by noise and vibration; and contamination and damage to plants and wildlife.
- *Reduced impact on natural resources* – through reductions in the use of non-renewable and non-sustainable natural resources such as coal, gas, oil and minerals.
- *Reduced impact on communities* – through reductions in the disruption of and disturbance to the local environs, neighbourhoods and inhabitants.

2 Economic:

- *Reduced risk of environmental penalties* – through reductions in fines and the costs associated with environment-related legal involvements.
- *Reduced costs of remediation activities* – through reductions in costs associated with clean-up and remediation works following environment-related incidents.
- *Reduced costs of resources and waste materials* – through reductions in cost of purchasing materials which are wasted during production and reductions in costs of waste disposal.

3 Organisational:

- *Improved opportunities* – improvement in tendering opportunities with clients which environmentally pre-qualify service companies and product suppliers.
- *Improved marketplace* – improvement in market sector business due to perceptions of customers who expect good environmental performance from suppliers and retailers.
- *Improved relationships* – improvement to interactions with business regulators, environmental authorities, stakeholders and the public.
- *Improved internal operations* – improvements to company organisation and management leading to better environment-related operational efficiency and effectiveness and enhanced morale of personnel.

Fundamental obligations

Any and all companies have a number of fundamental obligations in respect of the environment in which they operate. These demand that environmentally responsible practices are followed to meet the requirements of four key areas:

1 *National legislation* – national legislation is applicable which seeks to protect the natural environment and influences the ways in which a company operates. The most recognised pieces of national legislation are the Environmental

Protection Act 1990 (Environment Agency, EA, 1990) and the Environmental Permitting Regulations 2010 (Department for Environment, Food and Rural Affairs, DEFRA, 2010), the latter arguably assuming greater importance today as much of EPA 1990 is now covered by these regulations.

2 *Local regulation* – devolved powers from national legislation allow local authorities to exercise regulation and control of company activities. There are many individual regulations, an example being the control of air emissions under The Clean Air Act 1993 (EA, 1993).

3 *Client/customer terms of contract* – within the arrangements between a client or customer and a provider of services or products specific terms may be imposed which have an environmental relevance and must be satisfied as part of the terms of contract.

4 *Company, or corporate, controls* – companies will set specific environmental agendas for business operations which are reflected in their policies, objectives and targets and which are measured by their environmental performance.

Note of the Environmental Permitting Regulations 2010:

The Environmental Permitting Regulations 2010 provides industry, regulators and others with a single extended permitting and compliance system and includes those systems for: discharge consenting; groundwater authorisations; and radioactive substances regulation . . . environmental permitting: cuts unnecessary red tape – bringing cost-savings to industry and allowing regulators to focus their resources on issues that matter; provides continued protection of the environment and human health – maintaining current standards; and increases clarity and certainty for everyone on how the regulations protect the environment – a clearer, simpler and quicker system allowing a better understanding of the law and its effects.

(DEFRA, 2010)

Environmental management

Environmental management is concerned specifically with optimising a company's environmental performance through managing the environmental effects of the company's business activities. The company's approach embraces its mission, policy, aims, objectives, strategy, procedures and practices which form the company's response to its environmental situation and circumstances. A carefully structured and formal, or documented, approach by any company is absolutely essential in demonstrating unequivocally that it is not only complying with its own environmental mission and policy, but also meeting current environmental legislation, recognised management standards and industry regulation. Furthermore, it demonstrates to a wider audience that the company is committed to wider environmental matters.

The most appropriate way for a company to address its environmental requirements is to establish a structured and systematic management approach. This provides the company with the formal approach needed to deliver the appropriate response and, moreover, prepares the company for environmental assessments, contract pre-qualification and independent auditing, all of which feature extensively in business today. With developing environmental awareness within almost all sectors of business, many organisations have undertaken an environmental review of their activities. These have led to various approaches to environmental management, including: the development of situation-specific environmental plans; intra-organisational environmental procedures; and comprehensively developed and implemented EMSs.

A systems approach to environmental management

The development and implementation of an EMS has been seen by many companies as a direct and active response to the environmental implications of their business activities. The international standard for EMSs, ISO 14001 (ISO, 2004), specifies that an EMS should be based on: a clear environmental policy; a robust environmental strategy; a targeted aim and objectives; and the establishment of structured mechanisms to evaluate and control the environmental effects of its business operations. In addition, the management system must comply with the relevant environmental legislation and industry regulation which influence the provision of its services and products. Through the use of an effective EMS, a company should be best placed to meet both the intra-organisational requirements of its activities and the pressures and demands of the external marketplace.

An EMS may be 'company based', where the system is developed at the heart of an organisation to structure, organise and manage the entire company including the points of delivery of its services and products. In the context of the construction industry this means that a company, normally head office, system is established and then applied to the delivery of services and products via the building and engineering projects the company undertakes. Alternatively, an EMS can be 'project based', which describes a management system specifically developed for and applied to projects without there being an overarching company system. Because of their limited scope and nature, these are differentiated, in practice, by being called a 'plan' rather than a 'system'.

An EMS can be freestanding, developed explicitly for the purpose of providing environmental management, or can be linked to an existing management system, for example a quality management system, or QMS. Furthermore, the systems for environment and quality management may share procedures and resources although operating independently, or can be integrated into a dual-application single system. The system does not have to be implemented in one fell swoop. Rather, an EMS can be developed and established gradually as the required documentation – system manual, management procedures, implementation plan and working procedures – is implemented and embedded. Envirowise, a managed programme for government which advises companies on improving business practices, profitability and competitiveness, suggests that:

> Companies can choose to either develop their own bespoke system, if their priority is to take steps towards improving their bottom line through resource efficiency, or they can follow the more formal requirements of a national or international standard such as ISO 14001 or the Eco-Management and Audit Scheme (EMAS).
>
> (Envirowise, 2009)

As with any management system, an EMS must be conceived, developed and applied within the context of the company and its business activities. Therefore, the nature and scope of application can be as limited or as extensive as that deemed necessary by the company. As a support process to a company's core business processes and outputs, an EMS must be directed to enhancing the core processes and to adding value to the holistic operation of the organisation. For some companies environmental management will lie on the periphery of their business, while for others it will exist at the centre. For those companies whose activities have the potential to affect the environment in demonstrable and potentially detrimental ways, environmental management should lie at the heart. For such organisations, environmental management, through the development and implementation of an EMS, will be fundamental to maintaining the effectiveness and success of its business:

> An EMS is increasingly valuable as a prerequisite for doing business and accredited certification helps to demonstrate a business's compliance with supply chain requirements. All businesses can benefit from a systematic approach to ensure their use of resources is well managed, to help achieve savings and ensure that the business is not breaching current or developing environmental legislation.
>
> (Envirowise, 2009)

Standards-based EMSs

Although a company's approach to environmental management can take many forms, the blueprint for effective response is a standards-based management system. ISO 14001 is the international standard guiding the development, implementation and maintenance of an EMS. This standard presents a specification for the development and implementation of an EMS and provides companies with guidance on environmental management application. The standard is structured in such a way that it allows any organisation to establish the appropriate system to meets its own needs as the basis for optimising the environmental performance of its business and for maintaining that performance in the future.

An important feature of any environmental management approach is that it should embrace the generic requirements of the company's industry sector while also meeting the company's business-specific needs. The standard is quite explicit in the context that a company should use the specification to establish a documented EMS which best fits the company. In this sense, the standard is not prescriptive. It understands that companies have different business agendas, marketplaces and pressures, and that they must have flexibility of systems development to respond to environmental matters and issues in different ways.

Relevance of parallel management systems

ISO 14001 recognises the need for different management systems to have a common frame of reference. It therefore presents the specification for an EMS with appropriate links to other systems such as ISO 9001 QMS. A company is therefore able to develop and implement an EMS based around similar concepts and within a similar context as an established quality system.

EMSs

Environmental management is:

- The fulfilment of policy, strategy, procedures and practice that form the company's response to its environmental situation in the course of undertaking its business.

An EMS is:

- The company's formal structure and organisation that implements environmental management.

Establishing an EMS involves:

- the development of a documented system of procedures for managing the environmental aspects of the company;
- the implementation of those environmental procedures in the company's business operations;
- the provision of objective evidence to ensure that the procedures have been applied to a company's operations;
- the maintenance of procedures and their upgrade to facilitate continual improvement to the management system and business processes.

In more simple terms, this system is:

- saying what the company does;
- doing what the company says it does;
- evidencing that what the company says it does is in fact done;
- improving what the company does.

A company implementing environmental management is required to evaluate its business activities critically in response to any potential threat that they may potentially present to the environment. It should do this within a framework of formality by establishing a documented management approach which meets the specification of a recognised management system standard and/or assessment scheme.

EMS standards

This section examines: the requirements, or clauses, set by the international standard ISO 14001 for the establishment of an effective EMS; the association of the BS 8555 guide for EMSs, particularly applicable to smaller organisations; and the use of EMS assessment schemes.

The EMS standard

ISO 14001

ISO 14001 is the international standard for EMSs. It provides a specification for system development and implementation such that a company can establish and maintain an EMS to ensure that its business activities conform to its environmental policy, strategy, aims and objectives and meet the requirements of business legislation and regulation. The standard is not prescriptive but rather it guides a company in the development of an EMS to suit the organisation's business needs, situation and circumstance.

The role of the standard

The standard provides a commonly understood basis and uniform approach to EMSs upon which assessment can be made of the adopted approach. The standard may be used by a company to self-assess compliance with environmental requirements set within the organisation. It may also be used by external bodies to independently assess environmental-related organisational practice. The standard provides a company with a useful, meaningful and objective tool to control its environment-related activities and performance. Equally, it provides the basis to facilitate external recognition of its management approach.

Status

ISO 14001 is approved by the European Committee for Standardization (CEN). It is therefore the European Standard – EN ISO 14001. The UK, as a CEN member, embeds the standard and as such BS EN ISO 14001 is the official English-language version. For ease the standard is simply referred to as ISO 14001.

Purpose

ISO 14001 is intended to present the fundamental elements of an appropriate EMS. The standard specifies the requirements for an EMS which can assist an organisation to fulfil its environmental responsibilities, to comply with legislation and to achieve environment-related business goals:

> A system of this kind enables an organisation to develop an environmental policy, establish objectives and processes to achieve the policy commitments, take action as needed to improve its performance and demonstrate the conformity of the system to the requirements of this international standard.
>
> (ISO, 2004)

The standard also enables an organisation to consider the wider issues associated with environmental management:

> The overall aim of this International Standard is to support environmental protection and prevention of pollution in balance with socio-economic needs ... Environmental management encompasses a full range of issues, including those with strategic and competitive implications ... Demonstration of successful implementation of this

International Standard can be used by an organisation to assure interested parties that an appropriate environmental management system is in place.

(ISO, 2004)

Methodology

ISO 14001 is based on the 'Deming methodology' – usually referred to as plan–do–check–act, or PDCA. This accommodates the 'process approach' – the management of activities using a system of processes. PDCA in application to environmental management involves the following (see Figure 3.1):

Plan – develop the processes needed to fulfil the business as specified by the organisation's environmental policy.

Do – implement the processes as required.

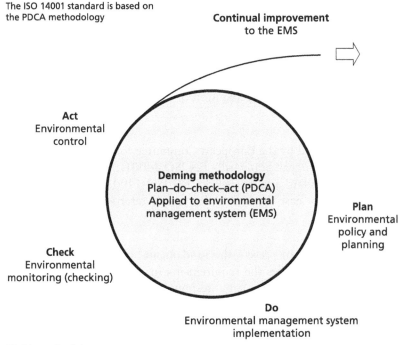

Figure 3.1 'Deming methodology', plan–do–check–act (PDCA), applied to EMS development and implementation processes

Check – monitor and measure the conduct of processes against the organisation's environmental policy, aims, objectives, goals, legal and wider responsibilities, and record and evaluate environmental performance.

Act – implement action to promote continual improvement of the EMS and business processes.

The way in which this methodology is applied will differ among organisations. An EMS, like a system applied to any other management function, will be configured to meet the requirements of the company, its business activities and its particular circumstance and market position. Its needs will determine the system's scope, level of complexity, detail, amount and type of documentation and resources used. In this way, organisations both small and large have the capability to develop and implement an EMS.

ISO 14001:2004 relationships and links to other management systems

ISO 14001 is a standard dedicated to presenting the requirements of EMSs. It does not include any requirements specific to other management systems. However, the standard does include elements which may be aligned or integrated with other systems and, therefore, an EMS may be configured around an existing organisational management system. Many companies establish an EMS based upon and linked to an existing QMS. The approach of aligning or integrating environmental management with quality management is often termed environmental quality management, or EQM.

Sections of ISO 14001

ISO 14001 presents a set of requirements with which the company's EMS must comply if it is to be recognised as meeting the standard (see Figure 3.2). The requirements are presented in the standard sectionalised and enumerated as follows, with key section 4 emphasised in bold:

1   Scope
2   Normative references
3   Terms and definitions
4   **Environmental management system requirements**
4.1 General requirements
4.2 Environmental policy
4.3 Planning
4.4 Implementation and operation
4.5 Checking and corrective actions
4.6 Management review

In addition, the standard includes two annexes for information:

Annex A   Guidance on the use of the International Standard
Annex B   Correspondence between ISO 14001:2004 and ISO 9001:2000

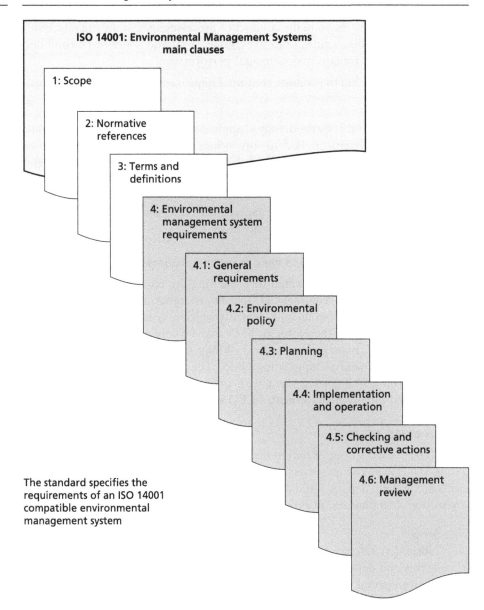

Figure 3.2 The main clauses of ISO 14001

It can be seen from the titles of the sections numbered 4.2 to 4.6 that the contents are commensurate with the PDCA methodology described earlier.

EMS requirements

The next section presents the requirements for the establishment of an EMS to the specification of ISO 14001.

1 Scope

The scope of the standard does not specify environmental performance criteria. Rather, the standard impacts on those environmental aspects of the organisation's business over which it can exert control and can influence. The standard states that it is applicable to organisations that want to:

a) establish, implement, maintain and improve an environmental management system,

b) assure itself of conformity with its stated environmental policy,

c) demonstrate conformity with this International Standard by

1 making a self-determination and self-declaration, or

2 seeking confirmation of its conformance by parties having an interest in the organisation, such as customers, or

3 seeking confirmation of its self-declaration by a party external to the organisation, or

4 seeking certification/registration of its environmental management system by an external organisation.

(ISO, 2004)

2 Normative references

There are no normative references cited in the standard. This standard states that this section is included so that the numbering of clauses is identical to previous editions.

3 Terms and definitions

Within the standard there are 20 terms and definitions. These describe specific wording used throughout the standard.

4 Environmental management system requirements

4.1 General requirements

The standard specifies that the organisation should develop, document, implement, maintain and continually improve its EMS. Also, it must explain how it will fulfil these requirements. The scope of the EMS in the context of the organisation and its activities must be defined and described.

4.2 Environmental policy

The standard requires that the EMS is:

• appropriate to the environmental effects of the organisation's business activities;

• committed to continual improvement;

• seeking to prevent pollution;

- directed towards compliance with legal and other requirements;
- focused on setting and reviewing environmental business objectives and targets;
- documented, implemented and maintained;
- communicated to employees and others working for the organisation;
- available for public scrutiny.

4.3 Planning

4.3.1 Environmental aspects
The standard requires that the organisation develops, implements and maintains environmental management procedures which:

- identify the environmental aspects of its business, whether they be linked to products or services;
- recognise those aspects which may have a significant effect on the environment.

4.3.2 Legal and other requirements
The standard requires the organisation to put in place procedures which:

- identify applicable legislation and other environment-related requirements;
- consider how legislation and other requirements apply to environmental aspects.

4.3.3 Objectives, targets and programme(s)
The standard requires that the organisation has:

- documented objectives and targets;
- objectives and targets which relate to functions and levels of the organisation;
- objectives and targets which are measurable and consistent with policy, legal and other requirements;
- considered its business requirements and views of interested parties;
- a programme for achieving objectives and targets;
- designated responsibility for achieving objectives and targets within the organisation's functions and levels;
- allowed a timeframe and means for achievement.

4.4 Implementation and operation

4.4.1 Resources, roles, responsibility and authority
The standard requires that:

- management should ensure the availability of resources which can operate and improve the EMS;
- roles, responsibilities and authorities are defined, documented, communicated and clear to all involved in implementing the system;

- a specific management representative is designated who takes responsibility for the EMS, who ensures that it is maintained and who reviews its performance and recommends improvements.

4.4.2 Competence, training and awareness
The standard requires that the organisation:

- ensures that personnel involved in system implementation are competent through appropriate education, training or experience;
- identifies training needs commensurate with the environmental aspects of their duties and tasks;
- maintains records of competence assessment and training.

In addition, the organisation should have procedures which make employees aware of the:

- need for conformance with the company's environmental policy and procedures;
- environmental effects and impacts associated with their work;
- benefits of environmental performance;
- roles and responsibilities they assume within the EMS;
- implications of nonconformance with procedures.

4.4.3 Communication
The standard requires that the organisation maintains procedures for:

- communication internally to functions and levels;
- receiving and responding to documentation from external sources.

4.4.4 Documentation
The standard requires that the documentation for the system should include details of the:

- environmental policy, objectives and targets;
- scope of the EMS;
- main elements, with references to related documentation;
- documents and records required by the standard;
- documents and records deemed necessary by the organisation to operate and maintain its business processes.

4.4.5 Control of documents
The standard requires that all documents are controlled, and therefore procedures should be implemented which:

- approve documents before they are issued;
- review, update and re-approve documents;

- clarify changes and revision status of documents;
- ensure availability of documents for operational use;
- ensure that documents remain legible;
- consider and control documents from external origins;
- prevent use of documents which are obsolete.

4.4.6 Operational control
The standard requires that the organisation should:

- plan operations with potential environmental effects with regard to its environmental policy, objectives and targets;
- develop, implement and maintain procedures to provide control where there might be deviation from the plan;
- specify operational criteria within the procedures;
- have procedures which control inputs from suppliers and contractors.

4.4.7 Emergency preparedness and response
The standard requires that the organisation has procedures to:

- identify potential emergency situations;
- recognise and respond to environmental accidents;
- mitigate environmental effects from situations and accidents;
- review and revise emergency preparedness and response mechanisms;
- test the effectiveness of these activities.

4.5 Checking and corrective actions

4.5.1 Monitoring and measurement
The standard requires that the organisation has procedures to:

- monitor and measure those operations which could have environmental impact;
- document environmental performance;
- compare performance with objectives and targets;
- ensure the efficacy of monitoring procedures and equipment;
- calibrate monitoring equipment to ensure accuracy of data;
- maintain environmental records.

4.5.2 Evaluation of compliance
The standard has two numbered clauses which require that the organisation has:

4.5.2.1: a mechanism to evaluate compliance with legal requirements and keep records

4.5.2.2: a mechanism to evaluate compliance with other requirements and keep records

4.5.3 Nonconformity, corrective action and preventive action

The standard requires that the organisation has procedures for handling nonconformities and for taking corrective and preventive action which:

- identify and correct nonconformities to mitigate environmental effects;
- investigate the cause of nonconformities to avoid recurrence;
- evaluate need for and implement action to prevent nonconformities;
- record results from preventive and corrective actions taken;
- review effectiveness of actions.

4.5.4 Control of records

The standard requires that the organisation has procedures for:

- maintaining records to demonstrate conformity to the EMS;
- identification, storage, protection, retrieval, retention and disposal of records;
- maintaining records which are legible, identifiable and traceable.

4.5.5 Internal audit

The standard requires that the organisation should ensure that internal audits are conducted on the environmental management system to:

- determine if the system conforms to the standard and other requirements;
- ensure that the system has been appropriately implemented and maintained;
- provide audit information to management of the organisation.

In addition, an audit programme should be established which:

- determine requirements and responsibilities for planning and conducting audits;
- establish criteria for audit frequency, scope and methods to be used;
- maintain records of audit results.

4.6 Management review

The standard requires that the organisation undertakes a review, by senior management, to ensure that the EMS maintains its suitability and effectiveness, is improved where required, and retains currency with environmental policy, objectives and targets. The standard specifies the following inputs to management review:

- results from audits and evaluations of compliance with legal and other requirements;
- communications and complaints from external parties;
- measures of environmental performance;
- evaluations of operations against objectives and targets;
- results from preventive and corrective actions taken;
- follow-up actions from previous reviews;
- recommendation for improvements.

Comment

ISO 14001 does not specify how a company should configure and present its EMS. The standard presents a set of elements for an EMS which forms the basis of an effective approach. The standard allows an organisation to address environmental matters in its own way as required by its particular business situation. The standard does require that an organisation has specific commitment to environmental policy, objectives and targets. It also requires the organisation to comply with environmental legislation and other environment-related requirements, to provide specified system documentation and records, and to conduct periodic management review of system implementation as applied to its business processes.

BS 8555 based EMSs

BS 8555

While ISO 14001 is arguably the most well-recognised specification for EMS development and implementation, BS 8555 is included here as it is particularly appropriate to the establishment of an EMS by small to medium enterprises (SMEs), although it can also be applied by larger organisations. This British Standard is entitled *BS 8555: Environmental Management Systems: Guide to the phased implementation of an environmental management system including the use of environmental performance evaluation* (BSI, 2003).

BS 8555 links the requirements of ISO 14001 with the guidelines for environmental performance evaluation presented in ISO 14031 (ISO, 1999). BS 8555 outlines how an organisation can implement an EMS which can subsequently be certificated to ISO 14001 and achieve recognition by the Eco-Management and Audit Scheme (EMAS), details of which follow subsequently. The inclusion of ISO 14031 in the system's development is apt as it appreciates the importance of organisational performance indicators that focus on company business needs such as competitive advantage and meeting customer requirements – those elements which drive any organisation and maintain its position within its business market and supply chain.

The BS 8555/Acorn Scheme Workbook

The Institute of Environmental Management & Assessment (IEMA), working in conjunction with the British Standards Institution (BSI), produced *The BS 8555/Acorn Scheme Workbook*. The IEMA states that 'this joint IEMA/BSI publication provides a user-friendly companion document for SMEs (small- and medium-sized enterprises) and larger organisations wishing to implement an Environmental Management System (EMS) using the process outlined in BS 8555' (IEMA, 2003).

Implementing the workbook

The IEMA Acorn Scheme to EMS establishment propounds a six-phase process of system development and implementation linked closely to an organisation's core activities and appreciating both the intra-organisational dimensions and wider external business environment. The six phases are:

1 Commitment and establishing the baseline.

2 Identifying and ensuring compliance with legal and other requirements.

3 Developing objectives, targets and programmes.

4 Implementation and operation of the environmental management system.

5 Checking, audit and review.

6 Environmental management system acknowledgement.

The rationale for each of the phases and the main requirements of each phase are as follows.

Phase 1: Commitment and establishing the baseline

This phase seeks to secure commitment to environmental management and performance improvement by the company or organisation. It focuses on the identification of those environmental issues which are or may be created in the course of undertaking business activities, and seeks to consider initiatives which deliver improvements to environmental performance. Further, this phase seeks to engender and embed the support of those people who participate in and interface with the EMS. The main requirements of Phase 1 are:

- to ensure commitment to EMS establishment and environmental performance improvement from top management;
- to plan for the establishment of the EMS;
- to undertake a baseline assessment of business processes, products and services to identify environmental aspects and impacts;
- to develop environmental policy and initiatives for environmental improvement based on the baseline assessment;
- to determine environmental performance indicators commensurate with key aspects and impacts;
- to generate organisational approaches to engender employee involvement with the EMS, its continual environmental improvement and training.

Phase 2: Identifying and ensuring compliance with legal and other requirements

This phase seeks to identify the environment-related legal, and other, requirements impinging upon the organisation's business activities and the operation of its EMS to ensure that control measures are established which maintain compliance with requirements. The main requirements of Phase 2 are:

- to identify the environmental legislation applicable to the organisation;
- to identify other requirements such as industry standards, codes of practice and contractual agreements;

- to identify where and how these apply to organisational business activities;
- to check levels of compliance/non-compliance and assess the effectiveness of control mechanisms to ensure compliance;
- to develop action plans for handling non-compliance and improvements to control mechanisms;
- to establish control procedures, emergency procedures and appropriate associated training programmes;
- to establish indicators of compliance;
- to embed environmental performance improvement throughout the organisation.

Phase 3: Developing objectives, targets and programmes

This phase seeks to consider the core elements of the EMS and establish a framework for delivering systematic environmental management and performance improvement. The main requirements of Phase 3 are:

- to consider in greater depth the environmental aspects and impacts identified in Phase 1;
- to develop an approach for assessing the significance of environmental aspects upon the organisation and its business activities;
- to confirm environmental policy;
- to develop objectives and targets for performance improvement which underpin the policy;
- to establish performance indicators to monitor and confirm progress against objectives and targets;
- to configure environmental management programmes for meeting objectives and targets;
- to establish environmental control procedures;
- to communicate policy, objectives, targets and programmes to employees.

Phase 4: Implementation and operation of the environmental management system

This phase seeks to implement the EMS within the organisation. The main requirements of Phase 4 are:

- to configure the structure for EMS implementation, assigning roles, duties and responsibilities to designated personnel;
- to define the relationships of the elements and personnel which operate within the EMS;
- to identify, deliver and review training programmes and improve as required;
- to establish clear lines of communication both intra-organisationally and externally;

- to deploy effective system documentation, control its use and maintain record keeping;
- to test emergency preparedness and response mechanisms;
- to establish performance indicators to assess the effectiveness of the EMS in use.

Phase 5: Checking, audit and review
This phase seeks to establish appropriate system auditing and management review to maintain a continual improvement loop in system operation and development. This is an important phase for those organisations seeking to meet the requirements of ISO 14001 and system assessment under EMAS. The main requirements of Phase 5 are:

- to establish internal system audit procedures;
- to ensure that audit findings are disseminated to management for review;
- to establish preventative and corrective actions for non-compliances identified by audits;
- to implement formal management review of system operation against declared objectives and targets and determine the environmental performance of the organisation in relation to its business activities;
- to focus on continual business performance improvement and system improvement to maintain environmental policy.

Phase 6: Environmental management system acknowledgement
This phase seeks to place the organisation in a position of readiness to submit the EMS for ISO 14001 certification or EMAS registration, should the organisation deem this appropriate as part of its business operations. Of course, not all organisations will be seeking to achieve independent recognition of their EMS for the business and commercial reasons discussed previously. The main requirements of Phase 6 are:

- to provide evidence that the EMS has been appropriately configured, implemented and operated effectively against criteria specified by the independent certification/registration body;
- to assess the mechanisms of environmental performance evaluation to ensure its rigour for independent assessment;
- to provide an environmental statement, a requirement of EMAS registration.

Comment
The advantage with the IEMA/BS 8555 approach is that the scheme recognises that many organisations do not wish to or cannot establish a grandiose EMS in one fell swoop as might be the case with ISO 14001 certification or EMAS registration. This is one reason why the scheme appeals to SMEs. Where first-party EMS recognition is required, BS 8555 is a most appropriate way for an organisation to develop its EMS. It also presents the opportunity for the organisation

to progress through Phase 6 of the workbook to establish a fully fledged EMS capable of third-party recognition. Therefore, this approach is particularly useful for small, medium and some larger organisations seeking to establish an EMS.

EMS assessment schemes

Assessment

Detailed description and information follow later in this section as to the method by which a company can actively promote its EMS through independent, or third-party, registration of its system – the process of system assessment, or certification. Systems are assessed under a number of government-accredited assessment schemes, the most prominent being those operated by the BSI. Before looking at such an assessment, it is worth noting that a number of other assessment schemes exist to assist a company with its EMS development and implementation, the most prominent of which is the Eco-Management and Audit Scheme (EMAS).

EMAS

The scheme

EMAS is a Europe-wide approach to the formal recognition of effective environmental management. EMAS assesses and registers companies which have implemented a programme of environmental management and which pursue continual improvements to their environmental business performance.

EMAS is ratified by European law with administration provided by the individual member states of the EU under European Council Regulation 93/1836. Originally designed as an 'audit' scheme where companies would undertake an annual environmental audit of their activities, EMAS was revised in 1993 with a focus on 'eco-management'. EMAS was introduced to encourage sustainable industry growth while maintaining environmental protection. In particular, EMAS is intended to reduce industry pollution through the 'polluter pays' principle, encourage clean technology and promote more environmentally efficient management of resources. In seeking to achieve these aspirations, EMAS requires:

- continual improvement of environmental performance by companies;
- EMS establishment and auditing to be implemented;
- development and public statement of environmental business activities;
- training of employees in environmental management aspects of business operations;
- dissemination of information concerning environmental performance.

EMAS requires companies to develop environmental policies and management systems which are site specific. The environmental efficacy of products and services is considered at the place of production or delivery. EMAS provides a customer with objective evidence of a supplier's involvement and commitment to environmental quality and improvement. It focuses on environmental perform-ance rather than a systems management approach, although the two generally go hand in hand.

Requirements

There are seven aspects of environmental management consideration in seeking to comply with EMAS registration:

1 *Environmental policy* – A company is required to develop a company-wide environmental policy. This must comply with current legislation, industry regulations and any other pertinent environment-related requirements. The policy must be adopted and supported by the company's senior management and should be continually reviewed and amended as required to maintain currency and efficacy with business activities.

2 *Environmental review* – A company is required to undertake a company-wide operational review of its activities. The review should consider all the environmental effects and impacts of its operations from pre-production to delivery. A company must ensure that all legislation and regulations are met and that any non-compliance is identified and listed for action. A register of environmental effects should be compiled.

3 *Environmental programme* – A company is required to produce an environ-mental programme. The programme describes how the company-wide environmental objectives, derived from policy, are to be met.

4 *Environmental management system* – A company is required to establish an effective management structure within which the environmental programme is delivered. An EMS is a prerequisite to this as it provides the appropriate structure to facilitate continual improvement in company business activities and management.

5 *Environmental audit* – A company is required to establish an environmental audit procedure and schedule. The audit seeks to facilitate the detailed measurement and evaluation of the company's environmental performance and management practices. Audits should be conducted every three years and inform the continual improvement to business processes, management proce-dures and systems.

6 *Environmental statement* – A company is required to produce an annual state-ment of its environmental performance. This should include information relating to its policy, objectives, programme and actions allowing environmental performance to be scrutinised externally. This statement becomes public-domain information allowing independent consideration of the company's environmental performance claims.

7 *Validation* – A company is required to have its environmental policy, programme, management system, audit procedure and environmental statement verified each year. This must be undertaken by an independent accredited verifier.

EMAS recognition

The validation of compliance to the environmental standard by EMAS is third party and therefore independent. To obtain EMAS accreditation, a company is required to develop and publish its environmental policy, strategy and quantifiable and measurable objectives. Subsequent environmental audits and business achievements are available for industry and public scrutiny.

Use of EMAS

Sound and effective eco-management requires that a company implements a well-conceived and structured approach. In meeting the requirements of EMAS this involves the development and implementation of appropriate environmental policy, review and programme followed by the establishment of a management system and environmental statement, all of which may be evaluated through annual audit. In principle, there is much in common between the requirements of EMAS and ISO 14001, both of which focus, in part, on the establishment of an effective EMS. It is the EMS which provides for the appropriate support of a company's core business processes. Many organisations which pursue ISO 14001 based EMS certification will also seek EMAS accreditation. This allows for the wider recognition of a company's products or services and an opportunity to expand its business base to customers who demand greater environmental performance from their suppliers.

ISO 14001: EMS development and implementation

This section examines the process of EMS development and implementation, presented in a sequence of eight key stages following the requirements, or clauses, of ISO 14001.

Key stages

In association with the requirements of the standard, the process of system development and implementation has a sequence of the initial decision on the type and scope of the system plus eight distinct stages as follows:

Initial decision +

1 Preparation
2 Preliminary review
3 System planning and development
4 Implementation and operation
5 System monitoring
6 Management review
7 System registration (certification)
8 Continual improvement

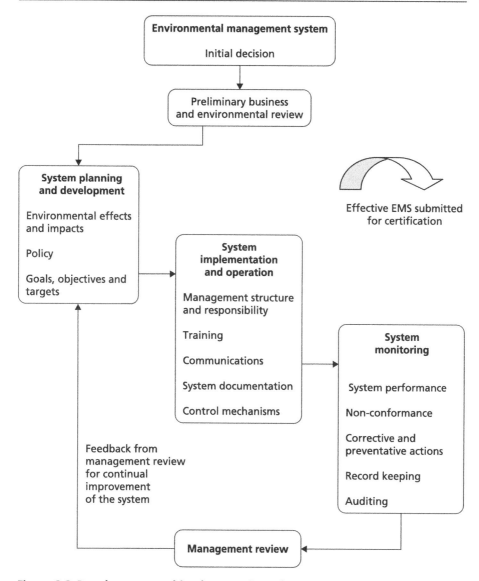

Figure 3.3 Development and implementation of an EMS

Where direct reference is made to particular sections and sub-sections of ISO 14001, these are indicated in the text. Figure 3.3 illustrates the various stages listed above.

Development and implementation of the key stages

The key stages can be established as follows.

Initial decision

The initial decision that a company has to make when considering the development and implementation of an EMS is the focus and scope of the intended system. There are two directions a company can take to creating and establishing an EMS:

1 A 'company-wide' EMS focusing on providing a corporate-based umbrella system implemented throughout the whole organisation which can be applied to each and every company activity.

2 A 'project-based' EMS focusing on providing a system developed for specific organisational activities or projects.

Most companies will lean towards the use of a company-wide EMS, centralising its system in its most prominent facility, normally its head office, and then cascade the system to its points of service, product or project delivery.

Advantages of the corporate-based system

The advantages of a corporate-based environmental management system are:

- a policy is applied to the whole organisation;
- a uniform system is used throughout the organisation;
- a set of standard management procedures and operational working instructions is used;
- one management system standard applies throughout;
- best practice can be shared across the organisation and to all its activities.

Disadvantages with the corporate-based system

There are a number of distinct disadvantages with the corporate-based system. These are:

- decisions taken at corporate level need to be cascaded down to projects, so focus and application can diminish;
- separation of head office staff and project staff can lead to lack of commitment, ownership, communication and support;
- lack of flexibility in a single system where projects require tailored systems to handle different environmental effects and impacts;
- extra staff are needed at corporate level to interface with project personnel at many locations.

Advantages of the project-based system

The main advantages with a project-based system are:

- dedicated focus on project-specific aspects where environmental effects actually impact;
- decision making is direct and applicable;
- procedures and instructions can be tailored to each individual project;
- ownership of the system is based within the project personnel.

Disadvantages of a project-based system

There are many disadvantages to a project-based system. These are:

- corporate policy and objectives may receive little attention;
- different environmental management procedures may be used across projects;
- varied standards of performance may be accepted from project to project;
- duplication of approach can readily occur;
- multiple project-based systems can become highly resource intensive;
- environmental knowledge and skills may differ across projects;
- the ISO 14001 standard will apply only to specific projects with certificated systems;
- lack of best practice dissemination across projects.

Preparation

Investment

The investment needs for establishing an EMS may be considerable. These include resource commitments in addition to financial support as follows:

- *Management* – executive and senior management will have an extensive and vital role to play in system establishment. They will oversee policy development and strategic considerations and will sanction the commitment of organisational resources. A key resource is that of the 'environmental systems manager', who takes responsibility for project managing the development, implementation and maintenance of the system. Middle and junior management will also be involved in implementing the system, and their activities will need to be structured, assigned and committed.

- *Time* – managers and employees will be required to devote their time to system establishment, perhaps in association with existing duties on other management systems and business processes. Personnel will likely need system training in operational procedures, documenting and record-keeping practices, and this will involve time allocation around existing work commitments.

- *Training* – managers and employees will likely need training to become more environmentally aware. Furthermore, they will require new knowledge and skills to develop, implement and maintain the environmental management system in application to the organisation's business processes. Training, on or off the job, will need to be structured and work planned around routine duties and tasks.

- *Finance* – money will need to be invested in system development and implementation, and this is particularly relevant where the company needs to introduce new activities or revise existing practices to effect improvements to the business processes. In addition, new staff may be required with specialist knowledge and capabilities, and there may be a requirement for inputs from external consultants. All of these will incur upfront investment costs.

Change management

The most challenging aspect of preparing for the EMS is likely to be organisational change. An understanding of change, capacity for change and pace of change is crucial to the effective establishment of the management system. Often, companies underestimate one or more dimensions of change, for example correctly determining the knowledge and resource capacity needed for change but miscalculating the pace of change. Inevitably, change involves people, and therefore getting individuals and groups on-board is essential. Determining the knowledge and skill base of personnel is relatively straightforward, but interpreting their perceptions, feelings and responsiveness to change is another matter. Change can be problematic because senior management fail to introduce change appropriately. Failing to explain the mission and its purpose, seeking to move matters too quickly, adopting an aggressive management style and undervaluing the contribution of staff are all symptoms of inappropriate approach and action. A company should seek to understand organisational inhibitors to change but must see them in context as understandable by-products of change and, moreover, the ways in which change is managed. In many organisations, employees say they do not fear or resist change per se but that change is badly handled by senior management, so creating fear and resistance.

A useful way for a company to understand change and embark upon change management is to utilise McKinsey's *7S Model* (Naoum, 2001). This is a model of seven business elements which need to be considered when formulating change because individually and collectively they are impacted by the change management process. These elements are: shared purpose; strategy; structure; systems; staff; skills; and style. Each element is required to meet the needs of the other elements, and all must satisfy the element of shared purpose. A common failure of change management is the lack of integration and unity of these organisational elements. While the model applies to understanding many aspects of organisational change, it can be applied to the concepts of environmental management as follows:

1 *Shared purpose* – embraces the mission, culture, policy and goals of the company (environmental management can be included as a goal to fulfil policy).

2 *Strategy* – includes the business plans of the company to achieve the shared purpose (environmental criteria can be included in the business plans).

3 *Structure* – configures roles and responsibilities within the company's organisation (environmental management roles, duties, tasks and responsibilities can be incorporated into the formal structure).

4 *Systems* – embraces the processes and management functions that deliver the business outputs (environmental performance standards can be specified within those functions which manage the processes).

5 *Staff* – are the human resources of the company that operate the organisation (staff appropriate to the needs of ensuring the environmental performance of the business can be incorporated into the company resource base).

6 *Skills* – capable, skilled and knowledgeable staff are needed to fulfil the shared purpose (those staff who are aware and informed in environmental management can be appropriately assigned by the company).

7 *Style* – involves the way that managers and employees behave and undertake their duties (environmental performance criteria can be specified as part of general working patterns and behaviour).

The consideration of organisational change using the 7S Model will allow the company to appreciate its current business activities, the potential implications of introducing the EMS, the availability and capability of its resource base and the impact of the proposed system on other organisational elements. This will provide a perspective on the strategic direction of the company, the position of the organisation in relation to its intentions, areas of strength and issues of concern. From this the company will be able to determine what it has to do, the resources required and the investment needed.

Outline programme

The timescale for developing and implementing an EMS will depend on a number of factors, the principal ones being the company's size, complexity, business activities and location. These factors exist in addition to the capacity for change and feasible pace of change. The company will need to develop an outline time-management programme to tentatively schedule activities against milestones. In addition to compiling the programme, the company will need to programme communications regarding system establishment to managers and the workforce. At this stage the programme would be presented in outline only since more detailed time-management programming will be carried out during Stage 2: Preliminary review. The outline programme will include a schedule for each main stage of system development from preliminary review through to management review. A separate programme would be compiled later to schedule those tasks involved in system registration with the chosen certification body.

Management responsibilities

All levels of management within an organisation have a part to play in establishing an effective EMS. The roles and responsibilities of management are differentiated by the level at which they operate within the company. They can be described as follows:

• *Executive management* – the senior management group of the company, or board of directors, who are responsible for the corporate management of the organisation and who develop and support the organisational culture, policy, objectives and programme which lay the foundation for EMS establishment.

- *Management representative* – the assigned member(s) of the senior management group, responsible for the oversight management of the implementation programme for the EMS. They liaise with the *environmental manager.*
- *Environmental manager* – the leader of the *environmental system team* responsible for implementing the programme. Accountable to the management representative, the project manager is line manager to the team.
- *Environmental system team* – the environmental system team is a group of managers, each representing a company department, process work section or management function, whose responsibility is to encourage liaison and co-operation throughout the various groups of the organisation.
- *Environmental co-ordinators* – represent departments, process work sections and management functions and oversee environmental management applied to their organisational area.
- *Line managers* – are responsible for supervising assigned activities within the business processes, the operations of employees who service those processes and the environmental performance within their assigned area of management.
- *Employees* – have a general duty for environmental management within their activities, overseen specifically by their line managers and generally by the environmental co-ordinators.
- *Consultants* – are specialists procured in-house or externally to provide particular knowledge and skills in environmental management, for example the development of systems and environmental auditing.

System team

Introducing and embedding the management responsibilities, and in particular establishing the environmental management team, are critical activities in creating a workable and effective EMS. The environmental manager is at the heart of developing and implementing an EMS. The environmental manager has total responsibility for the system and the personnel involved, and will be accountable to executive management through the management representative. The environmental system team is a vital part of the management structure as it is responsible for bringing the various sub-divisions of the organisation together. Encouraging close liaison and co-operation is fundamental to making employees aware of environmental management and in gaining their support and commitment for operating the system.

Getting people on board to challenge traditional thinking and conventional methods of working is needed if meaningful participation is to be secured. It is likely that most parts of a company will create environmental effects, and therefore everyone across the organisation needs to contribute to environmental management. Creating group and individual ownership and support for the environmental management programme is a key aspect of effective management. Employees will need to appreciate the programme and, moreover, to understand the part that they can play in its development and implementation. Developing a unified view of the environmental goals and targets is a principal function of managers so that everyone in the organisation is pulling in the same and appropriate direction. Perpetuating awareness and providing clear communication is fundamental to successful establishment of the EMS.

Employee awareness and communication

Raising awareness among and across all the managerial levels responsible for establishing an EMS and among those employees who will be involved in the system's operation is vital. To develop the awareness of managerial staff there are two main tasks. The first task is to ensure that managers appreciate the policy, objectives, goals, targets and programme and understand their responsibilities. In addition, managers should be fully empowered to make decisions in relation to system development and implementation. This can be achieved through employing staff information sessions and workshops. The second task is to cascade their knowledge and experience to supervisors and employees. This can be achieved through a formal awareness campaign involving presentations and on-the-job workshops at the various staff and operative levels in the company.

Senior managers will need to be aware not only of system establishment, but also of the implications if application is inefficient or ineffective, and the ramifications if things go wrong. Therefore, an important dimension of managerial awareness is a sound appreciation of environmental legislation, industry regulation, criminal and civil liabilities. In addition, they should be fully aware of the potential commercial implications of environmental infringement such as the effects on supplies to customer, business profitability and company profile. Middle and operational managers' and supervisors' awareness should focus on those matters which affect the running of the EMS which supports the business processes. This will involve the specific details of system operation together with a good appreciation of performance criteria. In addition, it will advise employees of their responsibilities for environmental management and in particular highlight areas of operations where environmental risk is present and accidents and incidents might occur. At an operational level an important dimension of awareness is the emergency action to be taken in the event of an incident.

As there could be environmental implications for individuals and groups of employees, an important focus of awareness should be given to occupational health, welfare and safety. There are strong overlaps between environmental legislation and industry regulation and health and safety. Managers have responsibility and a duty of care for the environment-related health and safety of employees, and therefore awareness initiatives must embrace these.

Communications throughout the company should be well structured and seek to ensure that clear and effective information is cascaded from executive management through to system operatives and all employees. Communication should be considered as a two-way process, both down and up the organisational structure. Management should be robust in terms of cascading information which structures and reinforces the messages underpinning policy and objectives, yet retain sufficient flexibility to respond to ideas from employees concerning the programme and methods of working. In this way, management will gain the respect they need to oversee system implementation, while employees will feel they can contribute and have ownership.

Communication mechanisms need to be established at all levels throughout the company if appropriate information is to be transmitted effectively. Methods to achieve this include the following:

- *Management meetings* – these involve a series of meetings at each organisational level to facilitate communication to and between managers. Executive management will focus on policy making, agreement of objectives, goals and targets and the programming of the establishment of the EMS. Directive managers will focus on translating corporate aspects into systems implementation through functional management while operational and line management will target performance planning and task objectives. The management meeting series should cover: organisational structure; resources and personnel; corporate business requirements; auditing; and management review.

- *Presentations* – will be given by various levels of management to communicate the requirements of the programme. Sessions will share corporate perspectives of system requirements, with a focus on the potential benefits of effective management and operations. They will also provide an all-important opportunity to clarify any ambiguities with system mechanisms and allay fears among managers and workforce of the changes to working principles and practices. The presentation series should cover: environmental effects; work objectives; performance setting; and documentation use.

- *Workshops* – will be conducted to convey the requirements of the environmental management system implementation. They will facilitate face-to-face interactive discussion between managers and employees, with a focus on the relationship between job tasks and business performance. The workshop series should cover: the programme; system documentation; operations and work instructions; record keeping; and system auditing.

- *Self-learning documentation* – should be provided to all managers and employees. This is fundamental to communication and information dissemination, as the EMS will be described in a suite of system documents including: a manual; management procedures; working instructions; and an implementation plan. Copies of these documents should be provided to all persons involved in system development and implementation.

Preliminary review

Definition

A preliminary environmental review is:

> The detailed consideration (review) of all aspects of an organisation's business processes with regard to its environmental situation as a basis for developing an environmental management system.

(Griffith, 1994)

Purpose

The purpose of a preliminary environmental review, sometimes termed an initial environmental review, is to determine the current environmental status of the organisation in preparation for developing and implementing an EMS. This approach should establish a benchmark for the company's environmental performance and lay the foundation for introducing systematic and procedure-based environmental management. The review is not conducted to test the company's current environmental approach against an environmental management standard. It is a structured and systematic fact-finding exercise to scope the stance of the organisation and the structure of its business processes in readiness for establishing an EMS. It will embrace activities throughout the whole organisation and should be capable of examining a company's outputs whether they are products or services.

Key activities

A preliminary business process environmental review should include the following key activities:

- a detailed examination of existing environmental policies, business processes, management procedures and working practices;
- an appraisal of current environmental performance;
- an evaluation of environmental aspects and their effects and implications resulting from company activities;
- the compilation of an environmental effects register;
- the identification of principal legislative, regulatory and other environment-related requirements (environmental legislation register);
- an analysis of organisational performance in relation to business processes to identify potential areas for improvement;
- the production of a list of recommendations.

The organisational activities involved in the preliminary environmental review are shown in Figure 3.4.

Approach

To carry out a preliminary environmental review a combination of quantitative and qualitative information will be needed. A systems review and a technical review of organisational processes will be needed. Information will be gathered from examining organisational structure, business processes, methods of working and personnel. Although the approach to preliminary environmental review will differ among companies to reflect their circumstances and requirements, the generic aspects will consider the following:

- organisational structure and formal hierarchy;
- management responsibilities and commitment;
- activities within company divisions, departments and sections;

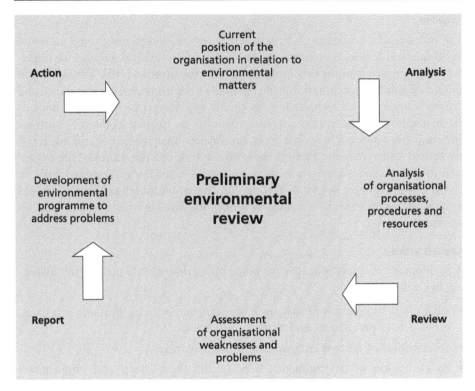

Figure 3.4 Preliminary environmental review

- compliance with legislation, regulation and other requirements;
- core business processes including design, production and distribution;
- technologies used in processes, and alternatives available;
- management procedures and working instructions;
- operations, methods of working and job tasks;
- deployment and utilisation of resources;
- environmental effects and impacts;
- current environmental management procedures and safeguards;
- existing links between environmental management and occupational health and safety;
- emergency procedures for environmental accidents/incidents;
- communication and dissemination of environmental information to personnel and external parties;
- current audit procedures;
- existing management review and improvement mechanisms.

The information required for these aspects may be acquired through a variety of mechanisms including:

- questionnaires (survey of personnel);
- interviews (with personnel);
- workshops (group personnel discussions);
- analysis methods (SWOT – Strengths, Weaknesses, Opportunities, Threats; PESTE – Political, Economic, Social, Technological, Ecological; PESTLE – as PESTE but including legal aspects);
- gap analysis (identification of procedures in place or in need of modification or development);
- inspections (scheduled and random);
- audits.

The collection and analysis of information should be co-ordinated by the environmental manager. Information may be acquired by one or more of three parties:

1 *Self-analysis* – conducted by in-company departments or work sections.
2 *Organisational in-house analysis* – conducted by the environmental management team.
3 *External analysis* – conducted by independent environmental consultants.

Information gathered through these data collection mechanisms may be structured into sections and sub-sections as follows:

- *Legislation and regulations:*
  - international and local legislation;
  - product or service regulation;
  - marketplace/industry legislation;
  - customer/client requirements.
- *Environmental effects and risks from the business:*
  - procurement of raw materials and suppliers' performance;
  - energy management and use;
  - emissions from processes;
  - sources of contamination;
  - waste products, sorting, storage and disposal (including special requirements for hazardous materials);
  - storage requirements of materials/outputs;
  - packaging, transport and distribution;
  - recycling and reuse;
  - neighbourhood/community issues.
- *Environmental procedures:*
  - policy;
  - strategy;
  - responsibilities;

- systems;
- resources;
- awareness, education and training;
- record keeping, audits, reporting and reviews;
- emergency planning and actions;
- accident/incident investigation and reporting;
- actions evaluation;
- insurance and indemnities;
- communications and public relations.

The aforementioned assumes that a company has some existing practices which are being structured into a formal and documented EMS. Where a company has no existing practices, an initial step towards effective preliminary environmental review will be to determine those organisational areas that require review. This would be achieved by conducting a scoping exercise. An outline list would be drawn up and through detailed consideration the list would be refined until the key areas were identified. These would then form the basis of the review.

SWOT analysis

The SWOT analysis technique is highlighted as an important method of gathering, synthesising and examining the organisational environmental situation and circumstance. The technique analyses the company's strengths, weaknesses, opportunities and threats. The technique is a basic management tool applicable to a wide variety of organisational analyses and is useful in the examination of environmental positioning. Strengths focus on those aspects of the organisation and its management which are positive in nature. These are robust areas upon which the company can build its system and improve environmental performance. Weaknesses are areas of management which place environmental performance at risk. They will require improved management control to ensure compliance with legislation, regulation and company policy and procedures. Opportunities represent areas where new or amended practice could bring benefits, for example in marketplace standing. Threats are areas which could damage the company if action is not taken to ensure effective environmental management and improved environmental performance.

Preliminary environmental review report

The analysis and findings from the preliminary environmental review should be contained in a report. This should focus on three principal issues as follows:

1 Determination of the nature and extent of the environmental issues identified and the organisational priority and timeframe assigned for rectification measures.

2 Development of an environmental management programme of actions needed to address effectively the issues of (1) above and how this will be resourced and implemented.

3 Specifications for the EMS given (1) and (2) above to ensure continual improvement.

System planning and development

[Refer to ISO 14001, section 4.3]

The EMS model

ISO 14001 propounds the PDCA methodology and EMS model. This is shown in Figure 3.5.

Environmental aspects (effects and impacts)

[Refer to ISO 14001, sub-section 4.3.1]

An important aspect of planning is to determine the environmental effects of the business and their potential impact. This is one of the most significant sections of planning to identify, quantify where possible and assess all the significant environmental effects from the company's business operations. Results from the identification of environmental effects and their impacts will link directly with the aims, objectives, goals and targets of the company and shape the development of the EMS. Where the EMS is configured appropriately to control the identified effects and impacts, it will be effective in ensuring, as far as practicable, the fulfilment of continual improvement.

The principal environmental effects can be grouped as follows:

- *Emissions to air* – include noise, vibration, fumes, vapours, gases, and particulates.
- *Releases to water* – include rainwater and sewerage discharges and controlled run-off to water courses.
- *Waste management* – includes non-hazardous and hazardous solid materials, particles and liquid wastes.
- *Land contamination* – includes discharge of any solid, particle or liquid which adversely affects the ground.
- *Use of raw materials* – includes natural resources, energy consumption, water use.
- *Local environment and community issues* – include disturbance/destruction of natural habitats, wildlife, flora and fauna, comfort and visual disturbance of local inhabitants.

All the potential environmental effects identified by the company as a result of its business activities should be analysed for their level of hazard, effect and severity on the local environment. A risk-based methodology, or 'risk assessment', is usually applied whereby a determination of high, medium or low risk is made based upon a quantitative scoring approach which considers severity and probability of the effect. A matrix can be developed to reflect the impact, source and calculated risk of the identified environmental effects. An example of an

---

The Environmental Management System Model

Plan–do–check–act (PDCA): is an ongoing, iterative model that enables an organisation to establish, implement and maintain its environmental policy based on top management's leadership and commitment to the environmental management system. After the organisation has evaluated its current position in relation to the environment, the steps of this ongoing process are the following:

(a) **Plan:**   Establish an ongoing planning process that enables the organisation to:

    (1)   identify environmental aspects and associated environmental impacts;
    (2)   identify and monitor applicable legal requirements and other requirements to which the organisation subscribes, and set internal performance criteria where appropriate;
    (3)   set environmental objectives and targets and formulate programme(s) to achieve them; and
    (4)   develop and use performance indicators.

(b) **Do:**   Implement and operate the environmental management system:

    (1)   create management structures, assign roles and responsibilities with sufficient authority;
    (2)   provide adequate resources;
    (3)   train persons working for or on behalf of the organisation and ensure their awareness and competence;
    (4)   establish processes for internal and external communication;
    (5)   establish and maintain documentation;
    (6)   establish and implement document control(s);
    (7)   establish and maintain operational control(s);
    (8)   ensure emergency preparedness and response.

(c) **Check:**   Assess environmental management system processes:

    (1)   conduct ongoing monitoring and measurement;
    (2)   evaluate status of compliance;
    (3)   identify non-conformity and take corrective and preventive actions;
    (4)   manage records;
    (5)   conduct periodic internal audits;

(d) **Act:**   Review and take action to improve the environmental management system:

    (1)   conduct management reviews of the environmental management system at appropriate intervals;
    (2)   identify areas for improvement.

This ongoing process enables the organisation to continually improve its environmental management system and its overall environmental performance.

Figure 3.5 Application of the PDCA methodology to the EMS model
Source: Adapted from ISO 14001

Environmental effect	Impact	Source	Severity	Probability	Risk
Air emissions • Dust	Nuisance to neighbours	Production site	1	3	3 low
Land contaminations • Storage of liquids	Ground pollution	Storage areas	1	1	1 low
Materials use	Hazard to people/ wildlife/ flora/fauna	Outside storage areas	3	3	9 high

Figure 3.6 Example section from an environmental effects and impacts analysis

environmental effects and impacts analysis is shown in Figure 3.6. Alternatively, a 'criteria-based' methodology can be applied. Questions are asked on environmental aspects to determine their potential impact and then mitigation measures developed accordingly. This method is essential where quantitative approaches such as risk assessment cannot be applied, as would be the case for low-risk projects or for activities taking place in an office environment.

When considering the potential environmental effects of business activities, a company may ask a considerable number of questions including the following:

- *Raw materials:*
  - Which materials are used, and is there an inventory?
  - From where are materials sourced or procured?
  - Are materials appropriately delivered, handled, stored and used?
  - Could alternative materials be used?
- *Pollution control:*
  - What processes discharge into the atmosphere, ground or water courses?
  - How much pollution is discharged?
  - Are discharges licensed?
  - Are limits placed on discharges?
  - Are environmental effects recognised from discharges?
- *Waste management:*
  - What waste products are produced from business processes?
  - How much waste is produced?
  - Are any wastes hazardous?
  - How is waste sorted, stockpiled, handled and disposed of?
  - Are there appropriate licences for disposal?

- Could waste be recycled or reused?
- Are accurate records kept of all wastes?
- *Energy use:*
  - What forms of energy are used?
  - How much energy is used?
  - Is energy use monitored and recorded?
  - Could alternative energy forms be used or existing use reduced?
- *Contamination:*
  - Can materials leak to the ground, and are they hazardous?
  - What are the materials that could cause contamination?
  - Are processes hazardous to local features and buildings?
- *Local environment and community:*
  - Which processes impact upon the comfort of local inhabitants?
  - Could there be visual disturbance to the neighbourhood?
  - Are processes/activities hazardous to the locality?
  - Could there be dangers to natural wildlife, flora and fauna?

Legislation

[Refer to ISO 14001, sub-section 4.3.2]

To develop and implement an effective EMS the company must have a working knowledge of environmental legislation. This should embrace both national and international perspectives. A company must ensure that it is aware of and complies with all legislation that affects its business operations. This is essential in meeting the requirements of ISO 14001 and the company's own environmental policy. The environmental manager should ensure that environmental legislation is monitored and tracked to identify legislative changes, new and proposed legislation.

Keeping abreast of environmental legislation is a task that should be undertaken annually as part of the management review. As legislation changes or new legislation is introduced, the company should examine its effects on business operations and initiate appropriate action to ensure compliance. It should be an intrinsic aspect of the environmental manager's role to keep executive management updated on current legislation and its implications. Likewise, the environmental manager should ensure that information is cascaded to managers and workforce so that the knowledge base of all employees remains current.

To formalise and structure the knowledge base of the company in environmental legislation it should compile an 'environmental legislation register'. This is a list of all environmental legislation applicable to the company and its business operations. It represents a good starting point for the appreciation and understanding of environmental legislation within the organisation. The register can be added to at any time as amended or new legislation is introduced or as company activities change and different legislation becomes applicable.

It is usual to register environmental legislation under categories applicable to particular groups of environmental aspects, for example legislation concerned with:

- general legislation;
- land use;
- air emissions;
- water releases;
- waste;
- ground contamination;
- noise and vibration;
- chemicals;
- materials;
- transport;
- community.

Environmental policy

[Refer to ISO 14001, section 4.2]

Definition
An environmental policy is:

A published statement of organisational intentions in relation to managing the environmental effects of the company's business activities.

Purpose
The purpose of an environmental policy is to:

- define the company's corporate philosophy towards environmental management in the context of its business activities;
- give authority to a written statement from executive management on the company's intentions for managing the environmental aspects of its business.

Requirements
To have relevance, currency and authority, an environmental policy should meet a number of requirements as follows:

- Be relevant to the core business, support and assurance processes.
- Identify the environmental effects and impacts of the business activities.
- Originate from executive management.
- Be supported by all management and workforce levels in the organisation.
- Encapsulate a culture of establishing, maintaining and improving the environmental performance of the company.
- Be amenable to scrutiny by company stakeholders, regulatory bodies and the public.

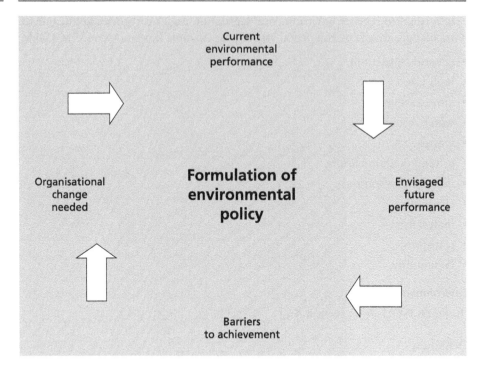

Figure 3.7 Organisational formulation of environmental policy

Approach
The approach to formulating an environmental policy involves a number of stages as shown in Figure 3.7. These are to determine:

- current environmental performance;
- envisaged future environmental performance;
- barriers to achieving the desired environmental performance;
- organisational changes required to ensure the desired environmental performance.

Policy content
The content of the environmental policy should incorporate the following aspects:

- *Corporate responsibility* – describes the range of responsibilities for the environmental effects of the business activities.
- *Legislative accountability* – describes the compliance with legislation and regulations associated with the business activities.
- *Performance expectations* – describe the intended environmental performance when carrying out business activities.

Statement
In many companies, the environmental policy will seek to address a broad range of aspects, interests and intentions. These will depend upon the nature of the

company and its business activities. Therefore, the policy will be as narrow or broad and as simple or detailed as the company needs to reflect its position and circumstances. As an environmental policy statement is, in reality, the public face of a company's response to, concern for and commitment to the environment, a statement should be made which is:

- clear and understandable to a wide audience;
- presented in an unambiguous format;
- a true reflection of organisational intentions;
- linked to aims, objectives, goals and targets for environmental performance;
- compliant with legislation;
- committed to continual improvement of environmental management;
- published with company identity;
- flexible for use in annual reports, publicity material and advertising media;
- authorised by signature of executive management.

As the environmental policy lays the foundation for development of an environmental programme and performance setting, it should be configured with measurable criteria in mind. This is important to developing goals and targets. Notwithstanding, the policy sets the overall ethos and philosophy towards environmental management and as such the statement which encapsulates the policy should avoid becoming over-detailed. The detail will be contained in the EMS programme, manual and associated documentation.

The appropriate formulation of environmental policy is fundamental to the effective establishment of an EMS, since policy shapes the organisation's philosophy. It also forms the basis for the environmental programme, the operational elements within the programme and the deployment of management and workforce. Commensurate with embedding leadership, policy must be developed at a strategic organisational level by executive management. Their perspective must then be cascaded throughout the whole organisation by each level of management to all employees. This is achieved through the implementation of EMS documentation – system manual, management procedures, implementation plan and working instructions – all of which are described subsequently. The environmental policy should be sufficiently robust to encourage the necessary change in working practices to meet the environmental demands placed on the organisation. Equally, the policy should be sufficiently flexible to accommodate fluctuations in organisational circumstances or to realign operations to comply with changing legislation.

The success of any management system is dependent on the acceptance, support and commitment of employees. Therefore, is it important that policy is seen as clear and sensible by both managers and workforce. For policy to be embedded and become meaningful to employees, its intentions must be capable of translation into understandable management procedures and working instructions. Although policy is a top-down management initiative, it should

---

ENVIRONMENTAL POLICY STATEMENT

1 The Company will ensure that, in the use of land and resources, its activities are commensurate with the needs of the environment.

2 The Company will, wherever practicable, use materials and products from sustainable sources.

3 The Company will pay particular attention to the environmental effects of its business including pollution, noise, dust and dirt and take measures to ensure that the impacts of these aspects are minimised.

4 The Company will pay careful attention to the procurement, storage and use of any potentially hazardous materials to avoid environmental impacts.

5 The Company will continue to develop an environmentally aware approach to its business activities and support the continued use of environmental management systems and procedures.

6 The Company will achieve this policy by establishing a clear set of environmental objectives and targets implemented by environmentally aware, trained and skilled managers and employees.

Signed .... Chief Executive Officer (CEO)

Figure 3.8 Example of a company's environmental policy statement

facilitate bottom-up involvement. Mechanisms will therefore be needed to cascade policy through levels of management to employees and to ensure that feedback, questions and ideas can be incorporated which may improve the policy or aid its implementation.

Environmental policy statement example
Figure 3.8 shows a typical environmental policy statement based on a composite of a number of policy statement examples from a range of organisations.

Environmental goals, objectives and targets
[Refer to ISO 14001, sub-section 4.3.3]

Definitions
Environmental goals are:

A set of objectives and targets which a company seeks to fulfil in relation to particular aspects of environmental management.

Environmental objectives are:

Measurable achievements in fulfilling environmental goals.

Environmental targets are:

The quantified performance specified to fulfil environmental objectives.

The environmental goals of a company are strategic in concept and fulfilled over extended periods of time. Goals therefore form a part of five-year, or longer, organisational business planning, although some goals may be short term to be fulfilled within an annual business plan. Objectives tend to be long term while targets are short term, yet both should be quantified where possible to facilitate the measurement of achievement or progress against planned performance. Objectives and targets should be challenging and demanding yet also motivational to employees. An example of this could be the following.

Environmental goals, objectives and targets should be set based on the environmental policy and the environmental effects and impacts analysis. The overall aim should be to set performance expectations which continually improve the environmental performance of the company. To ensure achievement, objectives must be linked to measurable targets and with the time and resources needed for their fulfilment appropriately included in the environmental programme. There would be little point in setting targets which were unachievable through inattention to resource requirements or inappropriate timescales for completion. Objectives and targets should be carefully and consistently monitored and reviewed to ensure that the company's environmental goals are fulfilled. If they are not meeting performance requirements or timeliness, the goals may have been too ambitious or, conversely, performance may be lacking. Management will then be required to take appropriate action to rectify the situation.

There are three broad groups of environmental objectives as follows:

1 *Improvement objectives* – those linked to environmental effects and impacts with high risk which require immediate action to ensure improvement.

2 *Monitoring objectives* – those linked to environmental effects and impacts with medium risk which are appropriately controlled yet require monitoring to ensure that control is maintained.

3 *Management objectives* – those of a general nature to meet longer-term goals.

Environmental goals, objectives and targets should be formalised through documentation and realistic timeframes set for their fulfilment. There should be direct links between aspects of environmental policy and objectives, and a target specified for each. For example, for the organisation policy aspect of waste management, an objective might be to reduce landfill waste by a target quantity of 5% per annum for five years.

Objectives should originate from executive management and be clearly seen to underpin environmental policy and goals. This is important in encouraging employees to see the relevance of their work in meeting the objectives and to support the holistic vision of the company. Objectives play a vital part in linking the strategic dimensions of the company to the operational aspects in the following ways:

- conversion of organisational environmental mission and purpose into principles for action;
- connection of environmental purpose and principles with the organisational structure;
- linkage of the organisational structure with environmental management functions and system;
- translation of environmental philosophy into applicable and meaningful work practices.

The *SMART* approach is a well-recognised set of considerations for organisational objective and target setting. Considerations should be:

S – Specific
M – Measurable
A – Achievable
R – Realistic
T – Time related

It is important in setting all objectives and targets that they are time related and can be tracked. Only in this way can improvements be demonstrated. Therefore, records should be kept which can provide objective evidence of environmental activities, performance and improvement.

Environmental programme

[Refer to ISO 14001:2004, sub-section 4.3.3]

Definition

An environmental programme is:

> A structured action plan(s) to formalise and implement the means of achieving the company's environmental objectives and targets.

An important aspect of environmental management is the transition from saying what the company seeks to achieve into actions which ensure that the intentions are fulfilled. This is achieved through the development and implementation of a programme of environmental activities. The programme should reflect a number of fundamental characteristics as follows:

- the specification of environmental objectives and targets and their relationship to goals and environmental policy;
- the assignment of responsibilities for environmental management functions throughout the organisation structure;
- the ways in which objectives and targets are to be met.

It is essential that the organisation's policy, goals, objectives and targets are supported by a realistic and workable programme for implementation. There must be a timescale in which the actions in the programme are to be carried out.

It should be challenging yet realistic. The programme should embrace any aspect where objectives and targets have been set. An attempt to incorporate all facets in one environmental programme may be too ambitious a challenge. Where there are many objectives to be planned and monitored, a master environmental programme will be compiled, supported by sub-programmes to reflect individual objectives and groups of objectives.

Programmes can take many forms but it is traditional convention to use Gantt, or bar, charts to reflect activities against timescales. These have the advantage of being simple to compile, and are generally well understood by users. They can facilitate the breakdown of a large and complex programme into sections, or sub-programmes, for providing greater insight into particular elements. The bar chart also facilitates the charting of progress where work on activities can be directly related to that which was planned. Other methods of planning can be used to support bar charts, such as network analysis. This examines the sequence, timing and interrelationship of the required actions in much greater detail, and with accurately quantifiable durations presents a useful numeric-based picture of the programme.

A key role of the environmental manager is to ensure that the implementation of the programme stays on track. This involves monitoring activities to set milestones and checking progress against those milestones. Sub-programmes will also feature activity milestones, and these must be monitored and assessed for progress in association with other aspects and against the master programme.

Much time will be devoted by management to programming and ensuring the progress of those activities which contribute to the development and implementation of an EMS. While the environmental manager assumes ultimate responsibility for the achievement of the programme, tasks will be delegated to other members of the environmental management team. To maintain clarity in managing sub-programmes which have been delegated, each sub-programme will be assigned to named personnel. This is important in creating ownership of the sub-programme and placing importance upon it as an integral part of the master programme. An environmental sub-programme should contain the following information:

- link to the master environmental programme;
- assignment of management responsibility;
- team members involved in activities;
- objectives and targets to be met;
- methods of fulfilling the programmed activities;
- criteria for measuring achievement;
- authority for completion;
- link back to the master environmental programme.

Each sub-programme should adopt the same format for description and mode of presentation to ensure consistency of approach. This is important as each sub-programme must link into the master environmental programme, and this will be easier if they are all alike. Similarly, progress review will be more meaningful as activities are co-ordinated. Managers should be able to see the progress of their sub-programme, its relationship to the progress of other sub-programmes and the position of all sub-programmes within the master programme.

Both the master programme and sub-programmes should be highly visible to management at all levels of the organisation. As the milestones set are crucial to system establishment, it is important that all personnel involved know what they have to do and when it must be achieved by. The company, therefore, should make programmes prominent and communicate them clearly and widely. Programmes should be linked clearly to methods and milestones for monitoring progress. Regular reviews should take place of each sub-programme to ensure that overall progress is maintained and that problems are identified, so that the environmental manager can assess the overall implications and devise the next actions.

Implementation and operation

[Refer to ISO 14001, section 4.4]

Management structure, responsibilities and co-ordination

[Refer to ISO 14001, sub-section 4.4.1]

Organisation chart

Establishing a workable and effective management structure and assigning responsibilities to personnel is fundamental to the successful establishment of an EMS. It is vital that everyone knows what part they play in operating the system and their responsibilities towards fulfilling the environmental objectives and targets. The structure sets out the formal arrangement of roles, responsibilities, authority, communication and interrelationships. It is necessary, therefore, to devise a document which carefully depicts and makes clear these arrangements. This document is the 'organisation chart'. The purpose of the chart is to illustrate the company's organisation of human resources and the assignment of responsibilities for the management of those resources which are necessary for the implementation of the EMS. See Figure 3.9.

In defining responsibilities and creating the environmental organisation chart, a number of fundamental principles are important, as follows:

- The executive directors of the company must be associated unequivocally with the chart so that support and commitment are seen to come from the top.
- Specified management functions should be clear to establish an organisation of champions for environmental matters within the total structure.
- All personnel involved with environmental management should be depicted in the chart.
- The organisation of management should be reflected in the company's environmental documentation – the manual and procedures.

Environmental management organisation chart

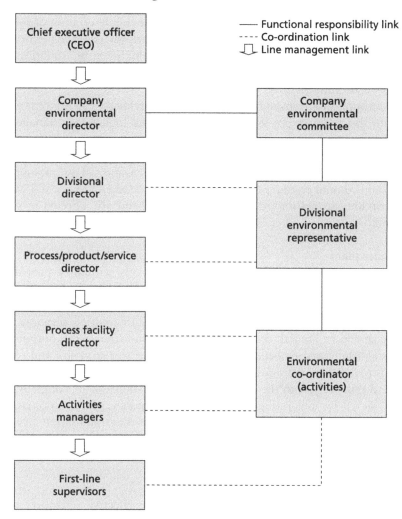

Figure 3.9 Simple organisation chart for environmental management structure

- The chart should show the relationship between managerial staff and operational personnel.
- Functional responsibilities and lines of co-ordination and communication should be shown.

Job descriptions

Each member of personnel needs to co-ordinate their position in the organisation with their environmental role, duties and tasks, and therefore 'job descriptions' should be provided to each individual to embed individuals and the tasks they undertake within the management structure. Job descriptions define the role of an individual within a company. It enables the individual to have a clear understanding of their job and the boundaries within which they function. Job descriptions comprise two elements. The first is a general description of role and responsibilities and the second is a specification for performance. The performance specification is the level of achievement needed to fulfil the responsibilities of the job. Within the job description there should be a statement relating to an individual's level of authority.

Environmental situations may arise where the individual must take emergency action and this requires pre-assigned authority from higher line management. Awaiting instructions from a line manager in an emergency may mean that appropriate action is not taken sufficiently swiftly to avoid a major environmental breach. Authority must be available to an individual to think and act autonomously if needs be and commit resources to mitigate any potential environmental impact. Job descriptions are an important part of an EMS and are reflected in system documentation.

Performance plans

In addition to the job descriptions, 'performance plans' should be provided so that each individual is informed of the environmental performance requirements of their job. Performance plans translate the role and responsibilities into measurable achievements. Job descriptions normally contain multiple responsibilities and these should be limited to a challenging yet feasible number of outputs. The key to establishing and embedding effective management structure and responsibilities is to link the organisation chart, job descriptions and performance plans closely and positively with the environmental aspirations of the company. This can be achieved by integrating the environmental policy, goals, objectives and targets with the activities of the individual through the organisation structure, their job description and their performance plan. The benefit of this will be to encourage ownership of and commitment to the corporate environmental aims.

Co-ordinating resources

The co-ordination of resources is essential to the effective operation of the EMS and fulfilment of the company's environmental policy. The environmental manager should review all managerial and operational functions within the organisational structure to ensure that all key aspects are resourced and that no gaps exist. In addition, it is important to ensure that functional overlaps, although often inevitable, are minimised to prevent duplication of tasks and inappropriate use of resources. The required degree of clarity depends very much upon developing a strong working relationship between the environmental manager and the environmental co-ordinators who oversee operations. Likewise, it is important that environmental co-ordinators liaise well with the first-line supervisors of the workforce.

An aspect sometimes overlooked is the need to co-ordinate resources that deliver products and services through different sections or departments. Although they may operate separately, they can also operate with mutual dependency through shared management. It is important therefore that cross-functional co-ordination is considered.

To achieve the necessary co-ordination of resources a number of mechanisms can be employed:

- *Service agreements* – specific arrangements between sections/departments to encourage co-operation in areas where there are likely to be mutually dependent interests.

- *Co-ordination meetings* – designated and regular cross-functional or inter-departmental meetings to review the management of common processes, procedures and practices.

- *A directory* – specifying the requirements for co-ordination throughout the organisation. This will feature in the environmental management manual, a key part of system documentation.

Training

[Refer to ISO 14001, sub-section 4.4.2]

To complement effective communication, education and training in environmental matters are essential to the company. First, they are fundamental to raising the awareness level among the workforce of environmental issues and expected operating standards within the organisation and, second, to alleviating fears of organisational change and management system implementation. An appropriate programme of education and training should be used which facilitates employee awareness, consciousness, understanding, implementation and commitment to the EMS. It is important to embed good procedures such that they become standard operating practices rather than the exception.

Education and training requirements do not simply appear but need to be identified. The identification of training needs for the organisation, its functions, management and individuals should be undertaken using knowledge base and skills analysis. This forms a part of role determination and job description mentioned previously.

Education and training programmes

There are essentially two related yet different knowledge and skill areas which need to be considered:

1 *Technical* – the knowledge and skills in environmental matters applied to the company's processes and operations.

2 *Personal* – the interpersonal capabilities needed to communicate and co-operate with other employees when undertaking processes and operations.

The programme for education and training can accommodate these needs through the variety of mechanisms mentioned previously: awareness sessions; seminars; workshops; on-the-job instruction; tool-box talks; and off-site courses. These should address not only the requirements for carrying out operations within production processes, but also the elements of EMS implementation. While not every employee will need an in-depth working knowledge of system implementation, understanding the function and contribution of the system and an empathy for the tasks of other employees will be useful.

Training will be required at different levels of the company both to accommodate the technical and human aspects and also to satisfy the different breadth and depth of knowledge needed to understand and implement the environmental system. This is likely to be as follows:

- Environmental awareness: *all personnel.*
- Policy and environmental management: *all personnel.*
- EMS: *senior managers.*
- Personal attributes and skills: *all managers and first-line supervisors.*
- Technical training: *operational personnel.*

Training must be commensurate with the aspirations of the company, and therefore a training programme should underpin the environmental policy, objectives and targets. In developing the knowledge and skills of the individual it should also reinforce the corporate culture and commitment to environmental dimensions of the business. It should achieve this by facilitating individual ownership of training rather than simply impose it. As the knowledge and skills of the individual develop, so the organisation gains from that engagement.

An effective training programme will maintain training records to ensure that knowledge and skill gaps are continually recognised and addressed. These are often held centrally by the human resources department and focus on individual employee development needs. While many training schemes are infrequent, loosely structured and static in practice, environmental training should be frequent and dynamic to meet the pace of change in the field. In fact, ongoing education and training are a requirement of management system standards and must be a part of system implementation and continuance. To meet both the technical and personal aspects of education and training, the programme should ensure that:

- the skill mix and balance of the workforce is appropriate;
- knowledge and skill needs are identified;
- appropriate mechanisms are used for delivery;
- there is ownership by the employee;
- assessment is regular;
- records are maintained.

Education and training are auditable aspects of systems implementation. Their presence demonstrates that the company's management and workforce are environmentally aware and implements knowledge and skills which are current and applicable. Auditors will seek to ensure that those personnel directly involved in system application are fully trained and able, and that all employees have a general awareness of system implementation. Overall, it is essential that a company demonstrates that its management and workforce take environmental matters seriously. This is founded in good awareness, understanding and practice, and this emanates from appropriate education and training.

Communications

[Refer to ISO 14001, sub-section 4.3.3]

Internal and external communications are an important aspect of EMS implementation and operation. Communication must be managed effectively to personnel within the company and to those outside who interact with or are influenced by the activities of the company. Internal communication will focus on the implementation of an appropriate 'communications plan' which, when implemented, continually informs and updates personnel on the management and operation of the EMS. In addition, it will assist and support training initiatives within the company which underpin the implementation and improvement of the management system. External communication will focus on the provision of environment-related information to company stakeholders, customers, suppliers, regulatory bodies and the public. This also should form a part of the communications plan.

Internal communications

Effective internal communications rely upon management understanding the following: the content and complexity of the information; the required degree of interaction with the recipient; and the desired response to the information. In relation to the developmental phase of an organisational system, much initial communication will be related to generating awareness among employees. In this situation, the degree of interaction between the sender and the recipient will be low and the content of the message will be simple to comprehend. As an organisational system moves into the implementation phase, the information to be disseminated to employees becomes more detailed and complex, whereupon the need for interaction rises and a greater response is expected. Because communication is dynamic and fluid, management must consider carefully what communications are made, how they are made and their effect upon the organisation.

Managers will need to adopt a range of communication mechanisms commensurate with the environment-related knowledge and skill needs of the workforce. These mechanisms may involve: general awareness campaigns; management team meetings; staff workshops; briefing papers; and newsletters. The overriding purpose of these mechanisms is to allow the company to develop and embed a culture of environmental empathy supported by a good general understanding of the organisational structure, the management function and implementation systems to deliver its outputs successfully. Detailed environmental knowledge and operational skill bases will be developed by appropriate training initiatives.

The primary difficulty encountered by managers when implementing any new management system is what is commonly termed the fear factor. This is an inherent aspect of all change management and often arises from misinformation, ill-timed information or poorly delivered information. Honest, timely and transparent communication is vital to combating potential dysfunctional effects. The establishment of an EMS, like any major initiative, will bring some organisational concerns and fears. Good communication will gain the confidence of staff, increase the knowledge and skills of the workforce, and can support challenge rather than breed reticence. An important dimension of achieving successful communication is linking it with education and training. Integrated communication and training can develop organisational culture by moving the mindset from passive awareness to comprehensive understanding and proactive participation.

External communications

With ever-increasing industry and public interest in the environmental performance of individual companies and business sectors, environment-related communication is more important than ever before. Companies must be clear and responsive to environmental issues, to the environmental effects of their business activities, and to safeguarding the environment. For almost all companies this means being open to the provision of environmental information, transparent in the reporting of environmental performance and amenable to environmental scrutiny by outside bodies. Although accepted and embraced by many companies, such requirements may sit uncomfortably with others. Notwithstanding, such demands are an inevitable requirement within many business sectors and cannot be ignored.

Environmental accountability is high on the political agenda of many countries within the EU and in other countries worldwide. Environmental management standards require that companies review their environmental performance, and while formal reporting is not a requirement many reputable companies see it as a way of evidencing their response and commitment to the environmental aspects of their business. Environmental reporting outside the company is, therefore, becoming an activity given greater consideration today by proactive organisations.

A formal environmental report, published annually, is a positive and useful method of external communication based on both quantitative and qualitative organisational performance data. Relating to the company's environmental policy, objectives and targets, it allows external parties to appreciate, in some detail, the environmental performance of the business. This can be compared with previous performance to measure improvements or serve as a benchmark for comparison with other companies. For smaller companies, it might be appropriate to include an environmental section within their annual financial report or merely issue a statement on environmental activities rather than produce a dedicated separate report. The essential point is that a company should seek to produce an environmental position document on a periodic basis for the interest of external parties.

In addition to an environmental report, a company should also establish mechanisms to handle a range of communication requirements. These might take the form of regular environmental newsletters, company magazines, networking circulars or public-relations events. These can assist the organisation to communicate directly with external parties including local communities, regulatory bodies, local authorities, industry professional groups and members of the public. Wider dissemination of environmental performance can be achieved through publishing material on the Internet, and while this may not be a necessity for all companies, those operating on an international level may find benefits through reaching a wider client audience.

System documentation

[Refer to ISO 14001, sub-section 4.4.4]

Purpose of documentation

Perhaps the most important dimension to the development, implementation and maintenance of an EMS is 'documentation'. Documentation is the formalisation of the system in writing within which efficient and effective management practices have been developed. Moreover, documentation provides the foundation on which environmental management application can demonstrate compliance with a specified environmental system standard. The standard requires that there is adequate and appropriate documentation to support the system in operation. Documentation and its use need to be managed to ensure correct application and so 'controlled documentation' is established. This ensures the effective operation of all those procedures necessary to maintaining the environmental safety and quality of the company's business activities. Documentation describes environmental management procedures and working instructions and then provides objective evidence that those procedures and instructions have been applied appropriately. Only in this way can a company be certain that it is implementing its EMS successfully.

The objectives of documentation

The documentation established for an EMS fulfils a number of fundamental and key objectives. These are to:

- present a well-defined written environmental management framework which configures organisational resources effectively;
- formalise environmental management through written documents based on uniform approach and operations;
- establish a written co-ordinating document – the EMS manual – for all other system documentation;
- provide a written set of management procedures and working instructions that meet the requirements of the EMS standard;
- provide a written basis against which actual practice can provide evidence, through written records, that the EMS has achieved what it has set out to achieve;
- support environmental performance improvement by the organisation through evaluating written records of performance against written goals and targets.

In general, documentation describes and explains how a company is structured, organised, resourced and operates. Therefore, EMS documents specifically detail how the environment-related dimension of the company is configured and functions. Without explaining what is to be done within the company and how things are to be done, the company could simply not function. It would, in fact, quickly become dysfunctional. Documentation is, therefore, essential. That said, the need for documentation can become synonymous with bureaucracy. Care needs to be taken to ensure that the EMS is documented to an appropriate level of detail to allow the system to be implemented efficiently and effectively while not becoming burdensome through written information overload. Likewise, the management procedures and working instructions that the documented system initiates must be sufficiently simple as to make them fully workable and achieve their purpose. Documentation must be accessible to those who need to use it while also being user-friendly. Although the volume of documentation needed to describe and explain environment-related company activities might be large, it can be broken down into sub-sections for ease of understanding and application. Documentation is an active constituent of the organisation and will change and evolve as the system matures through experience of use. Documentation, therefore, not only has to be developed and implemented, but must also be managed.

The documents

It is conventional to structure management system documentation in a hierarchy, often termed the document pyramid: one document at the top of the system is supported by an increasing number of documents throughout the lower levels with implementation of the system at the base. An EMS is no different to any other management system in respect of this fundamental framework. The framework is commensurate with the requirements for system development and implementation within ISO 14001.

The framework for an EMS incorporates four levels of documentation:

1 Environmental management manual (generic) – level 1.

2 Environmental management procedures (generic) – level 2.

3 Environmental implementation plan (situation specific) – level 3.

4 Environmental working instructions (generic with situation-specific elements) – level 4.

Many aspects of the system will be implemented throughout the company and its activities continually, and therefore documentation can be generic in nature. Where the system has a particular application, for example to a project, then aspects of the system will be situation specific and documentation must reflect this. As a consequence, the manual (level 1) and management procedures (level 2) are generic, the implementation plan (level 3), where required, is situational and working instructions (level 4) are generic but are influenced by situation-specific characteristics. The levels of documentation are also influenced by the level of organisation within which they are implemented. The manual and management

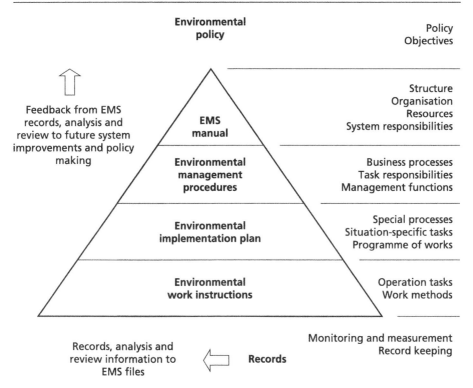

Figure 3.10 The EMS document pyramid

procedures are used predominantly to structure the corporate organisation of the company while the implementation plan and work instructions are used to structure the operational organisation of the company's activities. The framework, structure and levels of documentation for an EMS are shown in Figure 3.10.

1 *EMS manual (level 1)* – The system manual defines and describes: (a) the company's environmental policy and rationale; (b) the organisation structure; and (c) the management responsibilities to be followed to ensure that practice meets the desired and specified environmental objectives and goals of the business. The manual identifies and outlines the framework and structure through which the company will operate its EMS. It will also include structural links with related key aspects such as quality management and health and safety. The manual will include references to the management procedures (level 2) such that there is a document trail from level 1 to level 2.

2 *Environmental management procedures (level 2)* – The management procedures describe: (a) the processes to be managed; (b) how the processes are to be managed; and (c) the responsibilities for managing the processes. They are a set of procedures which are to apply throughout all the company's organisational sections and embrace all the company's activities. The management procedures

will include references to the implementation plan (level 3) to establish the document trail from level 2 to level 3.

3 *Environmental implementation plan (level 3)* – The environmental implementation plan describes the characteristics of any situation-specific application of the management procedures. This may be, for example, a service provided by the company within a project scenario. The implementation plan contains: (a) the situation-specific management plan; (b) any special processes to be implemented; (c) the programme for implementation; and (d) any special procedures needed to carry out the special processes. The plan must consider any requirements under current legislation and regulation applicable to specific management procedures. Prominent examples of this are particular restrictions to environment-related activity such as noise and pollution controls and health and safety provision of employees in specific situations. The plan will include references to work instructions (level 4) to ensure a documentation trail from level 3 to level 4. This is important at this level because working instructions are likely to be influenced by the specifics of the situation, such as site location and conditions.

4 *Environmental working instructions (level 4)* – The environmental working instructions: (a) describe tasks; and (b) describe how tasks are to be carried out. They should contain performance criteria reflecting the requirements of relevant standards and specifications, guidance on how to carry out the tasks and any training and instructions required in order to conduct specific tasks; an example might be working in an environmentally hazardous situation. To maintain the documentation trail, there should be references from level 4 to written record files.

What documents are required?
A fundamental and crucial question that must be asked is: what documents are required? It is worth restating here that the purpose of an EMS is to manage those environmental effects which result from the activities of the company's business. For each organisational activity that could impart an environmental effect, there should be a method for managing that activity which obviates or mitigates the environmental effect. Therefore, the documents required are those which:

- identify those activities which could create environmental effects;
- contextualise the environmental effects in relation to company environmental policy, objectives and targets and outline the framework for managing them (the manual);
- describe the procedures for managing the environmental effects (management procedures);
- consider any situation-specific aspects that could create environmental effects and how they may be managed (implementation plan);
- give instructions to personnel undertaking work tasks to alleviate environmental effects (working instructions).

Documenting activities

There are six key steps to identifying and documenting activities, as follows:

1 *Identify activities* – list organisational activities (product processes or services) and categorise into general and specific.

2 *Consider inputs* – list all resources (to processes or services) and consider the associated environmental aspects.

3 *Establish effects* – list all environmental effects associated with activities (processes or services).

4 *Evaluate impacts* – list all impacts, their scale, likelihood of occurrence and severity of consequence, through undertaking risk assessment.

5 *Determine actions* – list all actions that need to be taken at each management level throughout the organisation to manage the environmental effects involved with each organisational activity, describe how these actions will be taken and by whom.

6 *Write documents* – define, describe and explain the inputs, effects, impacts and actions for each organisational activity at each level of the EMS documentation.

Written document format

When the identification and documenting steps are complete, the potential environmental effects from each organisational activity will be known and the necessary actions will have been considered. Written management procedures and working instructions will have been produced for each. It is necessary now to collate these written documents into a co-ordinated and complete set of documentation which reflects the whole EMS. The particular type and format of documents for an EMS will, obviously, differ from one organisation to another depending upon their particular requirements. Notwithstanding, there are two main types of document format:

1 *Single document format* – where a manual includes all the levels of documentation required for the management system.

2 *Multiple document format* – where a manual is accompanied by sets of separate documents (management procedures, implementation plans and working instructions).

It is essential that the documents developed are presented in such a way that they are clear and also that they link from policy right through to post-implementation record keeping to operate completely. The documents will seek to achieve this by incorporating the following:

- a corporate flow diagram of organisational structure co-ordinating all of the system elements;
- sub-flow diagrams to define management procedures and work tasks within the system;

- detailed descriptions of the management procedures and work tasks;
- specification of procedures and tasks to guide personnel through their undertaking in a uniform and consistent way;
- job descriptions and performance specifications to define responsibilities and performance;
- monitoring mechanisms to ensure appropriate progress and performance;
- maintenance of operational records and filing for audit and review.

Document organisation and control
[Refer to ISO 14001, sub-section 4.4.5]

Document organisation should reflect the following:

- formal structure;
- the document development process;
- version and revision copies;
- ownership by managers;
- identification of documents to activities;
- a master file and master list of all documents.

The sensible creation and update of documentation is essential to ensuring the effective structure and implementation of an EMS. Documentation not only has to be developed and implemented, but must also be controlled. Control of system documentation is to know what documents exist, why they were developed, where they are, who is using them and whether they are effective in use. This can only be done if documentation is appropriately organised and managed. Every document used in the EMS must: (1) have a purpose; (2) have an owner (authorisation); (3) have currency (be the latest version); (4) be readily available to the user; (5) be traceable through the system; and (6) be archived for audit and review purposes.

System operation and control
[Refer to ISO 14001, sub-section 4.4.6]

Objective
The objective of operational control is:

> To implement the EMS's documented procedures, plans and instructions to ensure that all those activities of the company, which have a potential effect and impact on the environment, are carried out under controlled conditions.

To achieve this objective there are four key requirements, as follows:

1 Emergency preparedness plans and procedures must be in place to ensure that potential environmental emergencies can be handled quickly and effectively.

2 Control procedures must be in place to ensure that activities are carried out following the documented management procedures, working instructions and, where relevant, implementation plans.

3 Monitoring and measurement procedures must be in place to ensure that control procedures are implemented effectively.

4 Non-compliance and corrective action procedures must be in place to ensure that changes to control procedures are made when they become ineffective.

Emergency preparedness

[Refer to ISO 14001, sub-section 4.4.7]

Under the standard a company must have proactive written plans and procedures in place for managing any potential environment-related emergency. This requirement encompasses those activities undertaken within the corporate organisation and any and all activities carried out at other locations, for example within a project site situation. A company cannot rely solely upon the response and actions of the statutory emergency services to handle an environmental situation. They must have their own resources to respond quickly and effectively. In fact, most large companies will have such plans, procedures and resources in place as a requirement of their insurance policies. However, these may be somewhat general in nature, and specific measures may be needed to handle particular environmental effects and impacts.

An emergency preparedness plan will incorporate:

- identification of emergency situations;
- methods of alert and communication;
- evacuation processes;
- handling of injured persons;
- methods of assessing risk to people and property;
- procedures for containing and mitigating environmental effects and impacts at source;
- liaising with emergency services;
- incident investigation;
- inspection and testing for continued (post-emergency) risk;
- internal reporting and records;
- compulsory external reporting and records;
- public relations and media interfacing, where appropriate;
- feedback to facilitate improvements to procedures.

A further requirement is that of checking, or testing, the emergency preparedness plan. This should be done periodically and regularly. The most usual method is to carry out a simulation event. A simple but good example of this is a fire drill. While the nature of checking emergency preparedness will differ according to organisational activities and the overall types and levels of risk, an audit of the EMS will expect to see evidence of appropriate and competent checking.

Monitoring and measurement

[Refer to ISO 14001, sub-section 4.5.1]

It is conventional in management system terminology to refer to organisational activities being carried out under controlled conditions. By this is meant a state where processes and operations take place in ordered and predictable ways under the complete control of management and operating personnel. Essentially, if management procedures and working instructions are followed then the activities should be executed within those performance parameters expected. The output, whether it is a product or a service, should meet the requirements set by the organisational policy, objectives and targets time after time. This being so, the environmental management of the outputs by the system becomes second nature and management can focus more on improving the system, activities and its products and services.

Evaluation of compliance

[Refer to ISO 14001, sub-section 4.5.2]

To ensure that control mechanisms are working effectively, the EMS should have documented procedures to establish and maintain a mechanism that can confirm that controls are compliant or non-compliant. Moreover, the procedure will need to determine the level of compliance. Some activities may appear to be working effectively but not optimally, and in such instances management need to know just how compliant and effective the procedure is and which aspects might be improved. Sometimes, verification methods are superficial and merely observe that control procedures are working, irrespective of their level of efficiency or effectiveness. An approach is therefore needed which provides evidence of effectiveness. This can be achieved by measuring and testing the control procedures in use.

It would be unrealistic to expect perfect control procedures being applied consistently to all organisational activities. Conversely, zero non-compliance is virtually impossible. First, there is a need to identify unambiguously the level at which performance is non-compliant. Here there can be no doubt. A procedure fails, must not be allowed to continue and must be rectified. Second, a procedure operates but is not operating fully effectively. Here there is a range of operation from meeting a threshold standard up to fully effective operation. It is in this range that measurement needs to be applied. If the parameters of operation and a scale between them are set, management can measure where performance is on the scale and seek to improve the procedure. Over time, stepped improvements can be made until the control procedure operates at an optimum level of effectiveness.

There are a number of recognised monitoring and measurement techniques, of which the following are commonly used:

- *Control chart* – marks actual performance on a quantitative scale within set parameters, so producing a profile of performance over time.

- *Scatter diagram* – marks actual performance as points, or dots, which indicate normal or extreme performance.

- *Gantt chart and vertical Gantt chart* – mark actual performance as a bar against a set scale, producing a profile of performance over time.

- *Histogram* – marks actual performance as a bar (where the volume of the bar rather than its length is a unit) against a set scale, producing a profile of performance over time.

Testing in the context of measurement and testing is concerned with undertaking tests to ensure that measurement is accurate. This will involve checking the process and equipment involved with measurement to ensure their accuracy. In some situations scientific sampling may be undertaken to determine the effectiveness of control procedures. Sampling lends itself to environmental management as the effects of organisational activities, for example pollutants, can be measured and the impacts assessed. The main requirement of testing and sampling is to ensure that the procedures are appropriate and accurate, and provide consistent results. Testing can be absolute if so designed or can provide measures of performance within set parameters.

A fundamental requirement for the evaluation of compliance is to ensure that a structured, consistent and continuing regime is followed. The frequency of checking depends on the importance of the activity to be measured. Although all procedures should be routinely checked, those which are operationally critical to the business and those with significant environmental risk must be a priority. Any evaluation of compliance procedures must include recording, analysis and reporting methods. An appropriate recording procedure will gather, collate and analyse measurements and test results to provide written evidence of the performance of the environmental control procedures in operation. A written report of results should then be compiled which can be scrutinised by management with a view to sharing best practice and improving EMS procedures where required.

Non-conformity and actions

[Refer to ISO 14001, sub-section 4.5.3]

Where environmental performance falls outside the specified requirements of the control procedure as verified, measured or tested, it is non-compliant with system requirements. Mechanisms should be in place, first, to identify the problems which have caused the non-conformity. Second, the mechanism must allow the actions necessary to remedy the problem. Third, the mechanism must monitor the actions implemented to assess their effectiveness.

A routinely used approach to address non-compliance is the 'root cause' method. This is a structured methodology which does the following: (1) identifies the problem; (2) examines the causes; (3) generates potential solutions; (4) proposes the best solution; (5) implements the solution; and (6) reviews implementation. An effective EMS should not have to resort to the root cause method of analysing control procedures in all but the most difficult and complex situations. Rather, routine checks on all control procedures should identify issues for attention before they become problems with the potential to be dysfunctional to the system.

Preventive and corrective controls

Two types of controls are required within an EMS:

1 *Preventive controls* – those which seek to prevent environmental non-conformances.

2 *Corrective controls* – those which seek to rectify environmental non-conformance when it has occurred.

Prevention of environmental non-conformance will always be the desirable situation. Notwithstanding, the reality is that non-conformances will occur periodically and when they do the EMS must have controls in place to rectify non-conformance and invoke improvements to ensure their repetition. The system must therefore have both preventive and corrective control measures within its procedures and instructions. The standard does not expect a company to exhibit zero environmental effects and impacts, but it does require that environmental effects are prevented, or when they do occur that they are managed efficiently and effectively and, moreover, actions should serve as a catalyst for change and improvement to the EMS.

Environmentally critical activities

It is not necessary for a company to have controlled conditions for every single aspect of its business activities reflected in the system documentation. It is necessary, however, to have documented procedures and instructions in place where their absence would result in failure to fulfil the company's environmental policy, objectives and targets. The focus is therefore on ensuring that there are written and effective management procedures, working instructions and implementation plans for those activities which are environmentally critical.

The critical activities should be directly related to the critical success factors for the processes which deliver the products and services. These in turn will link to the objectives and targets which fulfil the company's policy. Critical activities should be assigned to particular personnel to ensure ownership, with each activity given key actions to be completed and criteria by which performance can be measured. Activities can be enumerated to reflect the degree of importance to a key process, for the level of environmental risk involved in their undertaking and resource priority. Where, for example, an activity is being monitored and controlled ineffectively, that activity may require greater resources to ensure that it is carried out appropriately. Where activities are continually completed successfully, they may require less routine management input, allowing resources to be redirected to activities where difficulties are perceived.

System monitoring (records management and auditing)

[Refer to ISO 14001, sub-section 4.5.4]

Purpose

An important attribute of the EMS is to establish and maintain a set of records which can provide evidence that the system has achieved the environmental objectives and targets which the company has set.

Record keeping and record management are requirements of the standard. Emphasis is placed upon establishing an audit trail linking objectives and targets to outputs, so determining the environmental performance of activities. The system requires procedures which maintain records in an orderly, accurate and consistent way and files these for subsequent use in auditing, management review and compliance with the standard, legislation and regulation. Records are not required from every organisational activity but rather from those areas of greatest environmental sensitivity and risk. Records will be read by a wide variety of company personnel and also third parties, and therefore the documents must be legible and understandable, sometimes by non-specialists. The rule of thumb with maintaining records is to keep them simple and systematic.

Records

Records should be maintained in the following key areas of environmental operation:

- planning consents;
- environmental impact assessments and statements;
- environmental management procedures and working instructions;
- control of Substances Hazardous to Health (COSHH) Regulations;
- licences and activities relating to waste management, emissions and discharges;
- consents for special processes and activities;
- reporting of Injuries, Diseases and Dangerous Occurrences Regulations (RIDDOR).

Records management

The standard requires that a documented procedure is established for the systematic management of records. This includes identification, cataloguing, filing, storage, retrieval, archiving and disposal. The records must be comprehensive and sufficiently detailed to allow informed use during management audit and review. The records should be related to organisational activities, processes, products or services such that the total production cycle can be tracked. Records should encompass all key aspects that impinge upon the company's environmental performance and link directly to objectives and targets. Records from different parts of the company will need to be collated and allow comparison where appropriate, and therefore the methods of collection and handling will need to be unified and consistent. In addition to providing depth of detail in specific organisational operations, records should provide breadth to facilitate the holistic review of the company.

The records management procedure will be specified within the EMS manual. It will likely require records to be maintained at a local level within designated boundaries of processes, products or services. This is useful in maintaining understanding and ownership of the record-keeping process. Information from different organisational areas will then be gathered and collated for wider analysis at corporate management level. The manual will provide details of the interrelationship of records and their co-ordination. It will also indicate the track of records for audit purposes.

The period of time for which records must be kept should be specified within the system's record-keeping procedures. While some records can be disposed of as they are updated and replaced, others must be kept for set periods of time. This would be the case with records connected with specific environmental legislation. Knowledge of the record type, use, storage and retrieval will allow management to establish an appropriate and effective procedure rather than a cumbersome and wasteful approach. Although many records are paper based as a result of the practical checking procedures used at the recording site, these can be transposed to electronic media for long-term and cost-effective storage. It is worth noting that some records must be retained as original documents, and this should be catered for by the system.

Auditing
[Refer to ISO 14001, sub-section 4.5.5]

Definition
An audit is:

> An independent examination of the EMS to provide assurance to management (and external parties) that the system is effective in implementation, fulfils policy, objectives and targets, and complies with the management system standard.

An audit can be one of two types:

1 *Internal* – a 'self-assessment' of the operation of the EMS undertaken by independent in-company staff or independent auditors employed by the company.

2 *External* – an independent 'third-party assessment' of the EMS undertaken by external auditors or certification body.

Although internal audits are a self-assessment by the company of its EMS, they should be undertaken by staff who are independent of the system's operation. This should ensure that the system is assessed thoroughly and objectively. Where the company does not have in-house staff capable of undertaking an audit, an independent auditor can be employed as a consultant. Where audits are external these are normally directed towards maintaining certification of the management system with the standard. Therefore, they are carried out by auditors from the certification body or an auditor appointed by the certification body to act on its behalf.

Audit considerations

There are some key considerations involved in managing audits for the EMS, as follows:

- *What is the standard against which the system is being assessed?* This can be ISO 14001, EMAS, or an intra-company standard of environmental performance.
- *Is there an internal capability to conduct an audit?* There must be knowledgeable, competent and experienced personnel to carry out the audit, or auditors will need to be brought in from outside the company. In any event, they should be independent of the system being audited.
- *Are audit capabilities matched to the technical characteristics of the processes and system?* Some audits may require particular knowledge and capabilities in complex technical situations.
- *Has an audit manager been appointed?* The environmental system manager should procure, arrange and manage any audit to ensure direction, co-ordination and effective procedure.
- *Is there an audit programme, or plan?* Audits should be periodic but frequent and with consistent method, and this can be assisted by a formal programme and itinerary in all situations.
- *Is there a method for evaluating the effectiveness of the auditors?* The company should ensure that audits are conducted efficiently and effectively and contribute to system improvement, and therefore a mechanism for assessing the auditors should be part of the adopted approach.

Audit process

There are a number of key steps in the audit process, as follows:

- *Determine the scope* – which activities are to be included in the audit and which are not.
- *State the standards* – which standards are to be used against which system compliance is to be assessed.
- *Choose the lead auditor and brief* – the internal or external auditor is appointed and briefed on the scope of the audit and standards applicable.
- *Review system documentation* – the system manual and associated documentation are scrutinised and the specific system requirements for the audit and audit team determined.
- *Specify the audit team* – the selected audit team is identified.
- *Plan the audit* – identify the dates, times and schedules for the audit activities.
- *Allocate audit duties* – specific duties of team members are discussed and assigned.
- *Confirm audit methods* – establish the audit procedures to be used and agree with auditees.

- *Preparatory meeting* – conduct a meeting of the audit team with the auditees to explain the scope, plan, method, procedures and schedule for the audit.
- *Gather evidence* – collect documented records of system operation and conduct detailed analysis of working system and processes for compliances and non-compliances.
- *Review evidence* – confirm areas of non-compliance.
- *Draft report* – document the findings from the audit with details of evidence.
- *Review* – discuss analysis with the auditees and agree areas of non-compliance.
- *Concluding meeting* – clarify the findings of the audit with the auditees and confirm recommendations.
- *Final report* – a final report is issued by the audit team to the auditees.

For *subsequent actions*, the auditees are responsible for determining and initiating any corrective actions for those non-compliant aspects identified in the audit. It is not the function of the auditors to be prescriptive with actions for remediation.

It should be remembered that an audit focuses on the performance of the EMS and not on the managers' performance. Although deficiencies in system procedures may be traced back to the management implementation of the system, it is appropriate to depersonalise this from individual personnel.

Frequency of audits

The frequency of auditing an EMS depends chiefly upon the type and complexity of business processes, the potential for environmental effects and possible impacts. Some aspects of operation will be very low risk, requiring little more than routine assessment as part of, say, a three-year audit plan. Other aspects may require auditing more frequently, possibly at annual or six-monthly intervals. It is commonplace for companies to adopt a three-year audit cycle routinely as this is commensurate with the requirements of assessment by system certification bodies.

Audit documentation

The written records, reports and reviews from the auditing procedure are an intrinsic part of the EMS. All documents associated with audits should be catalogued, filed and retained for future reference. Audit records will certainly be used by certification bodies for re-certification purposes, while information may be used by regulatory bodies and, indeed, the company itself in investigating environment-related incidents and reviewing company environmental performance.

Management review

[Refer to ISO 14001, section 4.6]

It is essential that the senior management of the company review the implementation and performance of the EMS at regular intervals. The frequency of review is not specified by the standard and depends upon the type, size, nature and activities of the company. Notwithstanding, a programme of regular, timely and frequent review is recommended.

There are two approaches to management review:

1 *Single annual review* – a comprehensive all-embracing event where the full environmental management system is reviewed.

2 *Multiple reviews* – a series of short events conducted throughout the year where all or parts of the EMS are reviewed.

For smaller companies the single annual review might be preferable. Such reviews take up a minimum of time and resources but may not optimise the potential benefits of review. They rely upon reviewing all aspects of the system throughout the organisation in one event. Where personnel are unable to take part, or many aspects are concentrated into a day's meeting, it is easy to miss important details and fail to capitalise on making improvements to the system. Multiple reviews conducted throughout the year can provide greater opportunities, particularly in large companies. It is likely that across a series of review events a higher proportion of staff will be able to attend, more issues will be picked up and overall more of the system will be reviewed in greater detail.

Any management review should reflect each section of the standard and all links between these sections. It is essential that environmental performance data is reviewed in context with the company's environmental objectives and specific targets and traced back to the original policy. If any objective or target is not being met the review must focus on the reasons for this together with potential remedies. If all objectives and targets are met and the policy is fulfilled then the review should examine the potential for improving further. This is likely to be accompanied by developing more stretching objectives and targets in subsequent years. It might also involve re-examining system procedures and working practices to encourage a little more effectiveness or efficiencies in day-to-day activities.

The following list contains key areas for environmental management review:

- environmental policy;
- environmental effects and impacts;
- environmental legislation and regulation;
- environmental goals, objectives and targets;
- environmental programme;
- organisational structure;
- information and communications;
- awareness and training;
- management procedures and working instructions;
- emergency procedures;
- monitoring methods, verification, testing and conformance/non-conformance;

- investigation of environmental incidents;
- records and reporting;
- audit results.

All reviews should examine in detail the reports generated through the audit process. Information should be available from each element of the EMS and should be reviewed objectively from the evidence gathered rather than from observation and conjecture. The focus of the review is to make improvements to the system and the activities of the organisation. Therefore, the review must address all aspects of non-conformance identified in the audit, what actions were taken and whether they were effective. It must also look at best practice with a view to sharing this throughout the organisation.

When carried out effectively, an environmental management review will ensure that:

- the EMS and its documentation remain appropriate and adequate for the company's environmental circumstances;
- appropriate actions are sanctioned to deal with any shortfalls in the system, documentation and practices;
- improvements are made to the system and documentation which improve operational processes, working practices and the company's environmental performance;
- the system continues to meet current standards and regulatory requirements;
- environmental awareness, information, communication and training are meeting the needs of managers and workforce;
- the system continues to provide the best approach to meeting the current and future challenge in environmental matters.

System registration (certification)

Assessment and registration, or 'certification', of an EMS by an independent and impartial organisation – a certification body – is essential in demonstrating that the system achieves the following:

- conformity with ISO 14001 and/or any other environmental management standard;
- compliance with environmental legislation and regulations;
- demonstration of the company's commitment to environmental safeguards in respect of the environmental effects and impacts of its business activities.

The assessment and subsequent award of a certificate to a company with an appropriate EMS demonstrates that it is aware of: environmental matters and environmental management concepts; the context of these within the scope and operation of its business activities; environmental risk and the need to take appropriate steps to manage these risks; and the requirement for active management of environmental effects and impacts of its business. Therefore, in addition to demonstrating adherence to system standards, legislation and regulation, certification confirms a company's wider perspective of environmental management needs. Details on EMS certification follow subsequently.

Continual improvement

[Refer to ISO 14001:2004, sub-section 4.6]

A key attribute of operating and maintaining an EMS is that it should remain current and applicable to changing environmental needs and the requirements of the company. It is therefore essential that the system is subject to a policy and practice of continual improvement. Continual improvement results from the fulfilment of challenging environmental goals, objectives and targets together with evolvement of the EMS and practices.

Identifying opportunities for improvement

The company should continually examine and evaluate its environmental performance through effective management review. An important aspect of review is the identification of opportunities for improvement. This takes place as elements of the business processes are scrutinised in detail. Actual or potential non-conformities with system standards for any processes will be highlighted for specific attention. Once deficiencies are highlighted, management can focus attention on the reasons for their manifestation and seek potential solutions. System improvements will be made to ensure that there is no reoccurrence of those deficiencies. The procedure for this is 'root cause analysis', which was explained earlier in the chapter.

Potential improvements

Improvements to an EMS can be many and varied depending upon a wide range of circumstances and situations. Examples of useful improvements which the organisation may seek are:

- process improvements to meet changing environmental legislation and regulation;
- use of alternative and/or new material resources to reduce environmental effects and impacts;
- changes to practices to encourage efficiency gains;
- modification to supervisory regimes to ensure fewer non-conformities;
- provision of further training in more efficient and effective working;
- streamlining the input of resources to processes.

Improvements programme

When seeking improvements to the EMS, it should be borne in mind that improvements do not have to be made to all aspects of the system concurrently.

Rather, the focus should be upon evolving particular aspects of the system to ensure the reliability of effectiveness before examining other parts of the system. This is important in ensuring that improvements made are well considered, implemented and embedded. The difficulty of attempting to improve the system in one go is that considerations are likely to be superficial and actions will lack robustness. A programme of improvements should be developed which identifies system aspects for attention, reviews difficulties and their reasons, prioritises their criticality to the system and considers actions to be taken. In this way, improvements are planned and implementation controlled.

Assessing improvements

It is important for the organisation to determine that the improvements it considers and implements are appropriate. It is all too easy for change to be made for the sake of change rather than adding value to the EMS. Such tweaking needs to be avoided while considered and meaningful modifications are made. Assessment should be made of any improvements to the system. The improvements programme should incorporate a period of time to allow modified practice to embed and then formally assess the effectiveness of that practice. It might be the case that further modification is required to make the improvements fully operational and effective.

Assessment of improvements can involve the following:

- simulating or trialling modified procedures within business processes;
- benchmarking environmental practice with sector practices;
- comparison of environmental performance with previous audit data;
- experience of improvements against preceding practice;
- monitoring the progress of modified performance against objectives and targets;
- gathering views and opinions of stakeholders.

EMS assessment and certification

This section examines: the types of independent, or third-party, assessment for EMSs; and the framework and process for certification of an EMS. System assessment and certification was presented in greater detail in Chapter 2 in relation to quality management systems, and as the approach is broadly similar for EMSs this section may be read in conjunction with the relevant section in Chapter 2.

Types of assessment

There are three types of assessment for an EMS, and these are identical to those explained in Chapter 2 when examining quality management systems:

- *First-party or self-assessment* – where an organisation has an established internal system and notifies those parties with which it seeks to work.

- *Second-party or collaborative assessment* – where an organisation has an established internal system which is open to the scrutiny of interested parties when seeking to work with them. In some business sectors companies may develop a system in collaboration with a key client or customer.

- *Third-party or independent assessment* – where an organisation has an established system awarded a certificate from a certification body.

A company must decide for itself which type of assessment it seeks for its EMS. That decision will be based on a wide range of factors surrounding the company's commercial marketplace, sector position, size, organisational characteristics and business activities. For a large company it may be quite obvious that a certificated EMS is required in meeting its environmental demands. However, for a small business sustained by a singular client base there may be no need to engage in an environmental system to support the delivery or its products or service. Therefore, the need for a formalised system is driven as much by the perceptions of clients, customers and other interested parties as it is by the perceptions of the company itself. For companies other than those which are very small, the trend towards pre-qualification of a company's capabilities across a range of management functions before engagement by clients and customers means that a formal EMS of some kind is virtually a prerequisite. The key point on system assessment is that recognition should be commensurate with company needs.

There can be no denying that the establishment of an EMS is advantageous to many companies and virtually a prerequisite for others. The view of Arque Construction, a UK-based small contractor working on private- and public-sector projects, sums up the importance of having an EMS in securing work from local and national public authorities: 'To get the council's highest accreditation level, you need an environmental management system' (Envirowise, 2009).

Certification

Framework

The same formal and strict framework for certification of management systems in the UK applies to an EMS. Certification bodies which assess the management systems of companies have their activities and conduct audited by accreditation bodies. In turn, the activities and conduct of accreditation bodies are monitored and controlled by government. In this way there is a hierarchal framework of accountability. This framework exists to ensure that every certification body acts professionally within codes of conduct, acts impartially in all its activities and exerts an appropriate level of competence to enable an effective assessment of management systems to be carried out. Internationally, the specifics of accreditation and certification frameworks have differences in other countries and continents to reflect local operating requirements and practice. Nevertheless, developing global harmonisation tries to ensure that the implementation of ISO 14001 is uniform throughout.

Within the UK, the United Kingdom Accreditation Service (UKAS) provides EMS accreditation for ISO 14001 and EMAS-compliant systems while EMAS itself provides EMS accreditation in European countries. Across European boundaries the European Co-operation for Accreditation (EA) ensures that an EMS certificate issued by a body in one country is recognised as equivalent to a certificate of another country: 'The European Co-operation for Accreditation (EA) Multilateral Recognition Arrangements (MLA) makes accreditation a "passport" which facilitates access to the EU and international markets through co-operation with the International Accreditation Forum (IAF)', IAF being the world association of Conformity Assessment Accreditation Bodies (IEMA, 2010).

The certification process

A company which wishes to obtain a certificate recognising that its EMS complies with ISO 14001 must apply for assessment of the EMS by an accredited certification body. The EMS will be assessed within a prescribed process of certification. The certification process for assessing an EMS is the same multi-stage process outlined earlier for a QMS. The six stages of assessment therefore remain as follows:

1 *Pre-audit* – initial assessment of the EMS.

2 *Desk-top study* – assessment of the EMS documentation.

3 *Certification audit* – assessment of the EMS in operation.

4 *Certification outcome* – results of the EMS assessment leading to the award of the certificate or alternative outcomes.

5 *Surveillance* – six-monthly checks on the operation of the EMS by sampling parts of the system.

6 *Certificate renewal* – three-yearly detailed scrutiny of the EMS by extended surveillance or full certification audit.

Stage 1: Pre-audit

The pre-audit stage ensures that:

- the organisation's processes and the EMS are sufficiently developed to engage in assessment;
- the documentation of the system is at the right level of development for assessment;
- the EMS has fitness for purpose based on the environmental management approach of the company;
- the internal audit procedures are appropriate to the operation of the system;
- the company is ready to engage in the subsequent stages of assessment;
- arrangements and resources for assessment have been considered and confirmed.

Stage 2: Desk-top study of documentation

The desk-top study stage ensures that:

- a detailed review of the company's EMS documents against ISO 14001 is carried out;
- the EMS documentation meets the requirements of the standard;
- any interpretations of the standard in relation to particular organisational implementation are grounded in the generic standard;
- the EMS documentation is arranged so that procedures are traceable against ISO 14001.

Documentation is normally divided into sections and sub-sections as follows:

- *Documents*:
  - Environmental policy, including statements of goals, objectives, targets, and likely environmental impacts of the business.
  - EMS manual.
  - Environmental management procedures.
  - Environmental implementation plans, where required.
  - Environmental working instructions.
  - Process, operation and activity guides for output products and services.
- *Legislative framework documents*:
  - National and local legislation.
  - Industry and business regulations and enforcement procedures.
  - Permits for business processes, such as generation of by-products.
  - Registration certificates for particular activities, such as waste management.
- *Background documents*:
  - Plans detailing activity sites, boundaries and environs.
  - Maps locating neighbours, communities, buildings and business facilities.
  - Survey information of natural, geological and topographic features.
  - Layouts of artificial features.
  - Protection orders on natural and built environment, flora, fauna and wildlife.
  - Forecasts of normal weather conditions in vicinity.

Stage 3: Certification audit

The certification audit ensures that:

- the EMS, in operation, complies with ISO 14001;
- the EMS is achieving legislative and regulatory compliance and making continual improvements to the environmental performance of the company.

The certification audit focuses on the following broad areas of assessment:

- management responsibility and activity in fulfilling the company's environmental policy;
- ongoing identification and evaluation of environmental aspects, effects and impacts;
- determination of environmental objectives and targets in relation to aspects, effects and impacts;
- monitoring of environmental performance against objectives and targets including measurement, reporting, record keeping and review;
- operational control measures implemented to safeguard against contravention of environmental legislation and regulation;
- internal audit plans, procedures and schedules to maintain the effectiveness of the EMS and perpetuate continual improvements to the system and processes.

The process of the certification audit normally involves the following:

- *The certification audit opening meeting* – where the lead auditor will: outline the assessment process; recap on the pre-audit and desk-top study stages; explain how the certification audit is to be conducted; agree an itinerary; assign audit activities; and formalise communication mechanisms between the audit team and the client.
- *The certification audit* – where the audit team undertakes the detailed assessment of the EMS in operation against the requirements of the standard.
- *The certification audit closing meeting* – where the lead auditor will report verbally on the findings of the audit team and give judgements on the efficacy of the EMS.

Stage 4: Certification outcome
The recommendation made by the lead auditor to the certification body results in one of the following outcomes:

1 Award certification
2 Award certification with minor corrective action(s)
3 Delay certification until improvements are made to the EMS in response to major corrective action(s)

The auditors are not empowered to award a certificate but they are empowered to make a recommendation based on the judgements from the certification audit. The certificate is awarded by the accredited certification body.

Stage 5: Surveillance
Surveillance ensures that:

- the certification body maintains over the period of the certificate a series of visits, which can be termed interim audits, to confirm that the EMS continues to operate effectively.

Surveillance visits are normally carried out at six-month intervals, with at least one visit taking place within any twelve-month period. The breadth and depth of examination carried out depend very much on the perceptions of the surveillance team but the focus is always on the continuing need to ensure that the EMS complies with the standard, legislation and environmental policy of the company. Surveillance will likely involve sampling selected parts of the EMS, and a number of key aspects will be examined:

- the effectiveness of the EMS;
- the efficacy of management and operations;
- activities directed to enhancing environmental performance through continual improvement to the EMS;
- response to corrective actions and observations made during the certification audit or previous surveillance visits;
- continuing procedures used for internal audit activities.

Stage 6: Certificate renewal
Certificate renewal ensures that:

- a thorough and detailed reassessment of the EMS is undertaken every three years.

The outcome from a certificate renewal assessment can be one of the following:

- renewal of the certificate and continued registration of the EMS to the standard;
- qualified renewal with a prescribed timescale for the implementation of specified corrective actions;
- recall of the certificate and discontinuation of registration.

For comprehensive description, explanation and discussion of the process of assessment and certification the reader is directed to Chapter 2. The process for the EMS is identical in concept, principals and application to the process implemented for the QMS.

The environmental supply chain

This section examines: the purpose of and approach to environmental supply chain management; preferred suppliers and partnering arrangements; and the requirements for effective site waste management plans for recycling, reuse and waste management applicable to organisations in the construction supply chain.

Supply chains

The purpose and function of an EMS is to ensure that a company can meet the level of environmental performance it has set. It can only achieve this if all the resource inputs that contribute to delivering its business outputs are also so committed. Therefore, the environmental performance of a company's suppliers has a direct effect on the environmental performance of the company. For this reason a vital aspect of environmental quality management is the procurement and management of materials, components, goods and services. This function is termed 'environmental supply chain management', or ESCM.

A supply chain is:

The flow of activities involved in providing materials, components, goods or services to customers.

An environmental supply chain is:

The flow of activities involved in providing appropriate materials, components, goods or services to customers commensurate with the customer's environmental requirements.

Environmental supply chain management is:

The control of or influence over the flow of activities involved in the supply chain.

Environmental supply chain management (ESCM)

Objectives

The objectives of ESCM are to:

- ensure the flow of environmentally appropriate inputs to business processes;
- simplify and make easier the procurement procedures for the inputs;
- minimise the costs of procurement, handling and usage;
- perpetuate benefits on both the input and output sides of business processes;
- add value to the business in holistic terms.

ESCM is sometimes thought of as a process which impacts on the input side of a business process alone. For example, raw materials go into manufacturing products which following dispatch present no further environmental implications. This is, of course, quite erroneous as control and influence can be exerted over, for example, transportation, handling and packaging. Such aspects can bring into being environmental effects and impacts. Therefore, successful ESCM is also concerned with the output side of business activities.

ESCM can play a significant part in the considerations made when a company develops its EMS. The way a supplier contributes to the business of a company can be an important influence in the development of its environmental goals, objectives and targets. It can also be significant in the configuration of organisation and procedures used to control process operations. The environmental policy may well be developed with the incorporation of commitments to supply chain management, in which case the approach will permeate throughout a company's activities.

Potential benefits

ESCM can assist a company in a number of ways:

- Ability to meet the environmental requirements of customers and other stakeholders.
- Maintain a competitive edge in the commercial marketplace.
- Meeting environmental standards and legislation.
- Ensuring the continuity of input resources.
- Protection against open-market disruption to supply.
- Establishment of an EMS that embraces both the supply (input) and demand (output) sides.
- Procurement to exact specifications and requirements.
- Developing closer and mutually beneficial business relationships.

ESCM evaluation

A company must evaluate two key dimensions of a supply chain:

1 *Product evaluation* – the detailed examination of the product to be procured, how it will be selected, the flow process to input, characteristics in use and output implications. This will be achieved by undertaking a risk assessment of the whole supply chain.

2 *Supplier evaluation* – the detailed examination of the characteristics of the supply company. This will be achieved by carrying out an assessment of the supplier to determine its risk potential.

A product evaluation determines whether any materials, components, goods or services would pose a risk to the total cycle of business activity of the company if they were to be used. This involves identifying potential environmental effects, assessing the levels of risk, determining ways to minimise such risk and suggesting alternative suppliers where appropriate and feasible to do so.

A supplier evaluation gathers information from the proposed supplier on all environmental aspects of its business. This can be used to determine the perceived level of risk in working with that supplier. Alternatively, the information can be used to make comparisons with other potential suppliers.

For both of these evaluations, the company is seeking to maximise its confidence in the ability of the supplier to meet its own business obligations and the responsibilities of the company in relation to environmental safeguards. A supplier with low environment-related risk would be one which has: a certificated EMS; a clear environmental policy statement towards customers; a good reputation for environmental performance within its marketplace; and a satisfied customer/client base. A supplier exhibiting a high risk would be one which has: no EMS; no track record, a poor reputation and bad press concerning its environmental performance; and a monopolistic market position. It could also be a new supplier or one with a short period of experience in the marketplace, or simply a new and unknown supplier to the company. There are many factors which must be taken into account in evaluating the risk of the product and its supplier.

Preferred suppliers and partnering arrangements

A customer does not always have to disregard a supplier because of a perceived environmental risk. It may choose to work with the supplier and actively seek to reduce or manage that risk. To retain the element of commercial competition while encouraging closer co-operation, a popular approach is to move away from open-market procurement to selective procurement, or what are usually termed 'preferred suppliers'. In this approach a customer maintains a shortlist of environmental pre-qualified suppliers which are invited to tender for contracts. Working more closely with a small number allows relationships to be developed advantageously over time with benefits to both customer and supplier. The *partnering* process, where the customer works in partnership with a single supplier over the long term through sequential contracts, is also finding greater favour within almost all sectors of business. Such an arrangement can lead to more highly effective relationships, allow customers to specify their environmental needs in much greater detail and help the supplier to meet customer requirements better.

Environmental position statement

A company may choose to publish an ESCM position statement. This is a statement from the executive management of the company declaring its position on supply chain matters and the contribution expected of its suppliers. This can be presented as a stand-alone statement or form a composite part of the company's environmental mission or policy. It is an important statement in that it commits both the customer and supplier to the concept and principles of supply chain relationships. Moreover, for both organisations it provides the foundation upon which environmental supply chain aspects can be incorporated in goals, objectives and targets, and become embedded through communication and training into management procedures and working practices. In this way, supply chain management can be absorbed into environmental management and evolve to become an intrinsic part of organisational culture.

Environmental procedures, the supply chain and waste management requirements

The Site Waste Management Plans Regulations

The Site Waste Management Plans (SWMP) Regulations came into force in 2008. The plans are a specific application of the duty of care provisions of the Environmental Protection Act 1990. The regulations are intended to encourage clients and principal contractors to estimate the nature and amount of waste that is likely to arise during a construction project and consider provisions that can be made to recycle, reuse or responsibly remove waste materials from site. In England, SWMP are required for all new construction projects exceeding a value of £300,000. In Northern Ireland, Scotland, and Wales, SWMP are not compulsory, although a voluntary initiative is recommended.

SWMP

A company that delivers its products or services at a project site is required to have in place environmental procedures for handling any waste resulting from its processes and operations. This aspect is particularly important within the construction industry, where construction work on the project site can give rise to a wide range and large quantity of waste. The management of waste can be achieved by developing and implementing procedures which are stand-alone in the form of SWMP, or specific procedures established as an embedded component within the EMS, in particular within the implementation plan which is situation, or site, specific.

SWMP are dynamic documents and should be configured to allow updates throughout a project. Such plans are developed at the pre-construction stage so that designers are able to consider how construction waste might be reduced and how materials can be recycled or reused and waste managed. Identifying potential waste materials at the design stage allows alternative materials to be considered, which might facilitate greater recycling and reuse and also minimise waste during the on-site construction stage. The regulations apply to sites where construction work is undertaken and therefore include: building; civil engineering; demolitions; and site clearance works.

SWMP content

A comprehensive approach to site waste management planning for major, or large, projects will consider specific aspects according to the value of the construction project. For projects with a value between £300,000 and £500,000 the content of the SWMP should include:

- the types of waste removed from the site;
- the identity of the persons (or company) removing the waste from the site;
- the name and location of the disposal site.

For projects exceeding £500,000 the SWMP should also include:

- a description of the waste;
- details of the destination site where the waste was transported to, that is, landfill site or incineration facility;
- the environmental permit (or exemption) held by the site where the waste was transported to.

Exclusions from SWMP

SWMP are not required on construction projects:

- where a construction site has a Part A Environmental Permit *(Part A permits control activities within a range of specified environmental impacts, including:*

emissions to air, land and water; energy efficiency; waste reduction; raw materials consumption; noise, vibration and heat; and accident prevention);

- where a nuclear-licensed site has an *integrated waste strategy (IWS)* in place that includes waste from on-site construction activities.

Responsibilities

A company acting in the capacity of a client organisation has the responsibilities:

- to produce the SWMP before any work begins;
- to appoint the principal contractor;
- to pass the SWMP to the contractor;
- to update the SWMP at least every three months (if self-managing a project).

A principal contractor has the responsibilities:

- to obtain information on waste and its management from its sub-contractors;
- to update the SWMP at least every three months;
- to keep the SWMP on site throughout the works;
- to ensure that other contractors know where the SWMP are kept and allow access to them;
- handing the SWMP to the client upon contract completion;
- to keep a copy of the plan for two years.

A contractor (traditionally denoted as sub-contractor) is responsible for specific aspects of the SWMP, and these will be reflected in the contract with the principal contractor as follows:

- upholding the duty of care under current legislation;
- purchasing strategies for materials;
- methods of work directed to reducing waste;
- on-site recycling and reuse of site-gained materials;
- disposal of waste;
- the provision and timing of information that should be reported to the principal contractor and/or client.

Review of waste management plans and procedures

The stand-alone SWMP should be reviewed at the end of a project to compare actual activities against those which were planned. Within an effective EMS the review of waste management forms an intrinsic element of system management. In both cases, detailed review will allow the lessons learned from the management of waste to be fed back into and inform revisions and improvements to the procedures used.

Infringements of SWMP Regulations

There may be a number of infringements of the SWMP Regulations leading to criminal offences. These are:

- failure to procure/prepare the SWMP (where the client and principal contractor would be found liable);
- failure to update the SWMP;
- failure to make the SWMP available on site;
- failure to file/store the SWMP for two years post-completion of contract;
- making false statements within the SWMP either knowingly or recklessly;
- hindering the implementation of SWMP legislation/regulation through failing to assist, failing to produce the SWMP or making misleading statements to those producing the SWMP.

The implications of infringing SWMP Regulations are serious as:

- infringements are offences in law;
- they can be prosecuted in a Magistrates' Court or in the Crown Court, depending on the seriousness of the offence;
- fines can be imposed up to £50,000 by a magistrate or unlimited in the Crown Court;
- 'strict liability' applies, which means that authorities do not have to prove intent to establish guilt but merely demonstrate contravention of the regulations.

Practical application of SWMP

Within the construction processes on site the most likely influence of the SWMP Regulations will be procedures established to identify, sort, contain and label waste materials for recycling or reuse and those directed to disposal off site. Specific consideration will need to be given to those materials which are hazardous, as distinct from inert, as they will require particular handling, storage and removal. Specific regulations may apply and involve specialised methods of dismantling, handling, storage and removal, a common example being asbestos-based materials. The differentiation between active and inert waste products is important as they will be subject to different handling and disposal costs. Examples of active wastes include timber, which may be reused, while inert materials like plastics may be recycled and hazardous materials like paints will require careful disposal. Effective SWMP implemented individually or as a composite part of an EMS will utilise on-site management procedures and working instructions to ensure that all site waste is carefully directed to storage containers, differentiated by type of waste and subsequent processing method.

Waste management supply chain issues

Any client company can play an important part in sensible and effective waste management by embedding the responsibilities for good practice within its supply chain procurement approach. All contractors and sub-contractors should be expected to adopt management procedures and working practices which aim to reduce waste, reuse and recycle waste materials, and dispose of waste responsibly. A company which implements effective SWMP or an EMS will build such measures into its approach to delivering its core business.

EMS: application to construction

This section examines the development and implementation of an effective framework for EMS application by a principal contracting organisation. It is based on case study material.

EMS case study example

Example system

The following example illustrates the considerations given and framework applied to developing and implementing an EMS. The approach was utilised by a medium–large-sized principal contracting company which provides services within the building and civil engineering sub-sectors of the construction industry.

System standards

This principal contractor had an existing QMS complying with ISO 9000 and adopted the generic approach to use that system as the basis for establishing an EMS.

General approach

The company adopted a 'corporate'-based EMS formed around its existing ISO 9000 compliant QMS. For 'corporate'-based and 'project'-based system terminology and distinctions, see earlier sections.

Structure

The example approach is presented as a series of diagrams, or figures, described in the text which follows. This is important because a company will develop its system through a series of diagrammatic reflections rather than just narrative-based documents. In addition, systems tend to be configured using template documents for implementation during system operation. It shows the thinking processes behind the considerations given to developing the EMS from the outline framework through to the establishment of procedures. The example illustrates both the corporate structure of the company's business and the system implementation plan embracing specific remote, or site, activities. An outline of the EMS manual is shown together with a detailed example of how supervisory and operational system procedures are configured. It must be remembered that it is not sensible to present a totally prescriptive approach. A company should develop an EMS to suit its own organisation, its business operations and its circumstances.

Elements of the system

These are as follows:

- Framework for EMS.

- Checklist for prioritising the company's environmental activities and target aspirations.

- Checklist of external and internal factors to be considered in positioning the company for environmental management.

- Consideration of main environmental effects of business activities and their implications.

- Checklist of key steps in developing the company's corporate and implementation approach to the EMS.

- Checklist for considering the phases involved in a preliminary environmental review of business activities.

- Outline statement of environmental policy.

- Corporate-level EMS development framework linked to activities and inputs.

- Key sections of the EMS manual.

- Site-specific application of EMS development.

- Key elements and sections in EMS implementation plan for site-specific activities.

- Supervisory and operational procedures involved in the company's waste management process (including site waste management plans, or SWMP).

- Schedule of company audits for the EMS.

- Review template for elements of the EMS.

- Archiving template for EMS documentation.

Development and implementation of system elements

Framework for EMS

Figure 3.11 shows three main sections, or groups of activities, that comprise the system framework, these being: (1) environmental assessment; (2) the system (with its standard-based key elements); and (3) application-specific implementation plan. The need for environmental management by the company is influenced by internal factors such as the environmental effect that its business creates and also external influences such as environmental legislation. The key elements of the system are commensurate with the requirements specified by ISO 14001. It should be noted that monitoring of the system and those business processes which the system oversees is facilitated through an audit loop to both the system and implementation plan, while a feedback loop provides information to the company's review mechanisms and environmental files.

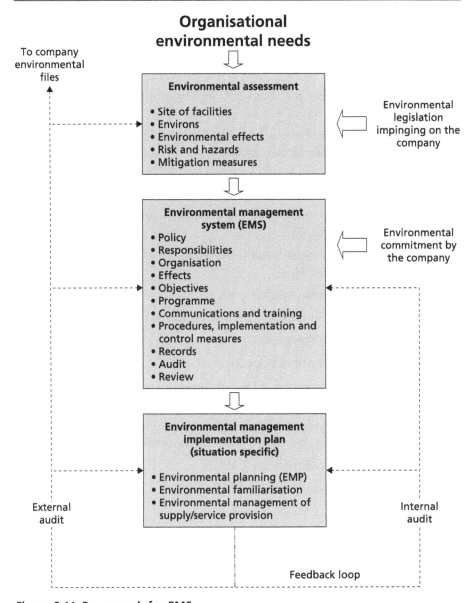

Figure 3.11 Framework for EMS

Checklist for prioritising the company's environmental activities and target aspirations

Figure 3.12 shows the ten key organisational activities involved in developing the EMS, and links these to the aspirations of the company. Consistent with the PDCA methodology recommended within the standard, the company has classified each activity with different target aspirations, namely: (1) a fundamental goal; (2) an improvement; (3) a greater improvement; and (4) ultimate goal.

		Activity	Target aspiration
	1	Develop environmental policy for the organisation	Fundamental goal
	2	Develop environmental policy statement for internal and external circulation	Fundamental goal
	3	Identify organisational areas for attention and focus for management systems	Fundamental goal
	4	Publish EMS manual, guidance notes and procedures for areas identified in 3, translate procedures into management assignments and action plans	Improvement
	5	Determine knowledge and skill base needs and implement training programme and review in relation to system procedures used	Improvement
	6	Evaluate system processes to provide better procedures, expand meaningful experience and feedback and review	Greater improvement
	7	Evaluate training programmes (in 5) and action plans (in 4) to determine improved performance	Greater improvement
	8	Produce report on systems operation and documentation	Greater improvement
	9	Develop and implement audit mechanisms	Greater improvement
	10	Review and restate policy, guidance notes and documentation for whole system to minimise detrimental environmental impact of organisational activities	Ultimate goal

Figure 3.12 Checklist for prioritising the company's environmental activities and target aspirations

Checklist of external and internal factors to be considered in positioning the company for EMS

Figure 3.13 lists the external factors and internal factors that the company needs to consider in positioning itself for establishing the EMS.

Consideration of main environmental effects of business activities and their implications

Figure 3.14 shows the principal potential environmental effects identified as a result of considering the physical requirements and business activities of the company. For each potential environmental effect the implications need to be considered. While some of the environmental effects may be quite obvious, for example air emissions and pollution as a result of business operations, others are less so, an example being existing site dereliction, where some companies leave

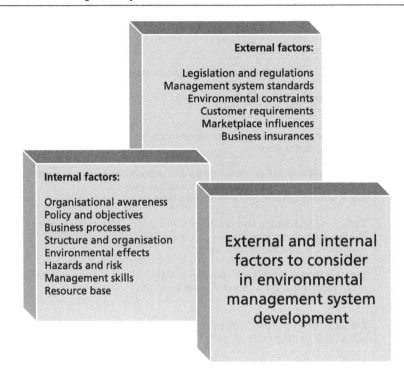

External factors:

Legislation and regulations
Management system standards
Environmental constraints
Customer requirements
Marketplace influences
Business insurances

Internal factors:

Organisational awareness
Policy and objectives
Business processes
Structure and organisation
Environmental effects
Hazards and risk
Management skills
Resource base

External and internal factors to consider in environmental management system development

Figure 3.13 Internal and external factors to consider in positioning the organisation for EMS development

redundant sites in a ruinous condition while relocating to new and improved business facilities.

Checklist of key steps in developing the company's corporate and implementation approach to environmental quality management

Figure 3.15 shows a checklist of the steps involved in determining the approach to the establishment of the general approach to system development.

Checklist for considering the phases involved in a preliminary environmental review of business activities

Figure 3.16 shows the four phases involved in considering a preliminary review of the company's business activities. These are important considerations and phases of development for system establishment. Information gathered and decisions taken at this stage do form the basis of subsequent system structure and task schedules.

Statement of environmental policy

Figure 3.17 shows the policy statement of the company as issued by the executive board of directors, pledging organisational commitment to all aspects of its business and its customers.

Environmental effect	Implications
Land use	Almost all business activities invariably occupy land, consume space above land and use areas below land. This is, without doubt, the most prominent environmental effect of business and commercial activities
Existing site dereliction	The preference to occupy greenfield sites leaves many localities with derelict sites and redundant buildings and structures in ruinous condition
Natural habitat destruction	Many activities leave their mark on the landscape, natural amenities and wildlife, much of which, once destroyed, can never be replaced
Use of natural resources	The use and destruction of natural resources is a severely detrimental by-product of business. For example, deforestation and quarrying leave their mark on the environment long after the raw materials have been extracted and used in the production of materials
Air emissions and pollution	Business activities give rise to many atmospheric pollutants in the form of toxic fumes from plant and equipment use and the production processes employed. In addition, in a completed building for example, there can be many sources of air emission such as chlorofluorocarbons (CFCs)
Use of water resources	As a by-product of business, increasing demands are placed upon water supplied. This may be overuse of existing supply points or the requirement to build further supply facilities, which themselves commence a cycle of environmental effects
Discharges and water pollution	Production processes frequently give rise to the spillage and discharge of contaminants through natural water courses and human-made drainage systems
Waste	Waste as a result of acquiring raw materials, refining raw materials for use in the construction process, delivering materials to site, and poor storage, handling and use of materials all give rise to the large amount of waste created
Comfort disturbance	Noise, dust, dirt, pollution and traffic are some of the most prominent comfort disturbances experienced by local inhabitants
Health and safety	By their very nature, some activities are accompanied by an inherent level of danger to both the employees working at a business facility and members of the local public community
Energy consumption	Businesses give rise to a considerable level of energy consumption both during the processes involved and in the use of the finished product

Figure 3.14 Consideration of main environmental effects of business activities and their implications

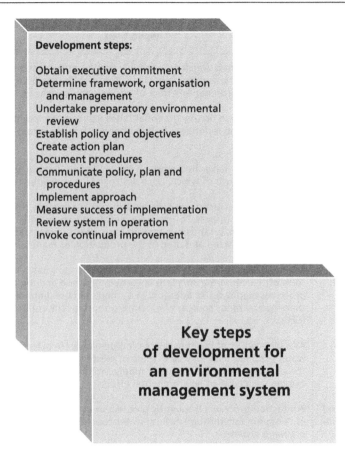

Figure 3.15 Checklist of key steps in the organisation's corporate approach to EMS development

Corporate-level EMS development framework: activities and inputs

Figure 3.18 shows the development framework for the system. It is a vital part of system development and implementation as it establishes the principal elements of the system, consistent with ISO 14001, and also relates these to external factors of influence and key activities involved in system establishment.

Key sections of the EMS manual

Figure 3.19 shows the outline structure of the company's EMS manual. It is structured into seven main sections with three detailed sections: environmental policy; company organisation and management structure; and review and audit procedures. While a company's manual can be configured in any way to suit the needs of the organisation and its business, the outline of this manual broadly follows the requirements of ISO 14001. It should be noted that in this case the company is presenting its management procedures, implementation plans and working instructions as separate but co-ordinated documents in a multiple document format to accompany the manual. This means that detailed descriptions of those

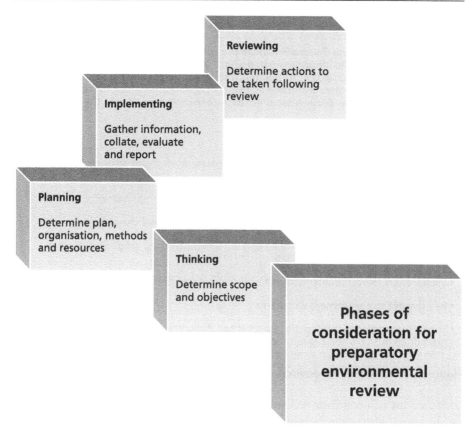

Figure 3.16 Phases of consideration for a preparatory environmental review of business activities

elements listed are not contained within the manual itself but in supplementary manuals.

Site-specific application of EMS development

Figure 3.20 shows the translation of the corporate system into site-specific application. As discussed, a company may undertake to deliver products or services at locations other than its home establishment. When that occurs the company needs a structure in place to ensure that the attributes of the corporate system apply to remote delivery. The structure mirrors that at corporate level, allowing for the specifics of the site situation, the site-based management team and the life-cycle of the provision. In essence, this figure is a sub-diagram of the corporate structure shown in Figure 3.18.

Key elements and sections in EMS implementation plan for site-specific activities

Figure 3.21 shows the key elements and sections of the company's implementation plan. The plan is a set of documents which include: general information,

---

Statement by Company Board of Directors

ABC Principal Contracting (International) Ltd provides a wide range of services to the manufacturing and allied industries. In so doing, it acknowledges the need to safeguard the environment in all aspects of its business. The overall aim of our business is to provide complete customer satisfaction and where possible ensure that the services we provide present minimal detrimental effect to the environment and both the customer and ourselves.

In all our activities, we commit to:

• Comply with all legal requirements and standards associated with environmental management.

• Review, through our EMS, all our activities and actively seek improvements.

• Develop and market services that have excellent environmental characteristics.

• Promote environmental excellence among our sub-contractors, suppliers and other employed organisations.

• Implement resources to safeguard the environment and its population.

• Be proactive to environmental concerns.

• Ensure that corporate policies are implemented through sound environmental procedures in all our sub-organisations.

Figure 3.17 Statement of environmental policy

such as a list of document revisions; schedules of information, such as supplier information, waste management and testing procedures; emergency provisions; and system audit, review and archive. Again, the structure follows the corporate system and meets the requirements of the management system standard.

Supervisory and operational procedures involved in the company's waste management process

Figure 3.22 shows an example of the development of a system procedure, in this case the procedure for managing the waste process resulting from organisational business activity. It is important to note that the system manual will contain such a procedure for each and every organisational activity that delivers the business process, whether it is a product or service. Also, there will be such a procedure to manage each and every potential environmental effect as identified in Figure 3.14. The manual therefore contains many such management and operational procedures, each reflected in a list such as the one shown in Figure 3.19, with each diagram supplemented by guidance and instructions provided in the written form of checklists or aides-memoires.

Schedule of company audits for the EMS

Figure 3.23 presents a pro forma developed by the company to reflect its schedule of company audit of the management system. It is worth noting that an audit of the full system is scheduled in each six-month period of operation

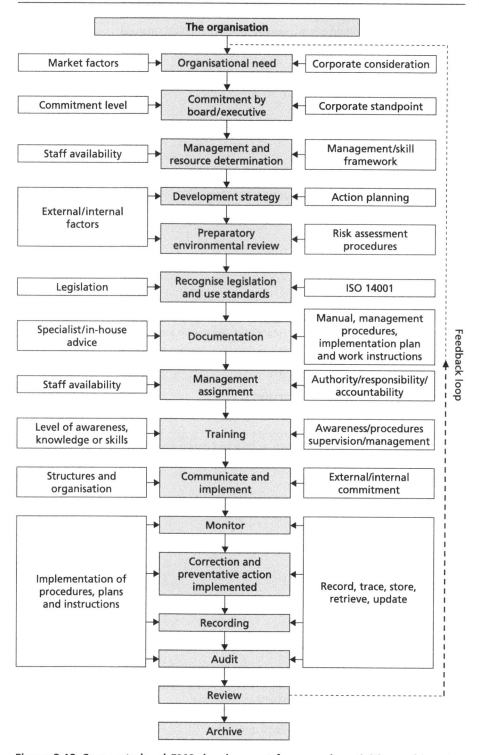

Figure 3.18 Corporate-level EMS development framework: activities and inputs

Environmental management system (EMS) manual
Outline sections: • Cover • Title page • Contents • Revision, or version, list • Company environmental statement • Summary of manual • Scope of use • Instructions for use • Environmental policy • Company organisation and management structure • Review and audit procedures • References to supporting documentation • Index

Detailed sections:
 	
• Environmental policy:
  – Introduction
  – Scope
  – Corporate environmental responsibility
  – Responsibility to environmental standards, legislation and regulations
  – Management of personnel
  – Responsibilities of employees
  – Environmental performance criteria
 	
• Company organisation and management structure:
  – The company
  – Organisation chart
  – Environmental management structure
  – Divisional, departmental and section environmental responsibilities
  – Environmental management programme
  – Communications
  – Training
  – Management procedures*
  – Implementation plan(s)*
  – Working instructions*
  – Implementation and control
  – Management of second organisations (sub-contractors) and third parties (e.g. suppliers)
 	
• Review and audit procedures:
  – Definition of audit
  – Internal audit
  – External audit
  – Audit procedures
  – Management review
  – Environmental files

Note: The sub-sections denoted * are presented as separate but co-ordinated documents in a multiple document format to accompany a manual.

Figure 3.19 Key sections of the EMS manual

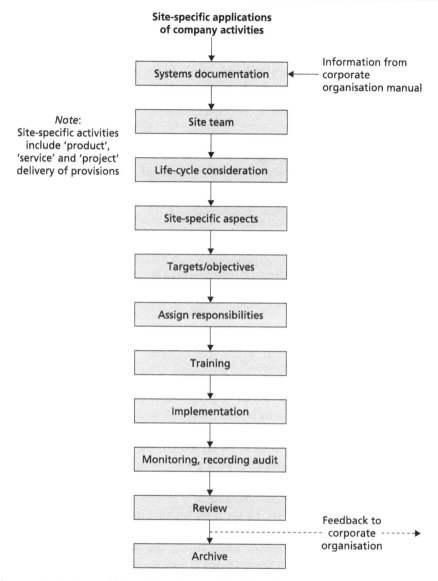

Figure 3.20 Site-specific applications for EMS development

with specific product, service and project provisions also audited to a six-month schedule. The full audit will be an interim six-month audit as part of the three-year system re-certification schedule. Therefore, the interim audits will check generally throughout the system and its operation while perhaps focusing on a small number of sub-system elements for closer scrutiny.

Review template for the EMS
Figure 3.24 shows the template or pro forma developed by the company for re-viewing the system during audit activities. It can be seen that as specific system

Environmental implementation plan for site-specific activities
Key elements and sections
(i) Approval statement by board of directors 1 List of amendments/versions 2 Controlled distribution list 3 Contents list 4 Provision information (product/service/project/site, etc.) 5 Schedule of documentation (procedures/instructions) 6 Specific requirements (contractual) 7 Schedule of suppliers (materials/components/service inputs/supply chain members) 8 Special requirements (procurement/handling/storage/disposal of particular process outputs, e.g. waste) 9 Schedule of client/customer requirements/inputs 10 Schedule of trial/testing requirements 11 Schedule of management documents (reference to corporate manual and other documents) 12 Supervisory/inspection (checking) programme 13 Emergency information (contacts and protocols) records 14 Audit schedule 15 System review 16 Records archiving

Figure 3.21 Key elements and sections in EMS implementation plan

elements are scrutinised, checked and reviewed, so information is gathered on: system evaluation in operation; aspects highlighted for further attention; proposed actions to be taken where matters arise; and follow-up actions to be taken to ensure that the system continues to operate effectively. It should be noted that each review sheet completed should be signed off and forwarded to corporate system files and archiving, consistent with good systems practice.

Archiving template for EMS documentation

The last figure, Figure 3.25, shows the template or pro forma developed by the company for archiving its system documents. It is important that documents are not only robust and completely and safely stored, but also retrievable if required at some time in the future. Therefore, the company in this case used a simple pro forma to sign off to its archive files each system document after use.

Summary

The case study presented is just one example of how a particular company decided to develop and implement an EMS. As with any and all management systems, it is essential that a company develops its systems to meet its own needs, industry sector requirements and business circumstances. Therefore, the example presented is not purported to be prescriptive or to reflect best practice. An appropriate management system is one which supports the company's core business processes and outputs while being efficient, cost effective and adding value to the holistic existence of the organisation. A company must determine this for itself.

Waste is anything a person produces or possesses which they intend or are required to discard, including scrap metal and products that are past their expiry dates

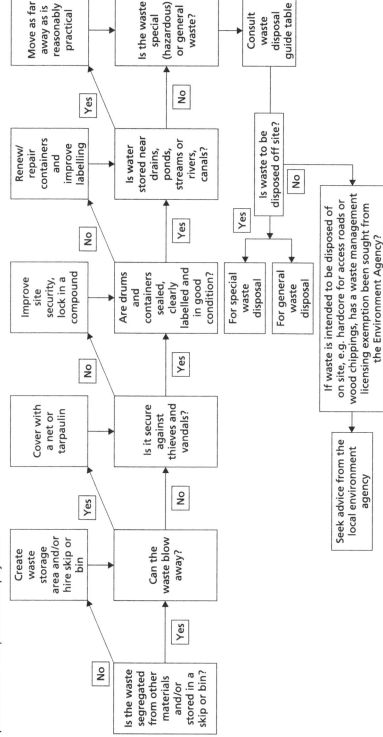

Figure 3.22 Supervisory and operational procedures involved in developing a waste management process

Application	Month											
Product/service/ project provision	J	F	M	A	M	J	J	A	S	O	N	D
1							/					
2							/					
3								/				
4								/				
5										/		
6											/	
7												/
8	/											
9		/										
10			/									
							/					
Full system												/
Interim audit												

Audit planned			/								Audit legend	
Audit undertaken			◣									
Audit closed out					■							

Figure 3.23 Schedule of audits for the EMS

A final thought

Strong environmental practices help us to differentiate ourselves in our industry. We are recognised as developing and delivering sustainable solutions and, in turn, we are seen by our employers, partners and customers as an organisation who truly wants to make a difference.

(Skanska, 2010)

Environmental Management System (EMS) Review	
Review date:	Manual ref.: ....
Review conducted by:	Other documents ref.: ....
Review elements: Product, service, project provision as required	
System process evaluation:	
Aspects requiring further attention:	
Proposed/identified actions:	
Follow-up on actions implemented:	
Approved:	Date: ....

Figure 3.24 Review document pro forma for the EMS

Provision completion:	Date:
Quality and environmental plan archived:	Date:
Signature:	(Approval manager)

Figure 3.25 Archiving document pro forma for the EMS

Environmental management: key points, overview and references

This section presents: a summary of the key points for the collective sections of Chapter 3; an overview of Chapter 3; and a list of references used in the compilation of Chapter 3.

Key points

The key points of this text are arranged under a number of sub-headings as follows.

Environmental management in general

- There is a widening expectation within business that companies should act in a more environmentally responsible way and improve the environmental performance of their activities.
- It is a prerequisite that companies meet increasingly stringent environmental legislation and closer industry regulation including performance assessment and auditing.
- Companies must respond to more demanding customer requirements in the provision of their goods and services, including pre-qualification during procurement.
- Many companies realise the need to respond to their environmental business situation in a positive and dynamic way through the rearrangement of their structure, organisation, resources, procedures and practice.
- Proactive companies have established a formal, or documented, EMS in response to their environment-related business needs.

EMS standards

- ISO 14001 is the international standard providing guidance to companies establishing an EMS.
- Companies with an EMS complying with ISO 14001 can apply to a government-accredited certification body for registration of their management system on a three-year renewable basis.
- ISO 14001 has a similar nomenclature to ISO 9001 QMS, facilitating correspondence between the standards and the management systems established to their specifications.
- The standard provides a framework within which a company can establish an EMS to meet the unique requirements of its business activities, situation and circumstances.
- ISO 14001 follows a 'process' approach and the Deming 'plan–do–check–act' (PDCA) methodology of controlling business activities.
- The European Eco-Management and Audit Scheme, or EMAS, is an alternative EMS registration scheme used widely throughout Europe and recognised and used in the UK.

EMS development and implementation

- ISO 14001 requires that a company develops, implements and maintains a formal EMS to ensure that its business activities conform with its environmental policy, current legislation and applicable industry regulation.
- A company may establish a dedicated stand-alone EMS or a quality-system-related approach which is one broadly based on, has similarities to and correspondence with a company's quality management system, often termed an environmental quality system, or EQS.
- EMSs are particularly pertinent to the nature and activities of the construction industry, where the potential of a project to affect the environment is ever present and its impacts can be considerable.
- EMSs are widely recognised and applied within the construction industry as a focused approach to maintaining environmental safeguards of construction developments and project sites.
- The EMS should follow the sections as detailed in ISO 14001, the system specification.
- To establish an EMS a company should: (1) develop a manual of procedures; (2) implement the procedures in the course of its business; (3) provide evidence of implementation through documentation; and (4) continually improve the system and its business processes.
- An EMS should have four levels of documentation: (1) a system manual; (2) a set of management procedures to guide supervision; (3) a situation-specific implementation plan, where necessary; and (4) a set of work instructions to guide process-based operations.

- A company may apply to a government-accredited certification body for registration of an EMS conforming to ISO 14001 where there is appropriate evidence of system implementation, and capitalise on the benefits of such recognition in the commercial and business marketplace.

- A company should ensure that its business supply chain is maintained and that all suppliers of goods and services meet the same environmental expectations that it does.

- The requirement for site waste management plans for construction projects of particular type and size means that environmental responsibilities are shared and must be managed both up and down the supply chain.

Overview

Companies in all sectors of industry, business and commerce have realised that there are growing expectations upon them to operate in a more environmentally friendly way and demonstrate improved environmental performance. This is no more so than in the construction industry where environmental issues are at the forefront of all its activities. Increasing pressure from customers, clients, industry regulation, national legislation and international standards of operation means that companies cannot turn a blind eye to the environment but must seek to understand the environment better and work with it and in its interests. Embracing environmental matters, issues and concerns is far from easy for some companies to accommodate as they tend to be overshadowed by the more pressing need to deliver quality products and services on time and to budget. Notwithstanding, many companies have seen the need to respond to the environmental dimension of their business through the establishment of a formal EMS. Such developments have tended to mirror QMSs, which have become an established and accepted part of modern business organisation and operation. Moreover, the presence of international standards for EMS development and implementation, following the traditions of quality management, means that there are considerable benefits to be accrued through proactive management of a company's interrelationship with the environment.

Chapter 3 has explained in detail the way in which any company can respond to the environment in the course of its business operations. It has described how ISO 14001, the international standard for an EMS, can assist a company to configure its policy, organisation, resources and procedures to best effect. Furthermore, it has also shown that a company should actively develop its relationships with input suppliers and output customers to manage the environment effectively within its supply chain. In this way, a company is best positioned to address the environment-related aspects of its business operations with a holistic perspective. This should ensure that the company not only recognises and responds to the environment in an effective way, but also absorbs its requirements and responsibilities within its organisational culture and practice.

References

BSI (2003). *BS 8555:2003: Environmental Management Systems: Guide to the phased implementation of an environmental management system including the use of environmental performance evaluation.* British Standards Institution (BSI), London.

DEFRA (2010). *The Environmental Permitting (England and Wales) Regulations 2010.* Department for Environment, Food and Rural Affairs (DEFRA). www.defra.gov.uk

EA (1990). *The Environmental Protection Act 1990.* Environment Agency, London.

EA (1993). *The Clean Air Act 1993.* Environment Agency, London.

Envirowise (2009). Eco management the easy way. *Construction Manager.* March edition. The Chartered Institute of Building (CIOB), Ascot.

Griffith, A. (1994). *Environmental Management for Construction.* Macmillan, Basingstoke.

IEMA (2003). *The BS 8555/Acorn Scheme Workbook.* The Institute of Environmental Management & Assessment (IEMA). www.iema.net

IEMA (2010). *Environmental Management Systems – Your A-Z Guide.* The Institute of Environmental Management & Assessment (IEMA). www.iema.net

ISO (1996). *ISO 14001: Specification for Environmental Management Systems.* International Organization for Standardization, Geneva.

ISO (1999). *ISO 14031: Environmental Performance Evaluation Guidelines.* International Organization for Standardization, Geneva.

ISO (2004). *ISO 14001:2004: Specification for Environmental Management Systems.* International Organization for Standardization, Geneva.

Naoum, S. (2001). *People and Organisational Management in Construction.* Thomas Telford, London.

Skanska (2010). *Sustainability.* www.skanska.co.uk/sustainability

Sunday Times (2009). *The Green List.* www.timesonline.co.uk/bestgreencompanies

WRAP (2010). *Sustainability in Construction.* Waste and Resources Action Programme. wrap.org.uk

CHAPTER 4

Safety management systems

Introduction

Chapter 4 focuses on *safety management systems*. In the context of this book the theme embraces the health and safety management system, or H&SMS, sometimes referred to as the occupational health and safety, or OH&S, system. The management of health and safety is relevant to all sectors of industry and, in particular, construction due to its inherent practical dangers. It is arguably the most important of all management disciplines since effective practices can directly safeguard lives. Chapter 4 examines the principles of safety management in the context of health and safety management system standards, and particularly OHSAS 18001. It progresses to look in detail at the characteristics and components of effective health and safety systems, together with an examination of system application to construction.

Safety management: fundamentals, principles and systems approach

This section examines: the need for health and safety management, focusing on its key dimensions and principles; the capability, competence and commitment required by employees; the influence of the international standard OHSAS 18001; and the benefits to a company or organisation of establishing an effective safety management system.

The need for health and safety management and health and safety management systems

The need for health and safety management

The provision of health and safety in the workplace is a paramount consideration for any organisation. Health and safety are relevant to all sectors of industry, business and commerce. Everyday activities in all types of companies and organisations are, quite ordinarily, surrounded by potential occupational hazards and dangers to safety, health and welfare. Clearly, some industries will be more prone to health and safety risk than others, as indeed some individuals and groups will be more susceptible than others. This results from the organisational business processes carried out and the human tasks that are engaged. To those who work on or interface with a construction site, the inherent levels of health and safety risk are all too clear. Construction remains one of the most dangerous of all occupations. The simple facts are that, first, accidents and dangerous incidents do occur in the workplace no matter how carefully one plans to obviate or mitigate their occurrence, and, second, all accidents are, conceivably, avoidable.

There are many obstacles and hindrances to the achievement of sensible, good and effective occupational health and safety. The time and cost demands from customers, pressures on production output targets, financial constraints, legislation and regulation are all external influences on the providers of goods and services. The size, complexity, resources, capability and competency of an organisation add a range of internal influences. Together, these factors conspire to make health and safety one of the most challenging aspects of organisational management. Notwithstanding, there are strong and powerful business incentives for companies to pursue and strive for better health and safety performance. Such incentives are, chiefly, legal and moral obligations, but in addition there are financial, commercial and marketplace incentives to an organisation which positively and avidly supports effective health and safety management.

Another simple fact is that good health and safety do not materialise by themselves – the organisation has to make them happen. Therefore, the company or organisation which successfully delivers good health and safety is one that recognises the business and organisational needs for occupational health and safety and employs a structured approach to their functional management, often through the establishment and use of a formal H&SMS.

Dimensions to health and safety

It is all too easy to dwell solely on the physical hazards and dangers to the individual occupying and using the workplace when, in fact, a number of dimensions are the purview of good all-round occupational health and safety:

1 *Safety* – the protection of persons from physical injury or harm at the workplace.
2 *Health* – the protection of the body and mind of persons from illness resulting from the processes, procedures or materials associated with the workplace.
3 *Welfare* – the provision of facilities which maintain the health of persons at the workplace.

It is important to note that the dividing line between 'health' and 'safety' is traditionally ambiguous and that health and safety are conventionally taken to include the physical and mental well-being of the person at the workplace. For all ordinary and sensible usage, health and safety embrace all aspects of protection of persons at their place of work while welfare embraces all aspects of facilities.

Health and safety management principles

The successful achievement of health and safety relies upon everyone in an organisation taking responsibility. As a consequence of this, a multitude of responsibilities follow:

- *An individual, in their own respect* – every person has a responsibility to ensure their own health and safety in the workplace.
- *An individual, in respect of others* – every person has a responsibility to ensure that their actions do not endanger the health and safety of others in the workplace.
- *An employer, in respect of employees and others deemed to be under their control* – every employer has a responsibility to ensure a safe and healthy workplace for their employees and provide required welfare facilities at the workplace.

So, an effective approach to health and safety is an organisational system which considers the responsibilities of the employer, all employees and any other party who interacts with or impacts upon the business processes and their management.

For effective health and safety management to be realised, its concepts and principles must be led from the top and permeate throughout the entire organisation, all of its business activities, processes and procedures and all employees. A number of principles support this aspiration:

1 Effective health and safety management evolves from a clear company policy and objectives for health and safety which the whole organisation can follow.

2 Effective health and safety management requires a robust management structure and organisation to deliver the policy and cascade its requirements throughout the organisation.

3 Effective health and safety management needs a planned and systematic approach to implementing the policy and objectives through a well-conceived health and safety management approach.

4 Effective health and safety management requires performance to be measured against set standards to determine the efficacy of approach and highlight where improvement is needed.

5 Effective health and safety management is perpetuated where the organisation reviews its performance and learns from the experience of its approach.

6 Effective health and safety management truly evolves when continual improvement is embraced as an inherent element of the organisation's approach.

Health and safety culture

A major challenge for any company is to establish and then, over time, embed their health and safety management approach within the day-to-day business such that it becomes an intrinsic component of organisational culture and operations. A company may need to achieve a real step change in its approach to fulfil this aspiration. The key elements to consider in the promotion of an active and positive health and safety culture are:

- the methods of control established for health and safety by and throughout the organisation;

- the means of eliciting co-operation to support health and safety between employees and between groups of employees as configured by the organisation;
- the routes and modes of health and safety related communication throughout the organisation;
- the capability, competence, attitude and behaviour of employees delivering the business processes and, in particular, the attributes of those employees with health and safety responsibilities.

The importance of embedding a structure and organisation for health and safety which includes these elements cannot be overstated. There need to be robust mechanisms with consistent application in each of these to gradually develop an ethos and climate where health and safety become the norm of operation rather than the exception. These important elements will be explored in later sections of this chapter.

Health and safety capability, competence and commitment

In response to increasingly stringent legislation, more demanding industry-driven regulation and the ever higher expectations of clients and customers, almost all companies will at some time experience the need to deliver better health and safety performance. This requirement comes in addition to and not instead of the need to provide products and services of good quality in a timely and cost-effective way – health and safety are yet another value-added dimension of successful business practice. In the same way that organisational performance and quality of business output have become a staple of customers, so health and safety are subject to a new breadth and depth of enquiry and evaluation by those same customers.

The pre-qualification of product and service providers is almost becoming the norm for private-sector contract arrangements as it is for those in the public sector. For any company to ignore or underestimate the need for considered and well-conceived health and safety management would be a business catastrophe in a marketplace where a structured health and safety approach is virtually a pre-requisite and an H&SMS is a genuine expectation. For companies to maintain competitiveness in health and safety management they must demonstrate that their organisation possesses the appropriate capabilities, necessary competencies and level of commitment to support effective health and safety performance. Moreover, they must provide evidence that their health and safety approach, through implementation, has supported a good safety record. These requirements form integral components of health and safety pre-qualification. A robust H&SMS can fulfil all of these requirements.

The capability, competence and commitment of the organisation are fundamentally based on the attributes of those persons who work within the organisation. For this reason, three key groups of factors are highly significant and must be considered:

1 *Organisational factors* – primarily the ethos, culture, procedures and practices which lay the company's ground rules for workplace health and safety.

2 *Job factors* – primarily the correspondence between the individual and the tasks they need to perform in the workplace.

3 *Personal factors* – primarily the knowledge, skills and physical and mental attributes that the individual brings to the workplace.

It is clear, therefore, that the organisation's health and safety approach is founded in and utilises extensively the people who work within the organisation. The structural arrangement of the health and safety management approach is of high importance, but of greater importance is the requirement to engender the support of and actively involve people. The most effective health and safety policy, objectives, procedures and practice are likely to be those which respect the requirements of both health and safety management and human resource management.

H&SMS

The system

The key organisational approach to pursuing and achieving a safe and healthy work environment is to ensure that all safety-related matters are considered, planned, organised, controlled, monitored, recorded, audited and reviewed in a structured, systematic and consistent way. The most appropriate way for a company to embrace this aspiration together with accommodating the requirements of legislation, industry regulation, customer and organisational requirements is through the establishment of an H&SMS. Although limited and targeted health & safety (H&S) management initiatives such as product-based production or service delivery H&S plans might satisfy the needs of the smaller company, for larger, more complex and business-diversified companies a company-based whole-organisation H&SMS is a preferable approach. Such an H&SMS will not only put in place the mechanisms for achieving effective H&S management throughout the organisation, but also add value by embedding appropriate internal culture and enhancing the capability to improve the core business.

H&SMS standards

The Occupational Health and Safety Assessment Series (OHSAS) Standard 18001: *Occupational health and safety management systems – Requirements* and OHSAS 18002: *Guidelines for the implementation of OHSAS 18001* were developed to meet the great demand from companies and organisations for an H&SMS standard by which they could assess their H&SMS. OHSAS 18001 presents a framework for a structured and formalised H&SMS which, in association with organisation-based needs and customer requirements, allows the organisation to achieve effective H&S management (BSI, 2007a; 2007b).

The OHSAS 18001 standard is based on the plan–do–check–act (PDCA) methodology described in the development of QMS and EMS earlier and restated in Figure 4.1. In addition, the H&SMS model is commensurate with the process model, also described earlier, where organisations manage their business

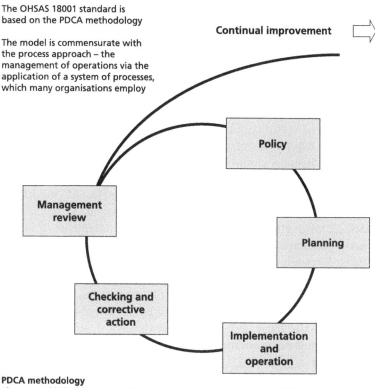

The OHSAS 18001 standard is based on the PDCA methodology

The model is commensurate with the process approach – the management of operations via the application of a system of processes, which many organisations employ

Continual improvement

Policy

Management review

Planning

Checking and corrective action

Implementation and operation

PDCA methodology
Plan: establish objectives and processes to deliver results to the H&S policy
Do: implement the processes
Check: monitor and measure progress against the H&S policy, objectives, legal and other requirements
Act: take actions to continually improve H&S performance

Figure 4.1 Health and safety management system (H&SMS) model suggested by OHSAS 18001

Source: Adapted from OHSAS 18001:2007

activities via systematic processes with specific management functions focusing on the undertaking of those processes and management of the interactions between those processes.

The standard specifies the requirements for an appropriate H&SMS within which the organisation can develop an appropriate H&S policy, objectives and procedures that accommodate legal, regulatory, customer and marketplace requirements. OHSAS 18001 is compatible with ISO standards for QMS and EMS approaches. The standard has broad application across and throughout industry sectors and embraces all types of companies and organisations from small to large and from simple to complex. OHSAS 18001 does not contain requirements that are specific to other management systems, but the elements which make up its system framework can be aligned with elements of ISO 9001, ISO 14001 and other management system standards. As outlined earlier and described in

Chapter 2, an organisation can adapt its existing management systems, such as a QMS, to develop and implement an appropriate H&SMS.

When combining or integrating H&S management with other management systems, it must be remembered that particular elements will differ according to the level of influence of other requirements. This is true in particular with an H&SMS where in addition to the OHSAS requirements the system is heavily influenced by legislation and industry-based regulation. In addition, the H&SMS, like all management systems, must be configured and applied to suit the needs, requirements and aspirations of the company within the context of its culture, organisation and business circumstances.

The effectiveness of the H&SMS and ultimately the success of the organisation in H&S management depend crucially upon how the H&SMS model is applied. The key elements depicted in Figure 4.1 must be perpetuated, directed and supported from senior-management levels through clear statements of H&S policy and organisational objectives. These must be embedded in sound and robust management procedures which focus on the core business processes. Then they must be translated into deliverable work instructions which encourage safe practices at the workplace. Moreover, the H&SMS must be managed as an active and evolving creation with continual improvement as an embedded objective. The OHSAS standard provides the basis and the wherewithal for a company to establish such an H&SMS and the promotion of improved organisational performance in H&S management.

Configuring the H&SMS

As with the development and implementation of the QMS and EMS addressed earlier, the H&SMS concept can be applied to an organisation in various ways. Specific application will be determined by a company according to its own requirements. There are essentially two routes for a company to consider:

1 A company-wide corporate-based umbrella system where the comprehensive H&SMS applies to the whole organisation and its business activities, processes and management.

2 A project-based system where the H&SMS applies to specific business activities and processes or the business outputs (products or services) based, normally, on a targeted H&S plan rather than a comprehensive H&SMS.

For the smaller company with limited organisation, defined resources and business activity focused on a single production or delivery site, a location-specific project-based H&SMS may fulfil all its requirements. Similarly, a company undertaking broadly similar and repeated work for a long-standing customer may have an arrangement to operate within a defined and agreed project-based H&SMS where this has been seen to meet the requirements of both the provider and customer. In contrast, a large company operating within the open marketplace where discerning customers expect and demand evidence of systematic, reliable and consistent H&S performance may need to implement and maintain an effective, recognised and certificated H&SMS.

An effective H&SMS can contribute to the acquisition of whole-organisation benefits including:

- better human resource utilisation;
- minimisation of H&S incidents, accidents and associated financial loss;
- recognition that H&S shortfalls are failings in organisational management which can addressed;
- awareness that H&S practice can be embedded in the core business of the organisation to deliver added value;
- systematic identification of H&S risk and the effective means to manage those risks;
- perpetuation of management directed at continual improvement of the organisation.

A traditionally held belief is that H&S are another one of those functional management disciplines which when addressed through a formalised management system become bureaucratic and laborious. As such, it is easy to adopt an attitude of minimum compliance. In reality, there is much routine to everyday H&S practice but it is also clear that there are sound business and organisational reasons to justify the use of the H&SMS approach.

Occupational H&SMSs – OHSAS 18001 requirements

This section examines the requirements for H&SMSs specified by the international standard OHSAS 18001.

Requirements

The requirements for H&SMSs under the Occupational Health and Safety Assessment Series (OHSAS) 18001 are presented as a set of clauses and sub-clauses (BSI, 2007a). For precise wording, terminology and specification refer to the standard itself. This section reflects the standard in the form of considerations and questions that an organisation should ask when developing, implementing and maintaining an H&SMS to the requirements of OHSAS 18001. For continuity with the standard the same enumeration is used to differentiate the various clauses, or elements of the system. OHSAS 18001 also presents two informative annex sections and a bibliography.

The clauses and sub-clauses of OHSAS 18001 (see Figure 4.2) are as follows:

1 Scope
2 Reference publications
3 Terms and definitions
4 Occupational health and safety (OH&S) management system requirements
   4.1 General requirements

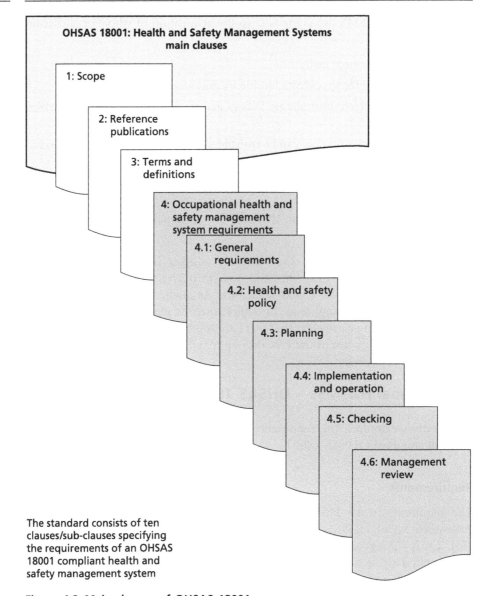

The standard consists of ten clauses/sub-clauses specifying the requirements of an OHSAS 18001 compliant health and safety management system

Figure 4.2 Main clauses of OHSAS 18001

    4.2 OH&S policy
    4.3 Planning
        4.3.1 Hazard identification, risk assessment and determining controls
        4.3.2 Legal and other requirements
        4.3.3 Objectives and programme
    4.4 Implementation and operation
        4.4.1 Resources, roles, responsibility, accountability and authority
        4.4.2 Competence, training and awareness

Consideration of the clauses

In relation to the clauses presented, a company wishing to develop, implement and maintain an H&SMS to the OHSAS 18001 requirements should consider the following aspects.

1 Scope

[Refer to clause 1]

The OHSAS standard specifies requirements for an occupational health and safety (OH&S) management system with the focus on the company controlling its health and safety risks and improving its health and safety performance. Consistent with other management system standards, OHSAS 18001 does not specify health and safety performance criteria or prescribe specifications for the development and implementation of a H&SMS.

The standard is available to any company which seeks to achieve the following:

a) Develop and implement an H&SMS to remove or reduce the risk to personnel and other stakeholders who might be exposed to health and safety related dangers in association with the organisation's business activities.

b) Develop, implement and maintain an H&SMS and seek to continually improve that system.

c) Ensure that the organisation's performance conforms to the organisation's H&SMS policy and underpins its goals, objectives and targets.

d) Demonstrate that the H&SMS conforms to the standard by:

1 self-declaring its H&SMS policy and management approach; or

2 obtaining confirmation of organisational performance from business stakeholders – clients, customers and output end users; or

3 independent, or third-party, confirmation of the establishment of the H&SMS; or

4 obtaining registration, or certification, of the H&SMS by an appropriate certification body.

OHSAS 18001 states that *all* the requirements of the standard should be embraced by the H&SMS, but that their precise inclusion will depend upon the type of company, its business activities and its outputs. As with other management system standards, OHSAS 18001 presents a set of requirements which a company should apply within the context of its own particular business and marketplace situation and circumstance. The health and safety focus of the standard is towards occupational health and safety and therefore it is quite explicit that it does not directly apply to the safety of property and business facilities nor the well-being and wellness expectations of company employees as other management approaches and systems are available for those purposes. So, OHSAS 18001 provides specifications for the sensible and appropriate establishment of an occupational health and safety management system but is not prescriptive in terms of the ways in which it should be applied by an organisation.

2 Reference publications

[Refer to clause 2]

The standard presents a short list of other publications in the public domain which provide further information or guidance on occupational health and safety management systems. These include a reference to the bibliography of the standard, which highlights associated ISO standards – ISO 9001 (QMS), ISO 14001 (EMS) and ISO 19001 – QMS and EMS auditing (ISO, 2000; 2002; 2004; 2005). In addition, specific reference is made to two publications: (i) OHSAS 18002, *Occupational health and safety management systems – Guidelines for the implementation of OHSAS 18001* (BS-OHSAS, 2007); and (ii) ILO, *Guidelines on Occupational Health and Safety Management Systems, or OSH-MS* (ILO, 2001).

3 Terms and definitions

[Refer to clause 3]

A set of twenty-three terms and definitions are presented within the standard. These are important to the understanding and application of H&SMS as they serve two purposes, first to explain the terminology used by the standard generally and second to explain terms included in the current version of the standard which have replaced those used in previous versions. The terms and definitions are cross-referenced to each other for continuity of explanation and understanding. For example, term 3.7: Hazard identification is defined as a 'process of recognising that a hazard (3.6) exists and defining its characteristics' – where term 3.6 defines a hazard.

The use of recognised and consistent system terminology is important to H&SMS development, as it is for any management system. It is not uncommon for companies, organisations, groups and individuals to interpret system elements in their own way and use their own vernacular. This can be unhelpful when a uniform understanding of the system is needed, so using specified terms and definitions should be favoured.

4 Occupational health and safety (OH&S) management system requirements

[Refer to clause 4]

4.1 General requirements

The standard specifies that an organisation should develop, implement, maintain and continually improve an OH&S management system conforming to OHSAS 18001 and describe how it will meet these requirements. Further, the organisation should seek to continually improve the system. The organisation is required to document the system and define its scope within that documentation (see Figure 4.3).

4.2 OH&S policy

'Top', or executive/senior, management are required to define, present and authorise the policy of the organisation towards health and safety and ensure that the policy:

a) Is commensurate with the nature and scale of the health and safety risks posed by the organisation's business activities;

b) Includes a commitment to prevent injury or ill health and seeks continual improvement in health and safety performance by the organisation;

c) Complies with legal requirements and other requirements of the business and its activities which involve health and safety hazard;

d) Presents an appropriate framework for setting and reviewing health and safety objectives;

e) Is established through documentation and implemented and maintained;

f) Is communicated to all organisational employees and controlled stakeholders including notification of personal H&S responsibilities;

g) Is available to any interested parties and stakeholders;

h) Is reviewed to maintain relevance and currency.

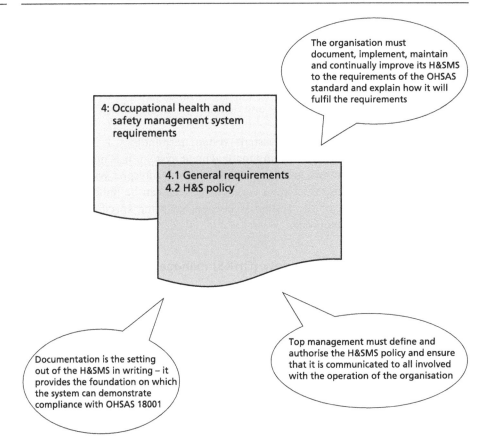

Figure 4.3 Key element – general requirements and H&SMS policy

4.3 Planning

4.3.1 Hazard identification, risk assessment and determining controls

The standard requires that the organisation establishes a procedure to identify hazards, assess their risk and determine appropriate controls (see Figure 4.4). Such a procedure should consider the following:

a) routine and non-routine activities;

b) the activities of all persons associated with the organisation's place of work;

c) human factors, capabilities and behaviours;

d) external hazards impinging upon the workplace and the H&S of those working there;

e) hazards created within the environs of the workplace under the organisation's control;

f) facilities, equipment and materials provided by the organisation or others present at the workplace;

g) changes to the organisation, its business activities or its resources;

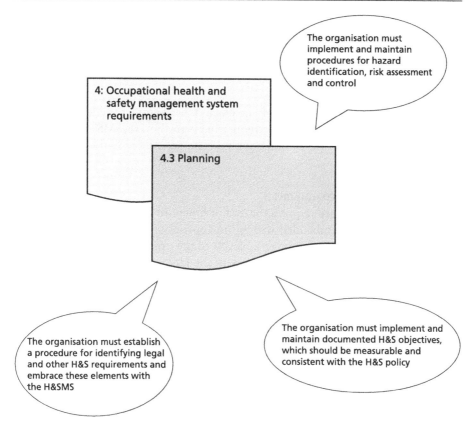

Figure 4.4 Key element – H&SMS planning

h) modifications to the H&SMS together with their effects on business processes and activities;

i) legal requirements relating to risk assessment and control mechanisms;

j) the configuration of workplaces, facilities, equipment, procedures and work organisation and their relationship to human resources.

The clause of the standard also calls for the organisation to consider its methodology for hazard identification and risk assessment. The methodology should:

a) be defined in terms of its scope, nature and timing such that it operates proactively;

b) provide for risk identification, prioritisation and documentation and the determination of control mechanisms.

When establishing control mechanisms the organisation is required to consider the reduction of risk in a hierarchy of priority as follows:

a) elimination;

b) substitution;

c) engineering controls;

d) signage/warnings and/or administrative controls;

e) personal protective equipment (PPE).

4.3.2 Legal and other requirements

The organisation is required to put in place a procedure for identifying and accessing legal and other requirements pertinent to the H&SMS. It must keep information up to date and should communicate such information to all interested parties.

4.3.3 Objectives and programme(s)

This clause of the standard requires the establishment of documented H&S objectives. These should be embedded within the appropriate management functions and throughout the various managerial levels of the organisation. The objectives should be measurable and be consistent with the H&S policy. Objectives should include commitment to prevention of injury and ill health, so they are concerned with human welfare in addition to occupational safety.

The organisation is required to implement and maintain a programme for achieving the H&S objectives, and such programmes should include:

a) the assignment and designation of responsibility and authority for achieving the objectives within the management functions and managerial levels identified by the organisation;

b) the wherewithal and time-frame by which to achieve the objectives set.

There is a requirement for the organisation to review the programme to ensure that the objectives can be achieved. Such programmes therefore need to be structured and robust to ensure that objectives are pursued, but the programme also needs to be flexible to accommodate necessary modification as circumstances change.

4.4 Implementation and operation

4.4.1 Resources, roles, responsibility, accountability and authority

The standard is clear that it expects top management to take responsibility for the establishment of the H&SMS (see Figure 4.5). It must demonstrate this by:

a) providing the organisational resources to sustain the system;

b) establishing the managerial responsibilities and accountabilities to facilitate effective H&S management.

There is a requirement to appoint a designated individual from executive management who will take specific responsibility for H&S management, irrespective of other organisational duties and responsibilities – the *H&S management representative*. Usually, this person will be designated the company H&S manager, or H&S director. This individual should have a defined role and authority for the following:

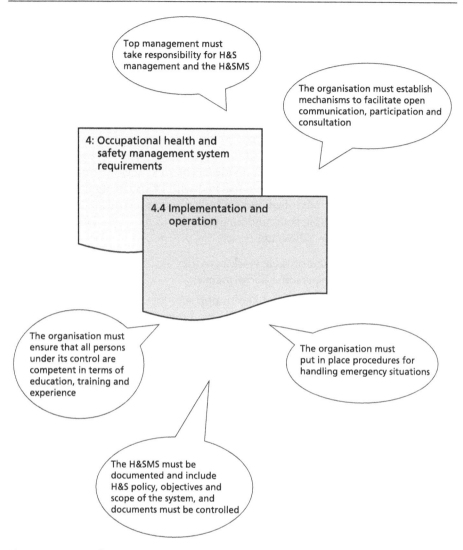

Figure 4.5 Key element – H&SMS implementation and operation

a) ensuring that the H&SMS is developed, implemented and maintained to the requirements of the standard;

b) ensuring that reports on the application and performance of the H&SMS are presented to executive management and utilised in association with the continual improvement of the system as required by the standard.

The H&S manager will have a pivotal role in ensuring that the H&S policy and objectives are applied to the business processes of the organisation through the establishment of management procedures and work instructions and, where appropriate, situation-specific implementation plans. This person will take single-point responsibility on behalf of the corporate organisation for perpetuating H&S

management and for overseeing the well-being of the H&SMS. Notwithstanding, the standard requires that all those with management responsibility shall demonstrate commitment for H&S performance within the organisation. Furthermore, it requires that all persons in the workplace also take responsibility for H&S where it comes within their sphere of control.

4.4.2 Competence, training and awareness

The organisation is required to ensure that any person over whom it has control and performs tasks that can impact on H&S is competent, based on their education, training and experience. Evidence of this must be maintained by appropriate records. Training needs associated with H&S risk and the operation of the H&SMS must be identified and suitable training provided. The organisation must establish, implement and maintain procedures to ensure that all persons are aware of the following:

a) the H&S consequences of their work activities and their behaviour and also the benefits of better personal performance;

b) their roles, responsibilities and the importance of achieving conformity to H&S policy and procedures and to the requirements of the H&SMS including emergency preparedness and response requirements;

c) the consequences of not following specified procedures.

Training should accommodate the differing levels of:

a) responsibility, ability and language capabilities;

b) risk.

4.4.3 Communication, participation and consultation

4.4.3.1 Communication

The standard requires that procedures should be employed which ensure:

a) internal communication among the various functions and levels within the organisation;

b) communication with contractors and other visitors to the organisation's workplace;

c) the receipt of, documentation of and response to communications from external parties.

4.4.3.2 Participation and consultation

The organisation is required to develop, implement and maintain procedures for:

a) the participation of workers by their involvement in:

- hazard identification, risk assessments and their control mechanisms;
- accident investigation;
- development and review of H&S policies and objectives;

- changes that affect H&S practice;
- H&S matters.

b) consultation with contractors where change affects their H&S.

The organisation should ensure that relevant external parties are fully consulted on H&S matters.

4.4.4 Documentation
The standard requires that the H&SMS shall include the following:

a) the organisational H&S policy and objectives;

b) a description of the scope of the H&SMS;

c) a description of the main elements of the H&SMS and their interaction with and reference to related H&S documents;

d) documents, including records, required by the standard;

e) documents, including records, identified by the organisation to ensure effective planning, operation and control of processes that relate to the management of H&S risks.

4.4.5 Control of documents
There is a requirement for all system documents to be controlled. The organisation must develop, implement and maintain procedures to:

a) approve documents before use;

b) review, update and re-approve documents;

c) ensure that changes and currency of revision are identified;

d) ensure that relevant versions of documents are available for use;

e) ensure that documents remain legible and identifiable;

f) ensure that documents of external origin used within the H&SMS are identified and controlled;

g) prevent unintended use of obsolete documents and identify them if retained.

4.4.6 Operational control
The standard requires that the organisation shall determine those operations and activities that can identify H&S hazards and configure the mechanisms of control to manage the associated H&S risks. For the operations and activities confirmed, the organisation is required to establish:

a) operational controls, and embed these within the H&SMS;

b) controls related to goods, services and equipment;

c) controls related to contractors and others visiting the workplace;

d) documented procedures for situations where their absence could lead to departure from the H&S policy and objectives;

e) operating criteria where their absence could lead to departure from the H&S policy and objectives.

4.4.7 Emergency preparedness and response

In meeting this clause the organisation should establish procedures which:

a) identify potential emergency situations;

b) respond to emergency situations.

The organisation has an obligation to respond to emergency situations and, moreover, combat the possible adverse effects of such instances. An emergency response must accommodate the needs of other parties such as emergency services. Furthermore, the response mechanisms should be tested periodically for their efficacy, and again this should involve other interested parties. As part of continual monitoring and testing, periodic review should take place to revise, update and improve the organisation's emergency preparedness.

4.5 Checking

4.5.1 Performance measurement and monitoring

The standard requires the organisation to support procedures which monitor and measure its H&S performance on a regular basis (see Figure 4.6). Such procedures must accommodate the following:

a) qualitative and quantitative measures;

b) monitoring of the extent to which H&S objectives are met;

c) monitoring the effectiveness of controls;

d) measures of performance that monitor ill health and attributes that highlight poor H&S performance;

e) recording of data and results from monitoring and measurement to allow appropriate analysis of corrective and preventive actions.

4.5.2 Evaluation of compliance

The organisation must have a procedure in place for evaluating compliance with legal requirements. It should maintain records of such procedures. In addition, compliance with other requirements should be determined, and again appropriate records should be kept.

4.5.3 Incident investigation, non-conformity, corrective action and preventive action

4.5.3.1 Incident investigation

The standard requires that the organisation should develop, implement and maintain procedures to record, investigate and analyse H&S incidents such that:

a) underlying deficiencies that contribute to incidents are identified;

b) the need for corrective action can be identified;

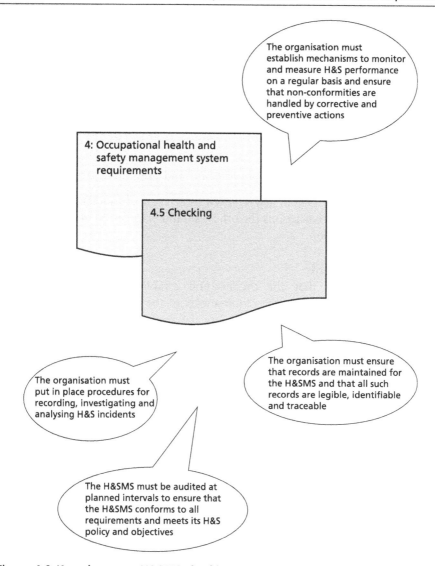

Figure 4.6 Key element – H&SMS checking

c) opportunities for preventive actions can be identified;

d) opportunities for continual improvements are determined;

e) results from investigations are communicated.

Again, the results of investigations should be documented and maintained.

4.5.3.2 Non-conformity, corrective action and preventive action
Procedures for handling potential non-conformities and associated corrective and preventive actions must be determined. The procedures should define the requirements for:

a) the identification and correction of non-conformities, and actions to reduce their consequences upon H&S;

b) the investigation of non-conformities, including determining their cause, and actions to avoid their recurrence;

c) the evaluation of the need for action to prevent non-conformities, and actions to avoid their recurrence;

d) the recording and communication of results from corrective and preventive actions;

e) the reviewing of effectiveness of corrective and preventive actions.

Where changes to the system are required as a result of modified or ungraded corrective and preventive actions then these should be reflected in updates to the H&SMS documentation.

4.5.4 Control of records

This clause requires that the organisation establish records which allow conformity of results against the requirements of the H&SMS and the OHSAS standard to be determined. Procedures are required which facilitate the identification, storage, protection, retrieval, retention and disposal of records.

4.5.5 Internal audit

The organisation must audit the H&SMS at planned intervals to:

a) determine if the H&SMS:

    1 conforms to the organisational arrangements for H&S management and the requirements of the standard;

    2 has been implemented and maintained appropriately;

    3 is effective in meeting policy and objectives.

b) provide information on the results of audits to corporate management.

Audits should be appropriately planned and based on objective evidence from risk assessments and outcomes from previous audits. A procedure must be established which embraces the following:

a) the responsibilities, competences and requirements for planning and carrying out audit events, for reporting the results and gathering records;

b) the determination of audit scope and criteria and frequency together with the methods to be used.

4.6 Management review

Executive and senior management of the organisation are required to review the H&SMS to ensure its continuing suitability, adequacy and effectiveness. Such review should consider the opportunities to improve the system and the underlying H&S policy and objectives (see Figure 4.7).

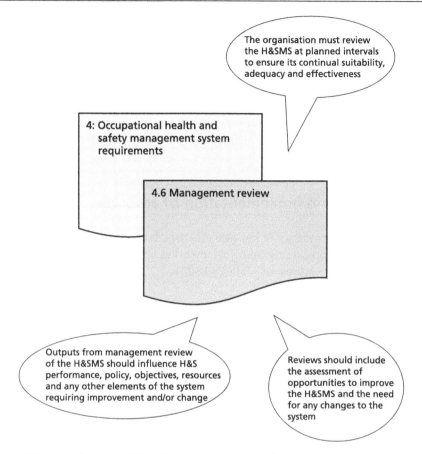

The organisation must review the H&SMS at planned intervals to ensure its continual suitability, adequacy and effectiveness

4: Occupational health and safety management system requirements

4.6 Management review

Outputs from management review of the H&SMS should influence H&S performance, policy, objectives, resources and any other elements of the system requiring improvement and/or change

Reviews should include the assessment of opportunities to improve the H&SMS and the need for any changes to the system

Figure 4.7 Key element – H&SMS management review

Management reviews should consider the following aspects:

a) results of internal audits and evaluations of compliance with legal and other requirements;

b) results of participation and consultation with other parties;

c) communications from external interested parties;

d) the H&S performance of the organisation;

e) the degree to which H&S objectives have been met;

f) status of H&S incident investigations and subsequent actions;

g) follow-up actions from previous reviews;

h) changing circumstances in all aspects affecting H&S management;

i) recommendations for improvement.

The outputs from these reviews should be commensurate with the commitment given to continually improve the H&SMS and include decisions and actions linked to potential changes to:

a) H&S performance;

b) H&S policy and objectives;

c) resources

d) other elements of the H&SMS.

This section of the chapter has presented the requirements of OHSAS 18001 – the requirements for the development, implementation and maintenance of and improvement to an H&SMS. The next section looks at the development and implementation of an H&SMS.

Development and implementation of H&SMSs

This section examines the set of key elements which need to be considered by the organisation when developing and implementing an effective H&SMS meeting the requirements, or clauses, of OHSAS 18001.

Key elements of an H&SMS

The OHSAS 18001 standard requires that five key elements are considered when configuring the H&SMS (see Figure 4.8). These are as follows:

1 *Policy* – this involves the declaration of a clear statement of the organisation's ethos and direction on H&S matters and their management.

2 *Planning* – this involves the identification and assessment of H&S risks and their mitigation in respect of the organisation's business activities.

3 *Implementation and operation* – this involves the arrangement of structure, organisation, resources and documentation for the H&SMS.

4 *Checking* – this involves the establishment of performance monitoring, measurement and auditing measures to alleviate and correct non-conformities in H&SMS operation.

5 *Management review* – this involves the evaluation and review of the H&SMS with a view to its continual improvement.

H&S policy

The H&S policy is a prerequisite to H&S management and lays the foundation for the conceptual development of the H&SMS. The H&S policy is a statement of the organisation's ethos, direction, intentions and commitment to occupational H&S. The policy is an expression of the organisation's attitude and behaviour towards H&S in the undertaking of its business activities and indeed the whole organisation. A carefully considered, well-structured and effective H&SMS can be a direct reflection of the organisation's policy. The H&S policy is not a safety document, as it does not present arrangements for H&S; rather it presents an umbrella of commitment to H&S under which all H&S matters are perceived, configured and managed (see Figure 4.9).

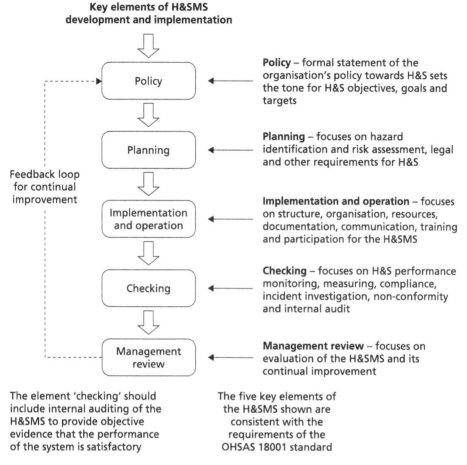

Key elements of H&SMS development and implementation

Policy – formal statement of the organisation's policy towards H&S sets the tone for H&S objectives, goals and targets

Planning – focuses on hazard identification and risk assessment, legal and other requirements for H&S

Implementation and operation – focuses on structure, organisation, resources, documentation, communication, training and participation for the H&SMS

Checking – focuses on H&S performance monitoring, measuring, compliance, incident investigation, non-conformity and internal audit

Management review – focuses on evaluation of the H&SMS and its continual improvement

The element 'checking' should include internal auditing of the H&SMS to provide objective evidence that the performance of the system is satisfactory

The five key elements of the H&SMS shown are consistent with the requirements of the OHSAS 18001 standard

Figure 4.8 Key elements of an OHSAS 18001 compatible H&SMS

As the H&S policy sets the tone for the entire organisation, it is essential that it emanates from and is seen to come from top, or executive, management. For this reason, the H&S policy statement will be written and agreed among a company's executive management board and presented and signed by the company's chief executive officer (CEO).

In addition to setting out the organisation's H&S perspective, the H&S policy must accommodate the requirements of H&S legislation. The Health and Safety at Work etc. Act 1974 (HSE, 1974) requires that an organisation employing more than four persons has a documented H&S policy statement. This is reiterated by the Management of Health and Safety at Work Regulations 1999 (HSE, 1999) which call for written H&S arrangements for organisations employing five or more persons. Essentially, legislation requires that a company should have a written H&S policy which includes or refers to the following:

- an H&S policy statement which includes organisational H&S aims and objectives;

The H&S **policy** is a published statement by the organisation of its intentions in relation to the H&S performance of its organisational activities and its services when providing business outputs

Health & Safety Policy

ABC Principal Contracting (International) Ltd is committed to ensuring safe and healthy workplaces and working practices. The Company meets its obligations under the Health and Safety at Work etc. Act 1974. In so doing, the Company will:

1) Ensure the safety, health and welfare of all employees;
2) Ensure the safety, health and welfare of all persons not under the control of but involved with the Company at its workplaces.

- The Company is committed to planning for and using procedures for hazard identification and risk assessment which form the basis for establishing control mechanisms applied to the whole organisation and its business processes.
- The Company supports the proactive improvement of its health and safety performance.
- The Company will maintain its H&SMS to the specification of OHSAS 18001 (refer to H&SMS Manual and Procedures).
- The Company has in place a designated H&S Management Representative whose single-point responsibility extends to all aspects of H&S planning, organisation, resourcing and management.
- The Company has in place a structure and organisation for health and safety which places a duty and responsibility on all employees for the safety of themselves and that of their colleagues (refer to H&SMS).
- The Company will ensure that all resources deployed to H&S management are capable and competent in H&S matters.
- The Company will monitor, measure, audit and evaluate its H&S performance to ensure that its requirements meet with its policy, legislation, regulation and any other requirements pertinent to H&S.
- The Company will engage in continual improvement to H&S and its H&SMS.

The Company is committed to whole organisation H&S management, applying its H&SMS to both its corporate activities and its output services in accordance with current management system standards and the requirements of legislation and industry regulation.

Signed on behalf of the Company
Chief Executive Officer (CEO) .... Date ....

The **purpose** of the H&S policy is to define a company's corporate philosophy towards H&S management and give authority from top management

The **requirements** of the H&S policy are to: be relevant to the core business; come from top management; be supported by all employees; be amenable to public scrutiny

The H&S policy is declared by a **policy statement** which should: be clear to a wide audience; be presented in a clear format; be a true reflection of intent; be linked to goals, objectives and targets; meet the OHSAS 18001 standard; be authorised by the CEO of the company

The **content** of a quality policy should include: corporate responsibility; customer accountability; performance expectations; communication requirements; and improvement

Figure 4.9 Typical health and safety policy statement – annotated for additional explanation

- an H&S organisation structure which includes duties and responsibilities;
- the organisation's H&S management arrangements.

An appropriate H&S policy will, first, set the direction of the organisation, and, second, provide a sound framework for H&S management by reflecting the following attributes:

- providing a clearly written and presented statement of the organisation's H&S policy;
- having the CEO sign off the written policy statement;
- demonstrating the commitment to H&S by top, or executive, management;
- giving H&S a whole-organisation context;
- matching H&S to business processes and outputs;
- charging H&S with the requirement for continual improvement;
- meeting the requirements of H&S legislation and regulation;
- desire for superior H&S performance;
- identifying the organisation's key H&S manager and giving single-point responsibility;
- explaining the H&S roles, duties and responsibilities of employees;
- outlining the routes of H&S-related communication;
- deploying appropriate resources to H&S activities;
- ensuring the H&S capabilities and competencies of employees;
- committing the organisation to reviewing H&S management and evolving future policy.

It can be seen from the many attributes stated that the H&S policy can be quite involved and multi-faceted. A policy may seek to include all attributes within the statement itself or make reference to links to other documents within the statement. For example, references may be made to the H&SMS manual and procedures or organisation charts contained in company handbooks to illustrate the structure of personnel and organisational resources. Given the extensive content of some policy statements, it can be seen just how important the statement is in establishing the underlying support to the organisation's H&S management approach.

Planning

In some respect, almost all business activities have a potential to generate hazards and risks to work methods and the workplace. At some time, most employees will become subject to an occupational hazard of some kind, and they could conceivably be harmed by that hazard – they are subject to risk. Hazards and their attendant risks can be foreseen and therefore can be systematically planned and managed.

Planning for H&S management focuses on three requirements:

1 Hazard identification.

2 Risk evaluation.

3 Control measures for prevention and protection.

Before these requirements can be considered in detail it is pertinent to understand the nature of hazards, the risks to safety resulting from such hazards and also the severity of harm that the hazard could inflict should it manifest as a health or safety occurrence.

Hazard

A hazard is 'something which presents the potential to cause harm'. It is a pre-existing condition that forms within the business processes and work tasks that need to be performed and which requires one or a set of variables to be present in order to occur. A hazard usually results from a precursor situation or chain of events and often manifests itself in the form of an accident. While an accident is generally considered to be 'an unplanned event that results in injury or damage', and is therefore unforeseen, a precursor or event is a contributing factor which should be recognised and therefore be considered. So, the identification of hazards within the business processes and their undertaking forms the starting point for H&S planning.

Risk

A risk to safety is 'the likelihood that harm could occur to a person due to the presence of a hazard and the level of harm that may occur'.

Harm

When a health or safety incident occurs, the level, or severity, of the harm inflicted influences a range of outcomes. This range spans from a near incident or minor injury to a major incident or fatal injury.

Therefore, for all hazards that are identified, the risk of occurrence (likelihood) and the level of harm (severity) must be determined. Planning for the H&SMS must develop and implement a procedure to evidence the consideration of: occupational hazards inherent to the organisation's activities; the likelihood of an occurrence of a health or safety incident; and the severity of harm that the occurrence might create. This procedure is termed *risk assessment*.

Duty to conduct risk assessment

OHSAS 18001 requires that an organisation develops and implements procedures to undertake risk assessment of its business activities. Likewise, the Management of Health and Safety at Work Regulations 1999 place a duty upon all employers and self-employed persons to conduct risk assessment. In addition to identifying possible hazards and evaluating their potential risks, the organisation is required to determine control of such risks.

Conducting risk assessment

Risk assessment incorporates the three requirements listed previously: (1) hazard identification; (2) risk evaluation; and (3) control measures for prevention and protection. Risk assessment is based on a simple methodology which uses qualitative and quantitative information obtained from factual experiences of safety incidents to develop a numeric figure to reflect the potential degree of risk. The outcome from such an approach is based on the statistical probability of a hazardous event occurring. Using evaluation criteria for likelihood of occurrence and severity of harm, simple calculations can be made to determine degree of risk in percentage terms – the higher the percentage, the greater the degree of risk. Alternatively, the degree of risk can be expressed as a priority rating of low, medium or high. Knowing the degree of risk enables management to focus attention on those activities, processes and work tasks where there is higher risk. Once the risk is identified in this way, the organisation can give detailed consideration to methods of control to mitigate it through the establishment of prevention and protection measures. This approach to risk assessment is detailed in Figure 4.10.

The precise approach to risk assessment using this methodology can vary. Some organisations may choose to use numeric values to represent risk while others prefer to give a rating. Likewise, the methods of presentation can vary, with tables, charts or matrices adopted. The important points are that the task of risk assessment is carried out irrespective of the precise approach, that it is based on a sound methodology that is satisfactory to the organisation and its activities, and that sound and robust outputs are achieved which meaningfully and usefully inform the establishment of H&S control measures.

Linking risk assessment to business activities, processes and tasks

Risk assessment must be applied to the business activities, processes and tasks undertaken by the organisation. Therefore, while the risk assessment methodology is generic, its application is organisation specific. Planning for H&S must consider all those business aspects which are routine and repetitive within the corporate organisation and also consider those which are specialist and vary with the nature of the business outputs and their delivery locations. The H&SMS documents will reflect both of these dimensions in the manual, procedures, implementation plans and work instructions. To carry out such risk assessment the organisation needs to link the risk assessment methodology to the business activities, and it can achieve this by developing an H&S method statement. These are explained in the next section of this chapter. In brief, the method statement is a two-dimensional spreadsheet which lists organisational activities on the Y-axis, and on the X-axis presents: the elements of work; the hazards; the persons at risk; the risk rating; and control measures. These can be developed for all or any part of the organisation and can reflect corporate or project perspectives and incorporate routine or specialist activities. H&S method statements are a key element of planning for H&SMS, as they make risk assessment organisation specific. Again, within the general concept of producing H&S method statements, a variety of precise approaches may be taken. The important point is that a

Methodology	Calculation tables

Methodology

Degree of risk =
likelihood of occurrence × severity of harm

So, suppose a situation is perceived where the likelihood of occurrence is '6' and the severity of harm is '4' where the likelihood of occurrence is an assigned value on a six-point scale from (1) remote to (6) highly probable (based on factual information) (see Table 1 opposite) and severity of harm is assigned a value on a six-point scale from (1) minor injury to (6) death (based on factual information) (see Table 2 opposite), then

Degree of risk =
likelihood of occurrence × severity of harm
or
$6 \times 4 = 24$

The degree of risk (24) is a numerical value which is the proportion of the possible maximum degree of risk. Maximum risk is 36 given by the likelihood of occurrence on the six-point scale multiplied by the severity of harm on the six-point scale.

The value (24) can be more usefully expressed as a percentage: 24 from 36 = 67%.

Degree of risk in percentage terms can be linked to a priority risk rating:

• low priority (L)
• medium priority (M)
• high priority (H)

where

(L) = 3 to 9%
(M) = 10 to 44%
(H) = 45 to 100%

So, a risk of 67% would lie in the medium (H) band.

The greater the value, the higher the priority and therefore greater consideration should be given to the risk and its control.

Alternative approach:
An alternative could be to grade risk on a scale of 1 to 10 where 1 = 1% and 10 = 100%. A table to reflect degree of risk where the variables' likelihood of occurrence and severity of harm have been assigned ratings of low, medium and high can be established (see Table 3 opposite).

Calculation tables

Table 1: Assessment criteria for likelihood of occurrence

Value	Criterion
1	Remote – certain not to occur
2	Unlikely – in exceptional circumstances
3	Possible – in certain circumstances
4	Likely – ordinarily occur
5	Probable – high chance of occurrence
6	Highly probable – certain to occur

Table 2: Assessment criteria for severity of harm

Value	Criterion
1	Minor injury – no first aid required
2	Illness – chronic injury
3	Accident – first aid required
4	Reportable injury under RIDDOR*
5	Major injury – under RIDDOR*
6	Fatality

* RIDDOR: Reporting of Injuries, Diseases and Dangerous Occurrences Regulations 1995

Table 3: Determination of risk priority rating

	Likelihood of occurrence		
Severity of harm	H/L 10%	H/M 50%	H/H 100%
	M/L 5%	M/M 25%	M/H 50%
	L/L 1%	L/M 5%	L/H 10%

Control measures:
Once the risk priority rating has been determined, attention can then be given to configuring appropriate prevention and protection control measures.

Figure 4.10 Risk assessment methodology

method is used and that the method is simple and clear and leads to accurate assessment of the risks and the identification of suitable control measures.

Control measures

The principles of control for H&S risk are to: (1) avoid risk; (2) mitigate risk at source; and (3) control risk. Accordingly, control measures to manage risk focus on:

1 prevention;
2 protection.

Prevention focuses on measures to reduce the likelihood of occurrence while protection focuses on measures to reduce the severity of harm. It is clear that the simplest and most effective risk management measure is to avoid the risk in the first place, yet this is not as straightforward as it might appear. It very much depends upon the size and scope of the organisation, the nature of its outputs, how and where they are delivered, and the resources used in their provision. For example, a company involved in design, development and supply would need to go back to the design stage to see if the fundamental design of the product raised the potential for hazard and risk during the subsequent production processes. In contrast, a company providing component assembly services at a project site may identify risk only at the site of its provision in the work tasks undertaken. So, the rating of risk will vary among organisations, and the extent of risk may be limited to particular activities or extensive and widespread through many activities and processes.

Irrespective of the orientation of the organisation, it is essential, where practicable, that risk is assessed throughout the entire operation of the organisation with a view to avoiding risk being built in to any of its activities, processes or work tasks. Where actions to eliminate risk in a business process are not possible, then measures should be in place to combat the risk at its source. For example, administrative controls might be needed or personal protective equipment available such that risk is mitigated when and where it is encountered. Where risk is likely to manifest, irrespective of its careful consideration, then there must be measures in place to control the risk. For example, emergency preparedness and response to emergency situations would be pertinent to the control of risk.

Reporting on and learning from risk assessment and actions

The natural extension to risk assessment is the establishment of procedures designed to report on hazardous situations, safety incidents and accidents to people. These are a requirement for an OHSAS 18001 H&SMS. These important procedures are embraced by requirements under Clause 4 of OHSAS and will therefore be explained within system *implementation and operation.*

Implementation and operation

Establishing and maintaining an effective H&SMS relies to a great extent upon engendering and embedding a whole-organisation H&S culture and environment. The Health and Safety Executive, among many organisations, propounds that

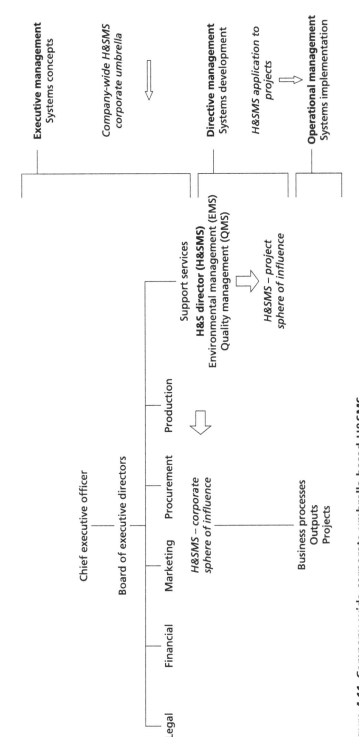

Figure 4.11 Company-wide corporate umbrella-based H&SMS

such a culture and environment is founded upon the four 'Cs': (1) communica-tion; (2) co-operation; (3) competence; and (4) control. Indeed, these elements embrace a great many organisational characteristics, attributes and variables which influence the approach to OH&S and the establishment of an H&SMS.

An organisation should consider the four 'Cs' along with other pertinent aspects but, moreover, must configure the method of H&SMS implementation and operation with particular regard to the requirements of OHSAS 18001. Essentially, the standard focuses on the organisation of the organisation and the requirements to establish H&S-related: structure; responsibilities; documents; com-munication; participation; and control.

Structure

A company needs to establish and maintain an organisation structure within which H&S is intrinsic. It has already been established that employers have a duty of care to ensure the safety, health and welfare of their employees. Also, that an individual has a duty for their own safety and well-being together with respons-ibilities for the safety and well-being of others. It is essential, therefore, that H&S is structured in ways that assign duties and responsibilities to the organisation, managers within the organisation and the individual employees who work within the organisation. Structure should focus on identifying and assigning management resources to the H&SMS with clear and defined roles and respons-ibilities and who have prescribed authority by the organisation and accountability to the organisation. Structure extends beyond the managerial framework, as every-one in the organisation has a responsibility for H&S matters as they influence their work. Consequently, the structure must reflect the intrinsic responsibility for H&S, from the chief executive at the top of the company to the operative at the workplace. This can be established, in concept and principle, by the com-pany organisation chart (see Figure 4.11). In practice, of course, embedding H&S within the organisation culture requires the support of hearts and minds within the total staff base and workforce. This important aspect will be examined subsequently.

The organisation chart, or what is often termed the company family tree, is a traditional method of illustrating the structure of roles, responsibilities, author-ities and accountabilities required by the H&SMS. Furthermore, the chart can be used to depict routes of communication and links of participation and con-sultation as required by the H&SMS standard. In practice, many organisations are large, complex and perhaps diverse, so much so that a sequence of organ-isation charts may be needed to illustrate and explain the structure for the entire organisation. It is common, therefore, for a set of charts to be configured focus-ing on the various managerial levels of the organisation, for example executive, directive, operations and first-line supervision. Alternatively, the company could be presented as organisational sub-divisions such as departments or work sections. The important point is that the structure of the organisation in respect of its position towards H&S is established and communicated to everyone under its control and to other interested parties.

Management representative

The management representative is a requirement of the OHSAS standard and a key individual within the organisation structure of the H&SMS. The management representative is a senior appointment within the organisation and often at executive level. Designated the company H&S director, or H&S manager, the management representative assumes total and single-point responsibility for all health, safety and welfare matters for the organisation. Although H&S throughout the operations of the organisation rely on a multitude of groups and individuals to deliver the safety agenda, the H&S director will take responsibility for oversight management of the H&SMS and be accountable to executive management. It is therefore a wide-ranging, onerous yet crucial managerial role within all companies and organisations. The H&S director will advise on and administer a plethora of activities at: (1) corporate level; (2) business output level; and (3) external level.

At corporate level, the H&S director's role entails:

- development of H&S policy and objectives;
- organisation structure for H&S management;
- engendering and embedding H&S culture within the organisation;
- ensuring that planning, based on H&S risk assessment, is applied to organisational activities, processes and work tasks;
- recruitment and/or assignment of capable and competent staff to H&S management;
- H&S induction and training for all organisational employees;
- implementing the report, investigation and analysis of all H&S-related incidents and accidents;
- review of H&S performance and audit of the H&SMS with a view to continual improvement;
- remaining cognisant of current and applicable H&S legislation, industry regulation and management system standards;
- advising on H&S matters as they affect the business outputs of the organisation at product, service or project delivery locations.

At output site level, for example in a project scenario, the H&S director's role entails:

- timely and effective liaison and interfacing between the corporate umbrella H&SMS and its applications to projects, contracts, service and product delivery sites;
- giving support to project site and first-line supervisors in their roles as H&S managers;
- overseeing monitoring, measurement, audit, analysis and review of all organisational projects and contracts;

- confirming effective liaison between all parties who interact with the organisation's H&SMS at project/contract level;

- ensuring site-based H&S induction and training for staff and workforce;

- developing effective co-operation and participation with H&S by other parties interfacing with the organisation and its business provision.

External to the organisation, the H&S director's role entails:

- interfacing with all outside bodies, institutions and organisations relating to H&S matters such as government departments, public authorities, regulatory bodies, customers, visitors to the organisation and the general public;

- interfacing with the media on matters relating to H&S and the organisation;

- reporting to external parties on matters of organisational H&S performance and operation of the H&SMS.

H&S management

In addition to the H&S director, the structure for H&S management will comprise a number of key H&S-related managerial and supervisory roles. For a large organisation there is likely to be an H&S management team responsible for designated sub-divisions of organisational responsibility. Managers and supervisors oversee the H&S activities of the team, and report to the H&S director. These positions may be supported by an H&S supervisor who oversees H&S at the project/contract level and then by first-line H&S supervisors who oversee work tasks. The exact configuration is unique to individual organisations and rightly is not prescriptive. The structure for H&S depends on many organisational characteristics. A company must determine for itself the extent, scope and orientation of management required. Further details on the roles of H&S managers and supervisors will follow subsequently.

H&S documents

As with all management systems the H&SMS needs to be defined, described and explained within the context of the business activities of the organisation. This formalisation of the system requires a set of written documents; indeed documentation is a requirement of OHSAS 18001. It was seen when examining quality management and environmental management in preceding chapters of this book that standards-based management systems are characterised chiefly by the documents which describe them. Consistent with the other management systems, H&SMSs are reflected in a set of documents which, normally, follow the convention of the document hierarchy, or pyramid (see Figure 4.12). The documents required are: (1) H&S manual; (2) H&S management procedures; (3) H&S implementation plan; and (4) H&S work instructions.

It is not intended here to divert into extended discussion of H&SMS documentation as the pertinent and significant aspects are commensurate with those

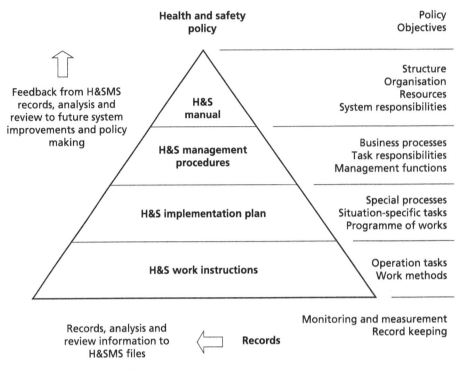

Health and safety
policy

Policy
Objectives

Feedback from H&SMS
records, analysis and
review to future system
improvements and policy
making

H&S
manual

Structure
Organisation
Resources
System responsibilities

H&S management
procedures

Business processes
Task responsibilities
Management functions

H&S implementation plan

Special processes
Situation-specific tasks
Programme of works

H&S work instructions

Operation tasks
Work methods

Records, analysis and
review information to
H&SMS files

Records

Monitoring and measurement
Record keeping

Figure 4.12 The H&SMS document pyramid

sections for quality management and environmental management presented earlier. It is essential to stress that H&S implementation plans and H&S work instructions are extremely important within an H&SMS since the majority of H&S matters are likely to occur at the business output delivery point, for example at a project site or service delivery location rather than in the corporate workplace. More will be said on this matter later when the unique characteristics of H&S on construction projects are examined.

With all management systems the relevant standard requires that documents be controlled. As such, OHSAS 18001 requires H&SMS documents to be subject to procedures which have been approved prior to use and are reviewed and updated. With an H&SMS the review and update of procedures is particularly important as legislation and industry regulation change frequently. These influences can have a profound effect on the way that business processes are conducted and on the ways in which they are managed, and the H&SMS needs to reflect these. The currency and appropriateness of the H&SMS are crucial to organisational performance. Any changes must be made first to the corporate-level umbrella system and then cascaded to implementation levels of the system. It may require some change to each level of the document pyramid and therefore can be complicated and time consuming. To ease the task of system update a company would find it best not simply to wait until change is invoked by, say, a variation of industry regulation but rather to accommodate systematic change through regular review and continual improvement to the H&SMS.

Management system documentation needs to be translated into workable practice documents for implementation at the workplace. As with other management systems, an appropriate way to achieve this is to use *administrative forms*, or what are sometimes referred to as *document pro formas*. These can be applied to all the key managerial and supervisory functions and used at the workplace to plan, monitor, control and report on all business activities, processes and work tasks. Examples of these forms will be presented and discussed subsequently.

Communication

Communication involves three distinct routes: (1) into the organisation; (2) within the organisation; and (3) from the organisation. While clear, transparent and reliable communications are required of all managers in all organisations and for all organisational activities, it is important that communications in relation to the specifics of the management function be appreciated also. Communications in respect of H&SMS, where applicable, often refer to the legislative and regulative requirements in addition to the requirements of the OHSAS standard. Communications with these and other specifics need to be correct and precise and above all managed.

Communication into the organisation is concerned with handling the plethora of information concerning H&S matters which is used to inform H&S policy, procedures and management practices. This information is likely to address current legislation, industry regulation and management system standards in addition to technical information influencing the business activities and processes and requirements of customers and the marketplace. It is important that all this information is managed and controlled in terms of its receipt and response. For this reason, all such information should be guided into the organisation at executive level and filtered and channelled to appropriate H&S managers within the organisation. Some information may be routine and repetitive while some may be irrelevant or redundant to ongoing practices. Only that information which is wholly applicable to current business activities needs to be circulated. The management of information into the organisation is one of the many important duties of the company H&S director. Information arriving at executive level can be passed to the H&S director, who considers the importance of the information and passes it, if required, to intra-organisational H&S managers.

Communication within the organisation is concerned with ensuring that information pertinent to H&S is structured and disseminated to managers and workforce throughout the organisation, sufficient for them to fulfil their roles, duties and responsibilities to H&S. This communication is absolutely essential to the effectiveness of the H&SMS. Communication must be directed to generating whole-organisation awareness of H&S and to information needed to deliver H&S management in the course of the everyday business. Awareness among employees is needed to develop the wider appreciation of H&S policy and objectives, engender support and commitment to H&S practices, and embed a positive H&S culture throughout the organisation. Communication of system-related information is needed to ensure that managers and supervisors have the wherewithal to fulfil their management responsibilities while operatives, who carry out the business processes, have the information needed to undertake their work tasks. Communication flows up, down and laterally within the structure of any organisation. Because this can be complex within a large organisation, the structure should formalise routes of communication within the organisation charts. It is also beneficial to do this so that all employees can relate to their location within the organisation and understand the routes of information both to and from their position.

Communication flow from the organisation is perhaps a less overt but important channel. H&S management requires in some circumstances that information is conveyed to outside bodies and organisations. An example of this might be a report on an accident to government authorities. Such information is obviously important but can also be sensitive to the corporate organisation. Therefore, outward information flow must be managed appropriately through messages that are correct and precise. Again, this is an aspect of communication best handled at executive level by the company H&S director.

In extensive and complex organisations it is not uncommon to find that external, or outward, communications are managed by a corporate communications department. This tends to assume a public-relations perspective in addition to information handling and communication management. This facility can be useful where communications need to be conveyed to the media or where technical information needs to be translated for public consumption. It should not, however, be charged with managing the flow of information into and within the organisation as this requires a technical and managerial understanding of H&S in the context of the business processes, and the communications department is unlikely to have such an understanding.

Participation

In many ways, communication is allied to participation in H&S matters. The participation of staff and workforce in resourcing and supporting the H&SMS is a prerequisite and quite obvious. The participation of any party or individual who interfaces with the H&SMS is less obvious but also essential to the effective implementation of the H&SMS. A working understanding and personal involvement with the H&SMS are key components of encouraging ownership of policy, shared objectives and effective managerial and work practices. An awareness of the H&S working practices of the organisation is useful among any interested parties who interface with it. This allows those who associate with the organisation to engage and operate within the same H&S framework. Indeed, contracted inputs to the organisation may need to demonstrate a commensurate H&S stance when pre-qualifying for a tendering opportunity.

Good communication is essential to eliciting participation in and support for the organisation's H&SMS. Where information is relevant to procurement and contracts with parties, then communication should be managed at a corporate level by the H&S director, as this involves corporate interfacing. Where information is relevant to work practices, then communication should be managed at operational level in the course of carrying out work practices.

Control

The output from risk assessment requires that those business activities, processes and tasks where risk has been identified are subject to operational controls. Operational controls apply to the organisation's business processes and procedures, the work of contractors and resource inputs from providers of goods, equipment and services. Effective control of H&S in all business processes and operational tasks is absolutely essential to the protection and well-being of employees and those that come into contact with the organisation's activities. Control in the form of prevention and protection measures is an important focus in H&S planning and risk assessment, as outlined earlier.

There are three principles of good control: (1) plan; (2) review; and (3) action. Every process and procedure needs to be carefully risk assessed and planned according to human vulnerability. They then need to be closely, regularly and consistently monitored and reviewed for conformity with planned performance. Where there is deviation from planned performance, then managerial action needs to be taken to restore the appropriate level of performance. This concept, sometimes referred to as plan, monitor and control, appears simple and straightforward. However, the application to practice in the busy workplace where there are many business processes being undertaken by a great many employees means that difficulties can easily arise. It is important to recognise that good control is a dynamic management tool. Actions implemented in response to a given circumstance may not resolve the problem within that one cycle of occurrence and following further review further action may be required. Control measures may appear to be static where preventive controls have been effective. They will appear active where protective controls are enacted to handle real occurrences such as an emergency response to an accident.

The effectiveness of H&S control measures depends very much upon the capability, competence and diligence of H&SMS managers and, in particular, first-line supervisors of work tasks. The focus of the supervisory role must be on active H&S monitoring, review and action taking. It might be easy to turn a blind eye to H&SMS non-conformance, and in the majority of occasions this may not lead to a catastrophic event. In limited situations, however, a simple lack of attention to duty or failure to meet a routine H&S responsibility could result in a major H&S incident or accident. To prevent such occurrence, a control measure must be sufficiently robust to ensure it fulfils the H&S requirement yet sufficiently simple and easy to administer to ensure it is enacted time and time again.

In support of effective H&S it is important that all staff understand the responsibilities that they assume in the workplace for both their own safety and the safety of others. This depends on the capability, competence and diligence of members of the workforce. Training needs to be provided which orientates an operative to the H&S risk and H&S responsibilities associated with the work tasks that they undertake. This is part of engendering an intrinsic organisational culture of H&S management.

Checking

The requirements for checking focus on: H&S performance measurement and monitoring; incident investigation; non-conformity; records; and auditing. Checking leads on naturally from the establishment of control measures but emphasises the gathering and analysis of evidence for the objective assessment rather than subjective assessment of system elements.

Performance measurement

OHSAS 18001 requires that the H&SMS establishes procedures to measure the organisation's performance for H&S on a regular and continuous basis. This is essential if the organisation is to achieve continual improvement in its H&SMS. The checking of organisational H&S performance on an ongoing basis requires that the principles of good control – plan, review and action – are effectively applied. Moreover, for observations to be meaningful, performance must be related back to the H&S objectives established during the H&SMS policy-making and planning stages. Qualitative and quantitative data will need to be acquired which can determine, by measurement, the extent to which the actual H&S performance has met the objectives set.

Key to data acquisition is a procedure which supports regular inspection of H&S in the context of application to the organisation's work processes and tasks. Administrative forms can be used to inspect activities at the workplace and gather information on the efficacy of H&S procedures as they are implemented. Active monitoring on an irregular basis through random sampling is also useful to spot-check that the system is not embedding complacency within activities as a result of routine scheduled monitoring. Data and information from inspections should always be retained in document form, and again the administrative forms can be used to action future attention and then filed for record-keeping purposes.

As inspection procedures often give rise to subsequent managerial actions, then checking must extend to monitoring and analysing any corrective and preventive actions implemented. This is essential to knowing and understanding just how effective management action has been in application. While highly successful H&S management will seek to obviate the need for any adjustment to planned procedures, the reality is that system analysis will throw up matters which do require modification to procedures and activities. The important point is that where actions have been taken these are checked, analysed and reviewed to ensure that optimum actions are absorbed into the system and that less successful actions are discarded. Likewise, the H&SMS should monitor and check ill health and well-being among employees in addition to safety aspects so that health, welfare and safety are holistically accommodated in the H&SMS checking processes.

All the information obtained by the checking processes will ultimately feed into management review such that long-term evaluation of the H&SMS takes place which informs and assists continual improvement of the system.

Incident investigation

OHSAS 18001 requires that the organisation implements procedures for recording, investigating and analysing H&S incidents. This supports the requirements of statutory legislation and regulation for reporting and investigating H&S accidents and major incidents. Deficiencies in the H&SMS which contribute to an incident need to be identified together with the appropriateness of any corrective and preventive actions used when an incident occurs. While some of the required information might be picked up during routine H&S monitoring, it is better practice for specific procedures to be in place. Such procedures will focus in detail on the particulars of an H&S incident. The activities leading up to the incident should be identified along with a determination of the root causes and effects. Once these are known, detailed analysis can take place to determine the underlying influences from which potential actions to prevent further occurrence can be established. The procedure for investigating H&S incidents must generate a detailed documented record of the incident, the investigation which takes place, the outcome of that investigation and the actions which ensue from it. Furthermore, subsequent monitoring must take place to follow the flow of the incident through to final resolution, for example the return to work of an employee following an accident.

Because of the inherent importance of matters concerning H&S, the outcomes of investigations should be used positively to inform the organisation and its employees of the lessons learned. A route of communication throughout the organisation needs to be established which disseminates such information. This can be achieved by regular generic and topic-specific electronic mail and internal newsletters to keep all staff and workforce informed.

In addition, there is a duty placed upon the organisation to report health, welfare and safety incidents and accidents to statutory parties outside the organisation. To meet this requirement duplicate records of any safety investigations and subsequent actions should be available to submit to outside bodies. These may, for example, contribute to independent investigations of major safety incidents by the Health and Safety Executive. Copies should be referenced and lodged in the appropriate organisational health and safety files.

Auditing the H&SMS

An important dimension to the requirements of checking is the internal audit procedure. The organisation should ensure that internal audits are carried out at scheduled intervals. These focus on determining if the H&SMS conforms to the organisational arrangements declared by the system and the requirements of OHSAS 18001. It is essential that the H&SMS operates as it should and, moreover, that objective evidence from monitoring confirms such effectiveness. The outcomes of internal audit should be fed back to corporate management to aid system review.

Only in this way can the performance of the H&SMS be assessed against the original performance expectations reflected by the organisational policy and objectives.

There should be a procedure, a schedule and plan of activities for auditing. These must establish that the approach is structured and robust and based on evidence from risk assessments, previous audit events and documented records of organisational H&S performance. Internal audit should be treated as a formal sub-process of monitoring and analysis which deploys a particular team to the particular task. It is important that the part of the H&SMS being audited is not assessed by its own staff – although an internal process, audits should be conducted by independent personnel.

Management review

Management review is conducted at executive, or senior, level with the express intent of critically evaluating the performance of the H&SMS. This is required to support the suitability and effectiveness of the H&SMS in the best and holistic interests of the organisation over the longer term. The anticipated outcome of management review is the assessment of performance of the H&SMS against organisational policy and objectives. This lays the foundations for developing and invoking improvements to the system.

Management review incorporates qualitative evaluation based upon experiences and observations and quantitative evaluation based on evidence obtained from internal audits, H&SMS performance records, H&S incident investigations, where applicable, and previous management reviews. Again, records should be kept of management review events, discussions and findings. Such information will be useful within the organisation for system improvement and outside the organisation in the support of independent assessments of the H&SMS, for example for the purposes of third-party assessment and system certification.

People and H&S: welfare and well-being

In the eyes of work-related legislation and in practice there is generally no differentiation between health and safety. Notwithstanding, the element that can be so conveniently omitted from the H&S perspective is welfare. What organisations really have to manage is occupational safety and occupational well-being. The Health and Safety at Work etc. Act 1974 imposes a duty on employers to ensure the health, safety and welfare of employees at work. The Management of Health and Safety at Work Regulations 1999 require the employer to ensure that employees are provided with health monitoring and management such that risks to health are identified and assessed in the same way that safety risk and assessment is provided. Consideration of the characteristics of well-being among its human resources is essential to any organisation. Safety-related incidents and personal accidents can have a profound impact on the individual long after recovery, while occupation-related ill health can impede an individual's capability to work in both the short and long term.

Welfare

An organisation is required to provide appropriate welfare facilities for its employees under statutory legislation and regulation. When one considers welfare one normally thinks of the provision of canteens, toilets and the like. However, welfare embraces the provision of facilities, equipment and associated procedures which are necessary as occupational safeguards against hazardous conditions and substances that may be encountered in the workplace. Where such safeguards are not taken, there would likely be resulting ill health and disease from exposure to hazardous materials. Therefore, the organisation must provide protection against such occurrences to its employees. This could include methods of working, the provision of personal protective equipment, workplace protection, post-work decontamination and cleaning, and washing facilities.

Ordinarily, welfare-related occurrences which could impact upon the health of employees would not pervade the whole organisation. Rather, they are likely to be confined to specific workplaces, for example a project site. Given this general characteristic, workplace-specific welfare provision will be planned and managed under the H&SMS implementation plan. This plan will, as previously explained, consider the particular risks attendant to the business output delivery site and put in place procedures to assess and mitigate the risk.

Well-being

Good health cannot be regarded merely as being without injury or disease, as that would be oversimplistic. Well-being is concerned with the all-round health and fitness of the individual. It incorporates the individual's physical condition and fitness to perform their work. It also includes their mental and emotional fitness to conduct their work in and around others and in the active workplace. An individual's state of mind – their psyche – is just as important as their physical condition, since almost all jobs require employees to think, rationalise, make decisions and interface and interact with other employees. So, well-being includes all the aspects that make the person fit, able and willing to engage in their work.

Occupational health management

The management of occupational health requires the organisation to establish procedures which plan, monitor and manage well-being among its employees. It must take all steps necessary to ensure that the physical, mental and emotional attributes of individuals are considered and safeguarded. An organisation should develop and implement an appropriate occupational health management strategy which embraces:

- planning for the obviation of health risks;
- monitoring of health among employees;
- communicating good personal health regimes for employees;
- self-management for good health;
- awareness education and training.

The monitoring of health among the staff base and workforce is an important management task. However, rather than adopting a top-down approach many organisations encourage self-monitoring and management. This entails the self-evaluation of occupational health and well-being by questionnaire surveys, self-directed testing and evaluation and self-arranged medical examination. Such regimes can be administered by the organisation or in conjunction with an appropriate outside body. Large organisations often support an occupational health clinic, or practice, where employees have easy access to medical, health and well-being facilities.

Reportable occupational health issues

An employer is required by the Reporting of Injuries, Diseases and Dangerous Occurrences Regulations 1995, otherwise known as RIDDOR, to report relevant incidents to the Health and Safety Executive (HSE) (HSE, 1995). Occurrences include accidents but also embrace aspects surrounding health and well-being, for example exposure to physical and biological agents and substances that make the person unfit to carry out their work. The occurrences which must be reported are:

- fatal and serious accidents;
- less serious accidents where a person is unfit for work for more than three consecutive days;
- dangerous incidents where persons are placed at risk;
- specified diseases associated with a person's job.

The H&SMS procedure required for RIDDOR requires the employer to inform HSE of the reportable occurrence. This is achieved by submitting a sequence of completed forms provided by HSE explicitly for the purposes of RIDDOR. These are submitted to HSE within specified timeframes according to the particular nature of the occurrence. A serious injury to a person or serious incident must be reported immediately, and this is done by contacting the Incident Contact Centre (ICC), a centralised national system. Reports can be submitted by telephone or by fax, e-mail or post. All RIDDOR-related incidents should be systematically and robustly investigated. A procedure to achieve this should be inherent in the H&SMS.

It is clear from the RIDDOR categorisations that many routine organisational activities undertaken on a day-to-day basis will fall outside these occurrences and therefore may not be reported so avidly, if at all. So, although the legal requirements place a duty on the employer to report RIDDOR-related incidents, many others may go unreported. To address this shortcoming and to ensure inclusivity in monitoring and reporting, the organisation requires a specific H&SMS procedure to pick up all health, welfare and well-being matters. While it would not be feasible to invoke the same formality for minor incidents as used for RIDDOR-related occurrences, it is still sensible to establish health-related record keeping in parallel with maintaining a safety 'accident book'. In fact, the

H&SMS procedure can use one or a set of document records differentiating safety-related accidents, ill health, diseases and well-being matters. These can be used in association with the HSE report forms necessary for supporting the RIDDOR procedures. In this way, all health and well-being matters in addition to safety matters will become part of good H&SMS monitoring, reporting and record keeping.

Hazardous substances in the workplace

Accompanying RIDDOR are the Control of Substances Hazardous to Health Regulations 2002 (COSHH) (HSE, 2002b). These regulations impose a duty on the employer to:

- assess health risk where an employee could be exposed to hazardous substances;
- prevent, or control, exposure to hazardous substances;
- take reasonable action to ensure the control of exposure;
- monitor employees' exposure to hazardous substances;
- ensure adequate information and training are provided to employees where there is potential exposure to hazardous substances.

COSHH applies to any and all substances that could cause adverse health effects. These can range from what might be considered to be quite innocuous but potentially hazardous substances, such as dust which can cause allergy, to highly dangerous biological agents capable of causing toxicity, infection or fatality.

The organisation's H&SMS must incorporate procedures for identifying and risk-assessing substances which could be hazardous to health. Moreover, procedures must be put in place to plan for the prevention, mitigation and control of substance-related occurrences. As with provision to combat RIDDOR-related effects to health, COSHH will need to consider methods of working, workplace protection, decontamination and personal cleanliness.

Personal protective equipment

Linked to welfare provision, the organisation may need to provide employees with personal protective equipment (PPE). The Personal Protective Equipment at Work Regulations 2002 (HSE, 2002a) require that employers provide their employees with personal protective equipment in the workplace where there is a risk to health. PPE ranges from the provision of basic equipment such as a safety helmet to a full air-respirator set for working in confined spaces with toxic vapours, for example.

Further details on statutory regulations

For further and precise details of the requirement for RIDDOR, COSHH and PPE the reader is directed to the respective regulations, which are available from HSE. The references section of this chapter provides details.

H&SMS: application to construction

This section examines an outline framework of considerations that need to be given by a principal contractor or contractor operating within the construction industry when establishing an effective safety management system meeting the requirements of OHSAS 18001. It is based on a composite of case studies.

H&SMS case study example

Introduction

This section presents an outline framework of consideration for the principal elements of an effective H&SMS meeting the requirements of OHSAS 18001. Examples for each element are provided as an indication of the considerations to be made and structures to be established when developing an appropriate system. Examples are also given of the administrative forms used in H&SMS implementation. As with other management systems presented and described earlier, readers who require complete sets of documents and template pro formas for system development are directed to private-sector consultancy practices which provide such documentation as part of their contracted services.

H&SMS – example based on case studies

This section focuses on an example H&SMS for application by the principal contractor within a construction-related practice. It is based on a composite of three case study applications of H&SMS, within principal contracting and one case study of a construction sub-contracting organisation. The example outlines a useful framework for H&S management by any principal contractor or sub-contractor.

H&SMS conceptual approach

The example describes a company-wide 'umbrella' H&SMS. This is appropriate to the corporate organisation and the many project-based construction projects that are undertaken by principal contractors and sub-contractors within construction.

H&SMS system standards

The companies contributing to this composite example of H&SMS establishment all utilise a system compliant with OHSAS 18001.

Relationship with other organisational systems

All of the companies contributing to this example also use ISO 9001 QMS and ISO 14001 EMS compatible management systems.

Influence of construction-related legislation and regulations

Unlike QMSs, which are ostensibly intra-organisationally based, H&SMSs are influenced fundamentally and importantly by generic and construction-industry-based legislation and regulations. These have a profound effect upon the way in which H&SMSs are developed and implemented. Legislation impacts directly on the duties and responsibilities of all employers towards their employees, while a plethora of construction-related health, safety and welfare regulations impinge upon the identification, assessment and mitigation of occupational hazards and risks and so fundamentally influence the planning of H&S management. In establishing the approach to H&SMS development and implementation the organisation must take account of these influences – they are statutory obligations, with implications and penalties for non-compliance based in both civil and criminal law.

In the example presented, the applicable legislation and regulations are intrinsically incorporated into the arrangements and structures shown. Comprehensive descriptions of specific legislation and individual regulations are not provided within the example as they would interrupt the narrative. However, the relevant generic and industry-specific legislation and regulations are outlined in accompanying appendices. Where elements of the H&SMS are required specifically to meet current legislation and regulations, then these are referred to.

It should be remembered that OHSAS 18001 specifies the requirements for an H&SMS in application to any industry sector. Particular industrial sectors will have their own requirements in terms of legislation and regulations in addition to generic health, safety and welfare requirements. The construction industry is influenced by a wide range of legislative and regulative requirements, all of which must be accommodated in terms of construction site practices and therefore incorporated into the H&SMS developed and applied.

Client and customer requirements

In addition to the requirements of legislation and regulations, there are client-based, or customer-based, related requirements to be met. These take the form of contractual conditions between the client and the contractors it engages to undertake the on-site construction works. Almost all public-sector clients and many private-sector clients make an H&SMS a pre-qualification requirement to tender opportunity or contractor selection. In addition, specific health, safety and welfare requirements may be identified by clients in association with particular types of works or location. For example, much construction work takes place in an active environment where the client's business processes are ongoing and must be sustained while construction work takes place. Such aspects can be accommodated within the contract's clauses themselves or in site rules governing the physical on-site construction practices of project participants.

System structure and main elements

OHSAS 18001 identifies the main elements of an appropriate H&SMS and the adoption of these featured throughout the systems used by the case study organisations. These were shown earlier in Figure 4.1 and are restated in Figure 4.13 as follows:

- policy;
- planning;

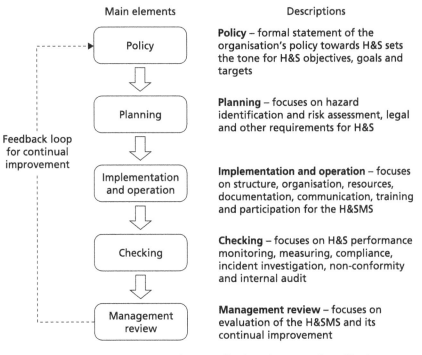

**Principal contractor's health and safety
management system**

Main elements

Descriptions

Policy

Policy – formal statement of the organisation's policy towards H&S sets the tone for H&S objectives, goals and targets

Planning

Planning – focuses on hazard identification and risk assessment, legal and other requirements for H&S

Feedback loop for continual improvement

Implementation and operation

Implementation and operation – focuses on structure, organisation, resources, documentation, communication, training and participation for the H&SMS

Checking

Checking – focuses on H&S performance monitoring, measuring, compliance, incident investigation, non-conformity and internal audit

Management review

Management review – focuses on evaluation of the H&SMS and its continual improvement

There are five key elements of an effective H&SMS meeting the requirements (clauses) of OHSAS 18001

In addition, the H&SMS must, within this structure, meet key industry-based regulation which requires the development and use of a *construction-phase health and safety plan*

Figure 4.13 Key elements of a structured H&SMS under OHSAS 18001

- implementation and operation;
- checking and corrective action;
- management review.

The main elements of the H&SMS given in OHSAS 18001 are well supported by the HSE publication *Successful Health and Safety Management* (HSE, 1997), which depicts an H&SMS comprising the following key components: policy; organisation; planning; measurement; audit and review. This publication is mentioned because it is often referred to by organisations seeking to develop their H&SMS, and all the case study organisations had referred to it. Indeed, many such publications are available and all, generally, follow the series of system elements presented in OHSAS 18001, although the wording or phraseology differs slightly among publications. The key point here is that the organisation has the flexibility within the standard to configure the H&SMS to suit its own requirements as long as the main system elements feature in its structure.

Policy

The contracting organisations all saw the overwhelming need to establish a clear and transparent H&S policy. These policies were seen not just to satisfy the requirements of the Health and Safety at Work etc. Act 1974 but also to establish the clear mood of the organisations in addressing the wider organisational management of H&S. A composite of the H&S policies enacted by the case study organisations was introduced and annotated in Figure 4.9 and is re-presented in Figure 4.14.

The main characteristics associated with the preparation of the H&S policy statement are as follows:

- To support a whole-organisation, or holistic, perspective to H&S management – for contractors this means that the H&SMS must be effectively applied to both the corporate family organisation and the many project-based temporary organisations established to deliver its construction services on site.

- To adopt and maintain a recognised standard of specification for its H&SMS such as that provided by OHSAS 18001 – for contractors this means that the H&SMS must embrace a set of structured documentation to describe the system (manual; management procedures; work instructions; and an implementation plan to accommodate the many project-specific applications to construction works on site).

- To appoint a designated H&S management representative for whole-organisation H&S matters – for contractors this means that the H&SMS must designate a director-level staff member who has single-point responsibility for the corporate umbrella system and application of that system to all its construction projects;

- To aspire to high levels of organisational H&S performance – for contractors this means that the H&SMS must ensure compliance with the legal and regulative requirements governing the on-site activities but furthermore seek to support continual improvement to H&S management on site.

- To recognise that occupational health, safety and welfare are important and integral dimensions of the effective delivery of the business output service – for contractors this means that the H&SMS must ensure the safeguard of all persons involved with the on-site construction processes and is achieved through effective risk assessment and control procedures.

- To provide appropriate resources to support whole-organisation H&S management – for contractors this means that the H&SMS must ensure that staff involved in H&S management are trained, capable and competent in H&S matters and that all employees based on site have the necessary knowledge, training, capabilities and competence to operate in what is, ordinarily, a hazardous working environment.

Principal Contractor's Health & Safety Policy

ABC Principal Contracting (International) Ltd is committed to ensuring safe and healthy work places and working practices. The Company meets its obligations under the Health and Safety at Work etc. Act 1974.

In so doing, the Company will:

1) Ensure the safety, health and welfare of all employees;
2) Ensure the safety, health and welfare of all persons not under the control of but involved with the Company at its workplaces.

- The Company is committed to planning for and using procedures for hazard identification and risk assessment which form the basis for establishing control mechanisms applied to the whole organisation and its business processes.
- The Company supports the proactive improvement of its health and safety performance.
- The Company will maintain its H&SMS to the specification of OHSAS 18001 (refer to H&SMS Manual and Procedures).
- The Company has in place a designated H&S Management Representative whose single-point responsibility extends to all aspects of H&S planning, organisation, resourcing and management.
- The Company has in place a structure and organisation for health and safety which places a duty and responsibility on all employees for the safety of themselves and that of their colleagues (refer to H&SMS).
- The Company will ensure that all resources deployed to H&S management are capable and competent in H&S matters.
- The Company will monitor, measure, audit and evaluate its H&S performance to ensure that its requirements meet with its policy, legislation, regulation and any other requirements pertinent to H&S.
- The Company will engage in continual improvement to H&S and its H&SMS.

The Company is committed to whole organisation H&S management, applying its H&SMS to both its corporate activities and its output services in accordance with current management system standards and the requirements of legislation and industry regulation.

Signed on behalf of the Company
Chief Executive Officer (CEO) .... Date ....

Figure 4.14 Typical health and safety policy statement

- To ensure that the organisation will periodically review and critically evaluate its H&S management approach – for contractors this means that the H&SMS must incorporate procedures which monitor, measure, audit and evaluate performance with a view to improvement of the H&SMS both within the corporate organisation and within its project-based organisations which carry out construction works on site.

It can be seen that all of these important characteristics feature within the H&S policy statement as shown in Figure 4.14.

Planning

Construction contractors, as major employer organisations, must adhere to general and industry-specific legislation and regulations. Such requirements are fundamental to the configuration of the H&SMS and therefore influence the process of planning. It was mentioned earlier that all employers must meet the requirements of the Health and Safety at Work etc. Act 1974. In addition, contractors must meet, among many requirements directly related to construction health, safety and welfare aspects, duties imposed by the Construction (Design and Management) Regulations 2007 (HSE, 2007). CDM 2007 is highly influential in the planning of construction works because it prescribes a particular approach to planning for H&S within the construction project environment.

Planning for H&S under CDM 2007 requires that particular activities are undertaken by designated participants to the construction processes as follows:

- *The notification of construction projects* – where construction works are proposed which involve non-domestic clients and will have a construction phase exceeding 30 days or 500 person-days then the client must notify HSE of the project (see Figure 4.15).

- *The provision of pre-construction information* – the client is required to provide designers and contractors associated with the construction works with information pertinent to the undertaking of the construction project. This project-specific information is essential to considering the hazards and risks that may arise during the works and which therefore impinge upon the planning of such works (see Figure 4.16).

- *The appointment of a CDM co-ordinator* – where construction works are notified, the client is required to appoint a CDM co-ordinator to oversee H&S management aspects relating to the construction project (see Figure 4.17).

- *The construction-phase plan* – the principal contractor is required to produce a construction-phase H&S-related plan prior to the commencement of the construction works on site. This plan describes how H&S will be managed throughout the construction phase and should be amenable to update and revision as the works on site progress (see Figure 4.18).

- *The H&S file* – where projects are notified, the client must maintain and retain upon completion of the construction works an H&S file for the project. The H&S file will comprise all H&S-related information relevant to the undertaking of the project, with information contributed by designers, consultants, the principal contractor and contractors/sub-contractors. The file is useful in informing similar construction works in the future but, moreover, is a requirement of CDM 2007 to assist in the provision of pre-construction information to those who may carry out subsequent works to the completed building or structure (see Figure 4.19).

- *The provision of welfare facilities* – in addition to H&S obligations the principal contractor has a duty under CDM 2007 to provide appropriate welfare facilities for persons on site during the construction phase (see Figure 4.20).

Notification of construction projects

Preamble:
The information presented below is typical of that provided by the CDM co-ordinator to the Health and Safety Executive (HSE) where construction projects must be notified under the Construction (Design and Management) Regulations 2007.

Information:
- Date of notification made to HSE
- Title name of project
- Construction site address
- Outline description of the project
- Outline description of the site works
- Client's name, address and contact details
- Designer's name, address and contact details
- Principal contractor's name, address and contact details
- Timeframe for principal contractor's planning for the project
- Planned start date for construction works on site
- Planned duration of construction-phase site works
- Anticipated maximum number of persons working at the project site
- Planned number of contractors working at the project site
- Details of contractors appointed to the project at date of notification
- Client declaration of obligations and responsibilities within the CDM Regulations

Notification:
Where construction works are proposed which involve non-domestic clients and will have a construction phase exceeding 30 days or 500 person-days, the client must notify HSE of the project.

Required documents:
Project notifications are made to HSE on standard Form F10.

Declaration:
In signing a notification to HSE the client acknowledges their role, duties and responsibilities within the requirements of the CDM Regulations.

Submission:
Notification is made by the CDM co-ordinator on behalf of the client organisation.

Figure 4.15 Notification of construction projects – typical information

A key component of construction H&S management which should be reflected within the H&SMS by appropriate planning procedures is a good understanding of and assessment of 'risk'. It was seen earlier that risk assessment involves three aspects:

1 *Hazard identification.*

2 *Evaluation of risk.*

3 *Prevention and protection measures.*

These aspects were described and discussed in the previous section, with the appropriate methodology for risk assessment presented in Figure 4.10.

Pre-construction information

Preamble:
The information presented below is typical of that provided by the client to designers and contractors associated with the project. This project-specific information is essential to the consideration of hazards and risks that may arise during the site works and so influences planning.

Information:
- Description of the project
  - Details of client, designers, consultants and CDM co-ordinator
  - Construction works start and completion dates
  - Minimum period from principal contractor appointment to start on site

- Client's project management requirements
  - Health and safety goals for the project
  - Site welfare provisions
  - Site security requirements
  - Client-specified safe working practices and procedures
  - Client-specified site rules and restrictions
  - Client-specified authorities for works
  - Project communications with project participants

- Project site/environmental aspects
  - Neighbourhood requirements, site boundaries and access
  - Statutory and local restrictions on site operation
  - Status/condition of existing structures affecting site works
  - Status/use of adjacent buildings/structures/land/facilities
  - Location and condition of existing services
  - Ground conditions, topography and structure
  - Storage on site of materials, components, hazardous materials
  - Waste products, handling, storage and removal

- Project site hazards and risks
  - Existing buildings/structures containing hazardous materials
  - Contaminated land
  - Existing storage of hazardous materials/substances
  - Hazards and risk from client's ongoing activities (if active project environment)

- Design and construction hazards
  - Design assumptions, suggested on-site working method and sequences
  - Co-ordination of design/design changes with construction phase
  - Design-identified hazards and risk to construction-phase works on site
  - Design-identified hazards from specified materials/components/substances

- Health and safety documentation
  - Specification of content and presentation of information for client's project health and safety file

Figure 4.16 Pre-construction information – typical information

CDM co-ordinator
Key duties

Preamble:
Where construction works are notified, the client is required to appoint a CDM co-ordinator to oversee health and safety management aspects related to the construction project.

Duties and responsibilities:

• To advise the client on its duties and responsibilities under the regulations

• To notify HSE where projects are required to be notified

• To co-ordinate health and safety aspects of design, and communicate with and co-ordinate other participants to the project

• To maintain effective communications between the client, designers, consultants, principal contractor, contractors and others associated with the project

• To maintain effective communications with external bodies and organisations involved with the project

• To liaise with the principal contractor concerning ongoing design matters

• To generate, co-ordinate and communicate pre-construction information

• To oversee all aspects of health and safety on behalf of the client

• To prepare, co-ordinate and submit the health and safety file

Figure 4.17 Key duties of the CDM co-ordinator

Risk assessment – hazard: categories; activities; types; and register
A clear and useful way in which the principal contractor can consider construction hazards and their risk is to gather information within selected categories, related to specific activities and to a particular type from which a register of site hazards can be compiled. To give continuity to the technological aspects of the work, the risk assessment may be directly related to the construction method statement which outlines the main elements of the works to be conducted within the project.

Hazards can be categorised in many ways but the case study contractors typically sorted their hazards into groups which reflect the particular nature of construction works as follows:

• hazards related to work activities;
• hazards related to work environment;
• hazards related to special (project-specific) processes;
• hazards related to work movement of persons and plant/equipment;

Construction-phase plan

Preamble:
The principal contractor is required to produce a construction-phase health and safety plan prior to the commencement of the construction works on site. The plan describes how health and safety will be managed throughout the construction phase. The construction phase plan can be developed from the H&SMS project-specific implementation plan.

The information presented below is typical of the key sections within which the development of the plan is considered.

Key sections:

Contents

- Introduction

- Project details

- Health and safety management system (H&SMS)

- Management system and communications

- Assessment and reporting

- The construction site

- Health and safety file

- Management review

- Performance monitoring

- Appendices
  - Organisation charts
  - Construction method statements
  - Health and safety method statements
  - Construction programmes
  - Site layouts
  - Legislative requirements (e.g. COSHH/RIDDOR)

Note:
Sub-sections of information within the sections noted above together with additional information necessary to the plan are presented in this section of Chapter 4.

Figure 4.18 Construction-phase plan for health and safety – typical information

- hazards related to use and movement of materials and components;
- hazards related to capabilities, competence, attitudes and behaviour of people;
- hazards related to works governed by specific regulations (substances, public safety and protective equipment).

Health and safety file

Preamble:
Where construction works are notified, the client is required to maintain and retain upon project completion a health and safety file for the project. The information presented below is typical of that provided.

Information:

- Title and particulars of the project

- Description of the construction works carried out

- Pre-construction information

- Design-related risk assessments

- Construction-phase plan

- Copies of hazard identification, risk assessments and control measures implemented

- Records of accidents and health and safety incidents, their investigation and analysis

- Details of actions taken and their effectiveness

- Records of performance evaluations and audits

- Details of hazardous materials/substances incorporated giving risk during future works

- Details of residual hazards and risk to future works on or use of site

- Details of structural solutions which impinge upon future works to the building product

- Details pertinent to the dismantling/removal of installed machinery/fittings/equipment

- Requirements for the maintenance and cleaning of the building product

- Details of the type, location, marking and protection of existing services and utilities

- Outputs from health and safety aspects of client's project progress meetings

- Outputs from health and safety meetings/committees and communications

- Health-and-safety-related details from project documents such as site diaries

- Authorisation, location and accessibility of the health and safety file

Note:
The above list is not exhaustive and will vary according to the type, nature and scope of the construction project.

Figure 4.19 Health and safety file – typical information

Provision of welfare facilities

Preamble:
The principal contractor has a duty under the CDM Regulations to provide appropriate welfare facilities for persons on site during the construction phase.

Key provisions:

- Site accommodation

- Sanitary conveniences

- Washing facilities

- Drinking water

- Rest facilities

- Storing and changing clothing

- Drying facilities

- Mess facilities

- Personal protective facilities (against materials/substances)

- First aid equipment

- Accident record book

- Emergency contact point (telephone)

- Personal secure storage

- Welfare provisions for contractors, client's representatives, consultants and site visitors

Notes:

The above list is not exhaustive but an indication of key areas of welfare provision.

Welfare facilities must be established in conjunction with other key aspects of project preliminary items including: site rules; safe working procedures; training; health and safety communications and co-ordination; emergency procedures; and procedures for handling accidents/dangerous occurrences on the site.

Figure 4.20 Welfare facilities – key provisions

Specific work activities and associated types of hazard and/or aspects to consider may be recognised, again, in many ways and the following listing is a composite of the way in which the case study respondents chose to consider hazards on their construction projects:

Demolitions:

- use of heavy plant/equipment
- building/structure collapse
- falling materials
- structural support/shoring
- work within buildings
- persons operating around demolition site
- noise, dust, pollution

Excavations:

- safe excavation practice
- earthwork shoring
- soil containment
- trench collapse
- falling materials
- visibility and obstructions
- presence of services
- removal of spoil
- backfill procedures

Working at height:

- use of ladders and hoists
- assembly of scaffolding
- safety screens
- scaffolding guardrails and boards
- work in stairwells and lift-shafts
- slippery surfaces
- openings in floors/roofs
- mobile work platforms
- use of safety harnesses and lifelines
- overhead power/service lines

Plant and equipment:

- appropriate selection
- responsible usage

- maintenance, service and repair
- supervision
- movement on/around site
- safe operation practice/speed

Manual handling:

- safe lifting practice
- excessive lifting of weight/size/awkward loads
- aids to move large/difficult loads
- team lifting of awkward/large components

Working in confined spaces:

- cramped working positions
- presence of toxic gases/vapours/substances
- working in standing/foul water
- presence of chemicals
- working in lift shafts
- operations in underground areas/drainage services

Use of electricity:

- faulty insulation of services
- presence of water
- humid environments
- working with transformers/circuit-breakers
- power supplies at height/in confined spaces/awkward access
- electric conductive/sparking clothing/boots

Materials and substances:

- chemicals
- petrol/diesel/oils/lubricants
- paints/cleaners/thinners
- adhesives
- removal/disposal of hazardous materials/chemicals

Fire:

- hot-work activities
- combustible materials
- flammable substances
- waste/rubbish storage/removal
- spillage of flammable liquids

Site housekeeping:

- workplace tidiness
- safe site access/egress
- plant/tool/equipment storage
- waste management
- warning signage

Natural and human influences:

- cold and heat extremes
- wind
- sun
- inclement weather
- physical distress
- work limits/rest periods
- welfare requirements

Public safety:

- site boundaries
- road/site traffic
- signage
- physical barriers
- walkways
- lighting
- unauthorised access
- environmental related issues (noise, dust, vibration)
- working periods/start-finish times

The types of hazard may be arranged, or grouped, into any configuration suitable to the needs of the organisation. The important point is that, irrespective of grouping, each of the potential hazards is identified and recorded so that appropriate consideration can be given to risk assessment and risk mitigation through the development of measures of prevention and/or protection. In this way an activity, taken from the construction method statement, is highlighted, the potential hazards associated with undertaking the activity are considered, the persons at risk identified, the level of risk determined and the control measures to mitigate the risk described. This approach can be used to consider every activity, or work task, in the method statement or for groups of activities, for example excavation-related items. The outcomes from risk-assessed work items are then used in the development of the H&S implementation plan to augment the generic manual, management procedures and work instructions of the documents which form the H&SMS.

Implementation and operation

Management structure for H&S

The management structure for H&S within principal contracting organisations must embrace the requirements of: (1) the company, or corporate, organisation; (2) the project organisation for undertaking construction works on project sites; (3) the contract arrangement with the client and other participants to the construction projects it undertakes; and (4) CDM 2007 and other key legislation and regulations governing construction industry business and activities.

The arrangement for H&S corporate governance was described earlier and presented in Figure 4.11. With appropriate links to the corporate organisation, this section focuses on the effective arrangement for H&S management at construction project level, incorporating the four requirements outlined above. An H&S organisation chart can reflect this arrangement as shown in Figure 4.21. The organisation chart reflects a hierarchy of H&S management from corporate level through to the undertaking of construction tasks on site. A configuration can take many forms. Conventional practice, exhibited by the case study organisations, is to support a structure where the management representative for whole-organisation H&S oversees the holistic H&SMS implementation, with delegated responsibilities given to project H&S supervisors, who then oversee first-line construction supervisors who manage the various teams of construction operatives.

The H&SMS is embedded within this structure, and culminates in the implementation of safe systems of work which merges H&S requirements with the technical aspects of carrying out construction activities and tasks. The H&S organisation chart can therefore be augmented by an accompanying chart which interlinks the management hierarchy with the key elements for H&S practice on site. This is shown in Figure 4.22. The key elements that contribute to effective H&S construction practices on site are: risk assessment; method statements; permits to work; training; safe working procedures; site rules; and safety monitoring and auditing. These aspects are described subsequently.

Responsibilities of H&S managers and site-based construction personnel

Within the H&S organisation chart the responsibilities of managers and construction personnel are as follows:

- *Health and safety management representative* – responsible for providing oversight management of all H&S matters for the organisation, and contributes to the development of: policy; culture; organisation structure; resource provision; planning; audit; and review.

- *Health and safety supervisor* – responsible for providing management of H&S on one or a number of projects, and contributes to the implementation of: H&S procedures, work instructions and site plans; communication and co-operation between project participants; performance standards setting, monitoring and review; and training for construction site personnel.

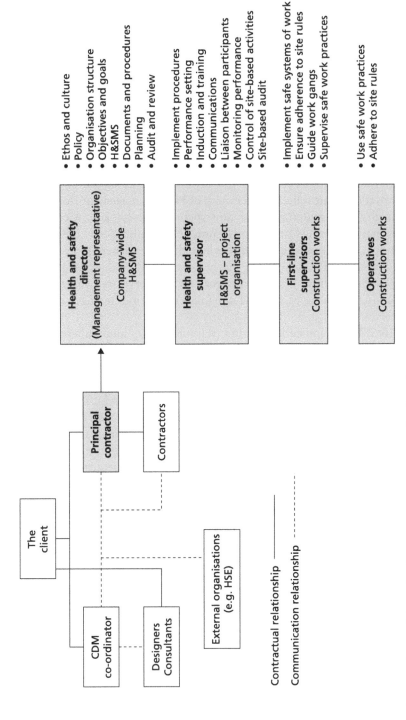

Figure 4.21 Health and safety management organisation chart

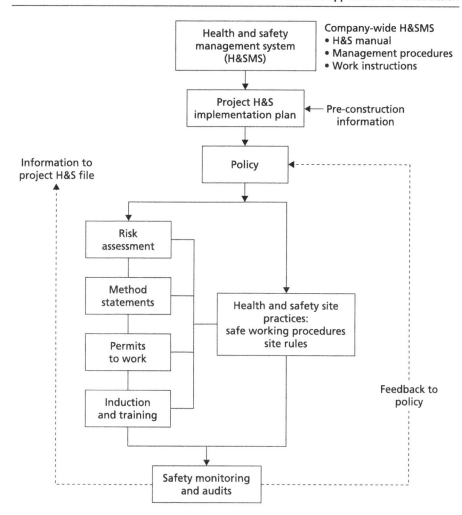

Figure 4.22 Key health and safety site practices

- *First-line supervisors* – responsible for providing management of H&S during the on-site activities and tasks undertaken by construction operatives, and contribute to the implementation of: H&S procedures; work instructions; and safe working practices;
- *Construction operatives* – responsible for observing H&S during on-site working by implementing safe working practices.

Implementation of the Construction Skills Certification Scheme

When considering the knowledge, skills, capabilities and competencies of the site workforce, appropriate training and certificates should be held by construction operatives. The Construction Skills Certification Scheme (CSCS) provides for persons who have successfully completely training and assessment to specified standards to become certificated and carry a CSCS card. The principal contractor may during pre-construction resource planning for the project choose its operative base from holders of CSCS cards. Likewise, the client may demand that the principal contractor employs such card holders, or the principal contractor may make this a requirement of the workforce of the contractors which it contracts.

The ownership of a CSCS card should not, however, obviate the need for the organisations based on site to provide additional general training, specific training, induction training and tool-box briefings. All of these are essential to meeting the requirements of OHSAS 18001.

H&S management system documentation

The case study contractors generally observed the conventional 'document pyramid' described earlier in this chapter. It should be noted that the variations in collating the H&SMS documents were apparent to accommodate the particular requirements of the CDM Regulations and to reflect the differences in the range of construction projects that were undertaken. Details of the documentation used are shown in Figure 4.23 and outlined as follows:

- *H&S manual* – incorporates: corporate policy; organisation structure; management responsibilities; and skeleton of general management procedures.

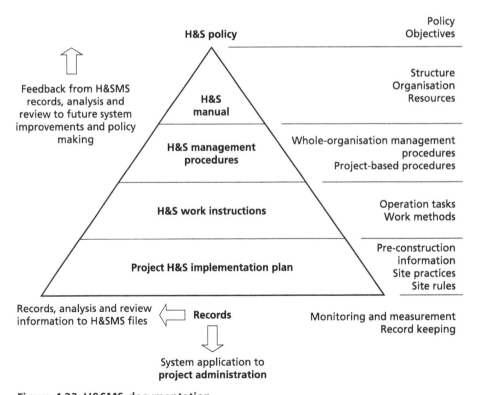

Figure 4.23 H&SMS documentation

- *Management procedures* – incorporate: (1) management procedures to be used in identified areas of organisational activity at (a) whole-organisation level and (b) project level; (2) organisational support services, for example personnel and training; (3) service management, for example finance, legal and marketing; (4) project management, for example client–contractor liaison; and (5) system management, for example document control.
- *Work instructions* – incorporate work instructions necessary to undertaking construction activities and tasks on site.
- *Project H&S plan* – incorporates: defined and described characteristics of any particular construction project; and revised and/or additional management procedures and work instructions to accommodate working on that project.

The H&SMS project H&S plan and the CDM Regulations requirements for the construction-phase H&S plan

The CDM Regulations require that the principal contractor completes a construction-phase H&S plan prior to the commencement of construction works on site. The construction-phase plan co-ordinates the 'pre-construction information' provided by the client and design participants with the H&SMS manual, management procedures, project-specific implementation plan, working instructions and the technical and managerial aspects of the construction project. The construction-phase H&S plan is intended to set out the general framework and arrangements for ensuring effective health, safety and welfare during the construction phase of the project. The CDM co-ordinator, appointed by the client, will confirm in writing that the plan satisfies the requirements of the regulations and that the documents have been developed sufficiently to allow works to commence on site. Possession of the construction site and authority to operate on the site are dependent upon this. The level of detail provided by the plan is determined by the H&S risks of the particular project. A construction project involving minimal risk will require a simple and straightforward plan while a large and complex project involving substantial risks will call for a more extensive plan.

The following list presents an actual example of the typical contents of such an H&S plan:

CONSTRUCTION-PHASE HEALTH AND SAFETY PLAN
CONTENTS
1 INTRODUCTION
2 PROJECT DETAILS
   2.1 Project Title
   2.2 Site Address
   2.3 Client
   2.4 Contract Administration
   2.5 CDM Co-ordinator
   2.6 Design Consultants
   2.7 Project Description and Scope of Works
   2.8 Extent of Principal Contractor's Design Responsibility

H&S practices

An effective H&SMS applied to the construction processes is centred on developing and implementing *safe systems of work*. A safe system of work is a healthy, safe and welfare-conscious approach to all construction activities and tasks, embedded in the project-empathic management procedures and work instructions of the organisation's H&SMS. The key effective H&S practices that deliver safe systems of work are: (1) risk assessment; (2) method statements; (3) permits to work; (4) training; (5) safe working procedures; and (6) site rules. In addition, inspection, monitoring and review of H&S performance are key practices and these are considered within the H&SMS elements of: checking and corrective action; and management review.

The organisation will need documents to define, describe and detail each H&S practice. The documents used for this purpose are a set of administrative forms, or system pro formas. These are also used as the basis for: checking H&S performance during the undertaking of construction activities and tasks; recording information during routine inspections; conducting audits of practices in application to work processes; and evaluation of system effectiveness to inform management review. The practices are as follows:

- *Risk assessment* – the function of this practice is to provide, for each construction activity or groups of activities, details of: potential hazards; persons at risk; level of risk; risk controls in the form of preventive and/or protective measures (see Figure 4.24).

- *Method statements* – the function of this practice is to provide, for each construction activity or groups of activities, details of: area of specialisation within the works; sequence relationship to other works; supervisory and monitoring arrangements; references to particular OH&S standards and requirements; control measures; response/emergency procedures (see Figure 4.25).

- *Permits to work* – the function of this practice is to provide, for identified construction activities, details of: works permitted; location; description of specific

Principal contractor's risk assessment					
Project title:					
Principal contractor (name):					
Assessor (name):					
Construction work element	Potential hazards	Persons at risk	Risk rating		Control measures
			L	S	R	
Sources of information					
Legend: L – Likelihood S – Severity R – Risk (likelihood × severity)			Document ref. no.:		
			Date:		

Figure 4.24 Principal contractor's risk assessment – suggested administrative form

Principal contractor's safety method statement			
Project title:			
Principal contractor (name):	Approved:	Yes	No
Construction method statement – link reference:	Document ref. no.:		
Assessor:	Date:		
Construction work element (description/details):			
Construction method (outline from construction method statement):			

Aspects assessed:	Requirements satisfied Yes / No / In-part* / NA (if In-part, provide details of actions required)
a. Potential hazards:	
b. Persons at risk:	
c. Supervisory arrangements:	
d. Monitoring arrangements:	
e. Requirements of regulations:	
f. Schedule of personal protective equipment:	
g. Schedule of plant/equipment/small tools:	
h. Workplace protection arrangements:	
i. First aid requirements:	
j. Emergency arrangements:	
k. Specialist arrangements (where required):	
l. Work approval and sign-off arrangements:	

Permit to work required: Yes   No
Requirements for training (brief details):

Figure 4.25 Principal contractor's safety method statement – suggested administrative form

hazards; precautions to be taken; reference to additional permits required (see Figure 4.26).

- *Training* – the function of this practice is to provide, for general or specific construction activities, details of: type of training required prior to under-taking construction activities, or induction training, and ongoing training; the trainee and training provider (see Figure 4.27).

- *Safe working procedures* – the function of this practice is to provide, for each construction activity or group of activities, details of: the work activity/task; location; description of work; safe working procedures, or methods, to be used (see Figure 4.28).

- *Site rules* – the function of this practice is to provide, for the coverage of all site activities, a set of rules for general site operation which all persons present on site must adhere to (see Figure 4.29).

Welfare

The H&S plan should reflect the general arrangements for the principal contractor's site establishment. Efficient and effective site establishment lays the management foundation for a successful construction project by configuring, structuring and organising those temporary welfare facilities needed to support the works on site. It is not intended within the scope of this book to describe and discuss the provision of welfare facilities, which are fundamental to all construction pro-jects, and include: preliminary items; site organisation; site layout; and provision of welfare facilities. Notwithstanding, while some aspects fall within the discre-tion of the principal contractor, the provision of welfare is a legal obligation under a number of significant regulations. These are presented in the appropriate appendix.

Planning for construction projects requires that:

- welfare arrangements are accommodated within the construction-phase health and safety plan;

- welfare facilities are provided, in accordance with applicable regulations, prior to commencement of works on site, and are maintained throughout the duration of the project;

- welfare facilities must reflect the requirements of and number of persons on site.

Communications and co-operation

General communications among the principal contractor's personnel based on site are essential to successful H&S management. To facilitate this, the case study contractors used a number of mechanisms, as follows:

- the establishment of an H&S committee with members drawn from the various management levels of the corporate and project organisations and site operatives;

Principal contractor's permit to work						
Project title:						
Principal contractor (name):						
Permit to work (issued for):			Scope of works:			
Issuing manager:			Date:		Document ref. no.:	
Location of the works:						
Description of works and identified hazards:						
Precautions required:						
Works conducted in active environments: (where plant, equipment or systems are in operation)						
Additional precautions required for active environment working:						
Additional permits required:	Yes	No	Hot work	Electrical	Confined spaces	Other
Authorisation: Issuing person .... Designation .... Date issued						

Figure 4.26 Principal contractor's permit to work – suggested administrative form

Principal contractor's induction and training requirements						
Project title:						
Principal contractor (name):				Approved:	Yes	No
Training category:	Induction	Routine	Specialist	Document ref. no.:		
Issuing manager:				Date:		
Reference to construction project/activities/works (description):						
Type of training (details):						
Trainee (name)		Trainer (name)		Date of training		
Follow-up actions/additional training needs identified:						

Figure 4.27 Principal contractor's induction and training requirements – suggested administrative form

Principal contractor's safe working procedures		
Project title:		
Principal contractor (name):		
Issuing manager:	Document ref. no.:	Date:
Reference to construction project/activities/works (description):		
Location of the works:		
Description of works:		
Safe working procedures/methods:		
Safety method statement reference (if applicable):		
Permit to work reference (if applicable):		
Training requirements reference (if applicable):		

Figure 4.28 Principal contractor's safe working procedures – suggested administrative form

Principal contractor's site rules
Project title:
Principal contractor (name):
The *site rules* specified below apply to ALL persons on site at ALL times: • Only those persons holding a Construction Skills Certification Scheme (CSCS) card will be permitted to work on site – cards should be carried at all times while on site and carried off site where works are undertaken in connection with the project. • Personnel must, on their own behalf only, sign in to this site and sign out from this site at the designated site access point. • Personnel must have received a site induction prior to commencement on site and commit to continuing routine and specialist training as required by the company. • Personnel must be present on site only within the periods specified by and authorised by the company. • Personal protective equipment (PPE), as prescribed by line managers, must be worn on site at all times. • The consumption of alcohol, intoxicating or narcotic substances is prohibited on site at all times. The site is a no smoking area. • The use of portable radios and similar personal electrical equipment is not allowed on site at any time. The use of mobile phones is not permitted on site except for use in designated areas such as site offices, accommodation and mess rooms. • Food and beverages should not be consumed on site except in designated welfare facilities. Personal litter should be disposed of responsibly in designated rubbish bins provided on site, and all personnel are charged at all times with maintaining a tidy site and tidy workplaces. • Personnel are responsible for appropriate and safe storage of small tools, equipment and work materials when not in use on site. • Personnel must observe the procedures and work instructions specified by the H&SMS. • Personnel must implement safe working practices on site. • Personnel are required to bring to the attention of managers and supervisors any matter arising on site in connection with the requirements of RIDDOR, COSHH and any other regulation in force. ALL personnel are reminded that they have a statutory duty of care for the health, welfare and safety of themselves and of others when working on site. Personnel found to be in breach of site rules will be subject to appropriate disciplinary action.

Figure 4.29 Principal contractor's site rules – suggested items

- the use of regular site-based H&S meetings to raise and discuss safety, health and welfare matters as they arose;
- the use of a central construction site safety notice board to display key health, safety and welfare information to site staff and workforce;
- the use of information boards around site to display specific H&S information, site rules and working practices.

Communication to other parties on site is also essential to achieving effective H&S practice. Therefore, clear, regular and detailed communications must be maintained with the client and with contractors, together with targeted communications to others visiting the site. Good communication is a prerequisite to perpetuating good co-operation between participants on site. In meeting these requirements, the case study contractors implemented the following:

- pre-start meetings with all project participants to highlight and discuss H&S matters in the context of undertaking the project on site;
- provision of pre-commencement H&S information highlighting identified hazards, risk assessments, site rules and safe working systems and practices;
- H&S routine items within the client's monthly project progress meetings;
- regular scheduled H&S management meetings;
- use of induction/awareness training and ongoing/update training in H&S as it affects on-site activities and tasks;
- use of tool-box talks to inform workforce of site rules and safe working practices in relation to specific site tasks.

Checking and corrective action

The checking, measurement and audit of H&S performance on site are features of an effective H&SMS. Each is essential to the critical evaluation of the H&SMS in use and to invoking continual improvement to the system. Additional good H&S practices on construction projects involve site inspection, investigation of incidents and audit of H&S performance, as follows:

- *Site inspection* – the function of this practice is to provide, in respect of construction processes and tasks, for: the identification of unsafe conditions of work; remedial actions taken; further actions to be taken (see Figure 4.30). Such inspections will, where appropriate, use quantitative measurement, testing and analytical methods to identify aspects of the H&SMS which do not conform to planned levels of performance.
- *Investigation of incidents* – the function of this practice is to provide for the detailed investigation of: construction activities where unsafe conditions of work are identified; H&S incidents and/or accidents that have occurred in relation to works on the construction site. Such reports are detailed and include: the type of incident/accident; persons involved; severity of harm; how the situation arose; causes; actions recommended; and follow-up actions/review of recommendations and actions taken (see Figure 4.31).

Principal contractor's site safety inspection		
Project title:		
Principal contractor (name):		
Issuing manager:	Document ref. no.:	Date:
Reference to construction project/activities/works (description):		
Location of the works:		
Status of the works (unsafe conditions):		
Remedial actions:		
Effect of remedial actions:		
Further actions required:		

Figure 4.30 Principal contractor's site safety inspection – suggested administrative form

Principal contractor's incident investigation report			
Project title:			
Principal contractor (name):			
Issuing manager:	Document ref. no.:	Date:	
Type of incident:			
Severity:	Minor	Serious	Major
Description of incident and its occurrence:			
Immediate causes: (Unsafe acts or conditions)			
Secondary causes: (Human, organisational or job factors)			
Recommended remedial actions: (Preventing recurrence)			
Follow-up actions (Review of recommendations and progress)			
Probability of recurrence	Low	Medium	High

Figure 4.31 Principal contractor's incident investigation report – suggested administrative form

- *Audit* – the function of this practice is to provide qualitative and quantitative information to evidence the performance of the H&SMS in operation as a basis for critical evaluation and management review. Audits can embrace the H&SMS as applied to construction projects and the corporate umbrella H&SMS. Audits can be wide ranging in scope to examine many aspects of the H&SMS in use. For example, an audit can focus on: safety policy; competence of employees; communication procedures; H&S planning; welfare facilities; or safety performance during construction activities (see Figure 4.32).

Management review

Management review is a procedure which focuses on making judgements about the performance of the H&SMS in the whole-organisation context and as applied to each construction project undertaken. On a project basis, the CDM Regulations require that the client maintains a project H&S file to which all participants to the construction project contribute. From the principal contractor's perspective all the information gathered from the H&SMS documents and records, such as site inspections and audits, will be submitted as evidence of its H&S performance on site. On an organisational basis, all information gathered during the undertaking of the construction works will be reviewed. This will be augmented by associated information from the organisation's support, service and system management sections such that a whole-organisation perspective of H&SMS operation may be assembled. The overall goal of management review is to provide evidenced judgements of potential improvements that may feed back into: future policy making; objective setting; organisational resourcing; and H&SMS documentation, procedures and site practices. In this way the requirements of OHSAS 18001 for H&SMS continual improvement can be encouraged and perpetuated.

A final thought

In summing up *Successful Health and Safety Management*, one of the first and foremost practical guides for managers developed by HSE, its deputy director general commented:

> The message it conveys is a simple one: organisations need to manage health and safety with the same degree of expertise and to the same standards as other core business activities, if they are effectively to control risks and prevent harm to people.
>
> (HSE, 1997)

Principal contractor's audit
Project title:
Principal contractor (name):
Issuing manager: Document ref. no.:   Date:
Aspect/area/activity audited:
Scope of the audit:
Audit planning and methods used:
Non-conformity/non-compliance identified:
Recommended actions:
Review of actions taken:
Further action required:

Figure 4.32 Principal contractor's audit – suggested administrative form

Safety management: key points, overview and references

This section presents: a summary of the key points from the collective sections of Chapter 4; an overview of Chapter 4; and a list of references used in the compilation of Chapter 4.

Key points

- Safety management is a fundamental and paramount management function within any organisation, and in particular within construction-industry-related organisations, where occupational hazards and risks are considerable and ever present on construction project sites.

- Although occupational accidents are by definition unforeseen events, the simple fact is that occupation-related accidents and incidents are avoidable and can therefore be managed in a positive and moreover proactive way.

- There are many obstacles to the achievement of good and effective occupational safety: these emanate from customers, clients, production targets, financial constraints, legislation and regulation, and all impinge upon the providers of products and services.

- There are strong and powerful business incentives for organisations to deliver better safety performance, and while these are chiefly legal and moral obligations there are also financial, commercial and marketplace benefits.

- The organisation which achieves effective safety performance is normally one which employs a structured approach to safety management through the implementation of a formalised H&SMS.

- Occupational safety includes three important dimensions: safety, health and welfare.

- Successful safety, health and welfare rely upon everyone in the organisation taking responsibility for themselves and others, and employers taking responsibility for all persons within their span of control.

- Effective H&SMSs are those which meet the management system standard OHSAS 18001.

- An OHSAS 18001 commensurate H&SMS is based on the plan–do–check–act (PDCA) methodology allied to adopting a process model approach to business activities and their management.

- The OHSAS 18001 standard is compatible for use alongside ISO 9001 QMS and ISO 14001 EMS.

- An effective H&SMS can be a company-wide umbrella system or a project-based system, or a combination of both, where in construction a company-wide H&SMS is generally adopted and then applied to the many individual construction projects which the company undertakes.

- An OHSAS 18001 compliant H&SMS must accommodate five key system elements: policy; planning; implementation and control; checking; and management review.

- The design, development and implementation of an appropriate H&SMS are reliant upon sound and robust system planning, within which hazard identification, risk assessment and control measures are central features of good management.

- In addition to meeting the requirements for effective planning within OHSAS 18001, an H&SMS must accommodate the requirements of key and significant construction-industry-based legislation and regulations which impinge directly upon the approach to and content of the planning process.

- The documentation requirements of an organisation's H&SMS should follow the traditional system document hierarchy – manual, management procedures, implementation plan and work instructions, allied to the requirements of industry regulation for a construction-phase health and safety plan to embrace the characteristics of a construction project's site works.

- An effective H&SMS applied to the construction processes on site centres on developing and implementing safe systems of work using safe working practices, the key aspects being: risk assessment; method statements; permits to work; training; safe working procedures; and site rules.

- The ongoing use and maintenance of the H&SMS must focus on the evaluation of performance of the system in operation, with a view to perpetuating continual improvement of the H&SMS in the long term.

Overview

While the effective undertaking of all key functional management disciplines within construction is imperative for the holistic success of the organisation, the management of health and safety is without doubt the most important. Construction works are inherently hazardous, and accidents to persons on construction project sites occur frequently and usually without warning. While many incidents are minor in nature, others are not. Where major accidents result in injury or fatality the effect and consequences for families, relatives, colleagues and organisations can be catastrophic. Health and safety managers have a continuous and onerous challenge to ensure a safe and healthy working environment. Principal contractors have an obligation to support their health and safety managers. The implementation of a structured and organised H&SMS is perhaps the most appropriate, best and effective way for the organisation to achieve this. The adoption of a structured H&SMS might be perceived as a legal and moral obligation, and indeed that is true. However, there are tangible financial, commercial and marketplace benefits for an organisation which positively and avidly supports good health, safety and welfare management. Moreover, and fundamentally, an organisation that pursues effective health and safety management has the capability to limit hazards, reduce occupational risk and save lives.

References

BSI (2007a). *BS OHSAS 18001:2007 Occupational health and safety management systems – Requirements*. British Standards Institution, London.

BSI (2007b). *OHSAS 18002 Occupational health and safety management systems – Guidelines for the implementation of OHSAS 18001*. British Standards Institution, London.

HSE (1974). *The Health and Safety at Work etc. Act 1974*. Health and Safety Executive, London.

HSE (1995). *The Reporting of Injuries, Diseases and Dangerous Occurrences Regulations 1995*. Health and Safety Executive, London.

HSE (1997). *Successful Health and Safety Management*. Health and Safety Executive, London.

HSE (1999). *The Management of Health and Safety at Work Regulations 1999*. Health and Safety Executive, London.

HSE (2002a). *The Personal Protective Equipment at Work Regulations 2002*. Health and Safety Executive, London.

HSE (2002b). *The Control of Substances Hazardous to Health Regulations 2002*. Health and Safety Executive, London.

HSE (2007). *The Construction (Design and Management) Regulations 2007*. Health and Safety Executive, London.

ILO (2001). *Guidelines on Occupational Health and Safety Management Systems, or OSH-MS*. International Labour Organization, Geneva.

CHAPTER 5

Integrated management systems

Introduction

Chapter 5 focuses on the integrated management system, or IMS. An IMS is a system which integrates the management functions of quality, environment and safety into one coherent management system to support the operation of an organisation's business processes and outputs. While integration presents any organisation with a tremendous challenge, dynamic and forward-thinking organisations within many business sectors including the construction sector are perpetuating the IMS. Chapter 5 examines standards for integrated systems, the requirements for the establishment of an effective system, and their application within the construction industry.

Integrated management and IMSs

This section examines: the concepts, definition and purpose of IMSs; the types of systems commonly used to configure functions of management; the advantages of IMSs; and the integrated management approach interpreted for practice within the construction industry.

Integrated management

An industry professional body perspective

In the past, many organisations managed OSH (occupational safety and health), environmental performance and quality reactively; they took few preventive measures until something went wrong. Subsequent action was limited to preventing a recurrence of that undesired event. The contemporary view is that organisations should take a pro-active approach; risks should be identified and controlled before the first adverse event. Such an approach is in principle more effective, but also more challenging. Success demands the design and implementation of robust management systems that incorporate, among other things, clear policies, procedures for planning and implementing risk assessment and control, and suitable arrangements for monitoring and reviewing performance leading to continuous improvement. . . . Organisations that have adopted separate management systems for OSH, environmental performance and quality may now be considering whether two or more of these systems might be brought together to form an IMS (integrated management system). . . . Integration appears to offer the prospects of substantial improvements in business efficiency and product/service quality, as well as in OSH and environmental performance.

(IOSH, 2010)

Industry perspective

Many companies and organisations throughout the energy, power, water, transport, manufacturing, and food production sectors are utilising IMSs, as are a number of prominent contractors and material suppliers within the construction industry. Within public-sector construction works, local authorities are beginning to introduce the consideration of integrated management initiatives in addition to traditional pre-qualification and pre-selection criteria of input resources. The IMS is being seen as an opportunity to gain additional marketplace profile and commercial advantage over business competitors, in addition to encouraging greater intra-organisational efficiency and effectiveness. A railway property management and construction organisation reported that an IMS was seen as an opportunity to improve its project-based processes with around a 50% reduction in the number of management procedures it had been using. Meanwhile, a large contracting organisation suggested that the IMS was appearing on the radar of its main clients and would in the future need to be considered equally as important as the single management systems it was using currently to secure projects.

Worldwide interest in IMS development and implementation

There is considerable interest worldwide in the development and implementation of the IMS. According to Arifin et al. (2009), the IMS is being applied in dual and triple applications within Malaysia, Singapore, Australia and New Zealand. This is perhaps not surprising given that the Asia-Pacific region is one of the fastest-growing regional markets for many industries including construction (BCC Research, 2010). Moreover, Arifin et al. report that ISO 9001 and ISO 14001 are dominant in quality management and environmental management applications respectively, while OHSAS 18001 informs health and safety management systems. With respect to practice specifically within Malaysia, they report that there is a very high level of awareness and understanding of the IMS and equally high levels of commitment by companies there to apply IMS principles. Their work suggests that the IMS is assuming considerable support in the Asia-Pacific region and will, likely, gain in popularity in the future. Such interest extends to Japan, where the Japan Quality Assurance Organization (JQA) suggests that there is good support for the use of individual and integrated systems for managing quality, environment and safety following ISO international standards (JQA, 2010). Throughout Europe, the European Business Network for CSR (Corporate Social Responsibility) (2010) supports the uptake of management systems meeting ISO standards across many business sectors including construction. Likewise, Bureau Veritas North America, one of the world's leading certification bodies, encourages the use of ISO standards-related quality, environmental, safety and integrated management systems. Collectively, there is considerable interest in the concept, principles and applications of the IMS worldwide.

Concept

The concept of integrated management is:

> To bring together individual organisational management systems to become structured, organised and implemented as a single management system.

Definition

In the context of application to quality, environment, and health and safety, integrated management can be defined as:

> The management and operation of an organisation's activities to ensure that its outputs, whether they be products or services, are provided to the required quality, with due regard to the environment and using safe and healthy working practices.

Purpose

The purpose of integrated management is to:

- improve the delivery of a product or service to the customer;
- impart greater efficiency and effectiveness in managing the company and its business activities;
- improve communication within the organisation by removing traditional management function system boundaries and barriers;
- reduce the duplication of bureaucracy, procedures and paperwork associated with using separate management systems;
- utilise the common and transferable elements of quality management systems (QMSs) to benefit the management of other organisational functions;
- enhance the input–conversion–output process model of delivering products and services;
- provide effective supporting and assurance processes which focus on the core business processes;
- capitalise on the synergistic effects of bringing organisational systems together;
- perpetuate a holistic, or whole-organisation, philosophy for the business.

Business expectation

The concept of integrated management is becoming more recognised as organisations within the construction industry strive to meet the requirements of clients across a multitude of business expectations and contractual requirements. For some of the largest international construction-related organisations these demands are at the forefront of their business vision and strategy and their operational management approach.

An example of such thought is exemplified in the business vision of Skanska UK. Skanska is one of the largest and most successful construction companies, both in the UK and internationally, and its activities are greatly underpinned by its clear and multi-focus business values as follows.

Skanska UK expresses its business strategy through its 'five zeros' vision:

- ZERO LOSS-MAKING PROJECTS – *loss makers not only destroy profitability, but also affect customer relationships.*
- ZERO ACCIDENTS – *whereby the safety of our personnel as well as subcontractors, suppliers and the general public is ensured.*
- ZERO ENVIRONMENTAL INCIDENTS – *which means our projects should be executed in a manner that minimizes environmental impact.*
- ZERO ETHICAL BREACHES – *meaning that we take a zero-tolerance approach to any form of bribery or corruption.*
- ZERO DEFECTS – *with the double aim of improving the bottom line and increasing customer satisfaction.*

The qualitative targets, as expressed in the five zeros, reflect our core values.

(www.skanska.co.uk)

It can be seen that the five zeros propounded by Skanska form the basis for a holistic organisational management approach that focuses on the requirements of corporate social responsibilities (CSRs), which include the three key management disciplines – quality, environment and safety – reflected in zero defects, zero environmental incidents and zero accidents.

IMS

Concept

An IMS is:

> The integration of one management system with at least one other management system where both systems operate concurrently, meet national and/or international standards, focus on the core business and its customers and are certificated by an independent external body.

Relating to the definition of management systems presented earlier, an IMS sets out and describes, for specified multiple management functions, the organisation's policies, strategies, structures, resources and procedures used, both within the company, or corporate, organisation and any sub-organisation, to manage the processes that delivers its products or services.

Definition

Therefore, an IMS can be defined as:

> The organisational structure, resources and procedures used simultaneously to plan, monitor and control the quality, environment and safety of the processes involved in providing a product or service in the marketplace.

Importance

The underlying use and importance of the IMS is propounded by Carillion, one of the largest construction service organisations in the UK. Carillion is an excellent example of commitment to an innovative management approach within construction.

Carillion was a pioneer of IMS development and implementation among construction-related companies following the introduction of the IMS in 2002. In delivering its core business to construction through building and infrastructure development using both traditional and non-traditional contractual arrangements Carillion has achieved the following:

- All construction-related business activities are certificated to ISO 9001 QMS standards.
- Around 99% of UK business is operating under an accredited ISO 14001 management system.
- Construction-related business activities are compliant with in-company safety management system standards (based on OHSAS 18001 principles).

This ensures that the performance of our business group in relation to areas of sustainability risk is controlled and audited both internally and externally by external certification bodies.

(www.carillion.com)

Types of system

There are three types of system commonly used by organisations to configure their functions of management. These are the:

- *separate* system, where individual policies for quality, environment, and health and safety are developed with separate corporate procedures and separate implementation plans;
- *semi-integrated* system, where separate corporate policies and procedures are developed and integrated into a single implementation plan;
- *fully integrated* system, where a single corporate policy for quality, environment, and health and safety is developed and integrated into a single implementation plan.

Examples of the three types of system are shown in Figures 5.1, 5.2 and 5.3. The corporate aspects of the systems are highlighted in the diagrams by shading while the implementation plans are shown as clear. In each case it can be seen how the systems evolve from policies into procedures through documentation which can then be applied to the organisation's operations through the implementation plan. Plans are applied to organisational processes through the application of working instructions. As activities are undertaken they are monitored and controlled, with feedback on performance gathered into records. These are channelled by an improvement loop back to inform future corporate policy making.

The principal shortcoming of the separate systems approach is that, traditionally, it has focused on complying only with external pressures and demands from the business world, rather than focusing on improving the company and its organisation and business activities. Conversely, separate systems can also be too inward looking, focusing on their specific interest rather than taking a wider organisational perspective. Where a number of implementation plans are used,

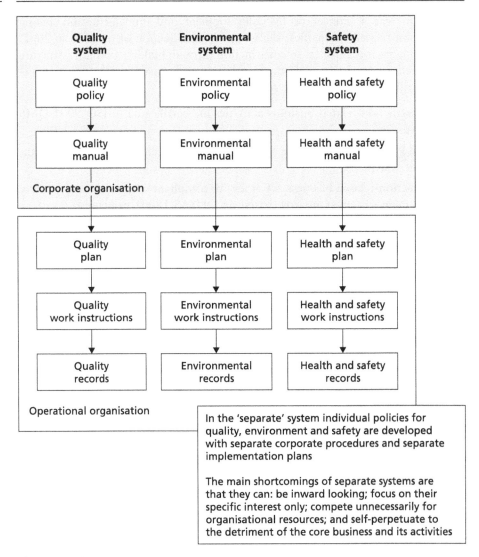

Figure 5.1 Separate management systems

as in the use of separate systems, their use and effectiveness are hindered by the multiple tasks involved in fulfilling each management function. With greater emphasis on each specific function rather than the multiplicity of functions, conflict can easily develop, particularly in competing for organisation interest and resources.

The main disadvantages of separate systems include the following:

- Communications are limited and knowledge is not shared.
- Decisions are made in isolation.
- Confusion can arise through the multiplicity of information.
- Bureaucracy, procedures and paperwork can perpetuate through duplication.

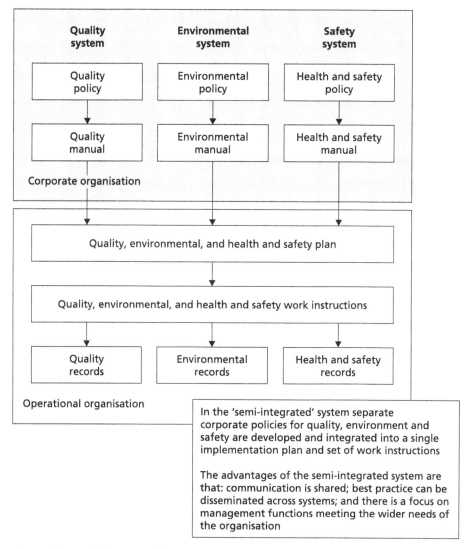

Figure 5.2 Semi-integrated management system

- Resources can be wasted through lack of direction.
- Self-perpetuation can arise, to the detriment of the wider organisation.
- Conflict among different management functions can arise.
- Complacency in conducting procedures can occur over the long term.

An effective system is one that enables information on management functions as they are applied to organisational processes to be relayed, acted upon and evaluated clearly and easily using short communication paths. Integration throughout the organisation from corporate policies to implementation plans shortens communication pathways. It also ensures that information is consistent

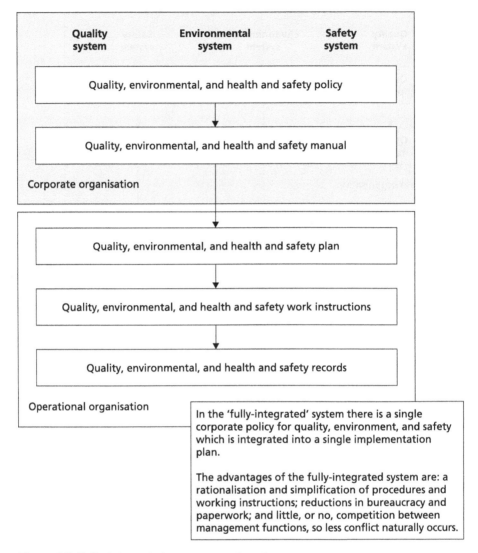

Figure 5.3 Fully integrated management system

and concise such that it is more efficient and effective in use. Bureaucracy and paperwork used in inspection procedures, record keeping and audits, always cited as obstacles to effective practice in traditional separate management systems, can be reduced. The integration of procedures and the associated reduced number of documents to be maintained have far greater effectiveness in use. Moreover, those people who implement an IMS and work within and around it will find it more focused, easier to use and with less repetition of tasks and paperwork.

The main advantages of semi-integrated and fully integrated systems include the following:

- Communications are open and knowledge is shared.
- Best practice can be disseminated easily.
- Decisions can be made appropriately and quickly.
- Resources can be allocated to all management functions as required and redeployed easily when necessary throughout the group of functions.
- Rationalisation and simplification of procedures and working instructions.
- Reductions in bureaucracy and paperwork.
- Focus on and among specific management functions while meeting the wider needs of the organisation.
- Little, or no, competition between management functions and less conflict.

Systems structure: vertical and horizontal

Separate systems tend to be 'vertical' systems (Griffith, 1999). Each system is a separate entity and operates in parallel to others with little or no communication and sharing of information across system boundaries. See Figure 5.4. Each management function has independent corporate system documentation and implementation plans. While each has an appropriate improvement loop, feedback is channelled

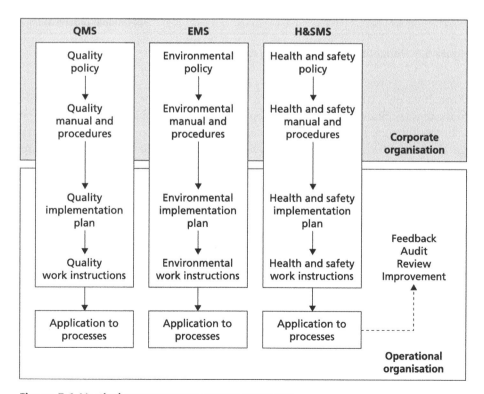

Figure 5.4 Vertical management system structure

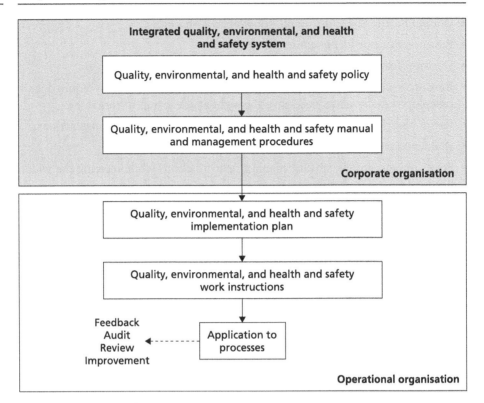

Figure 5.5 Horizontal management system structure

through its own structure to its individual separate corporate policy. This leads to those difficulties associated with separate systems highlighted previously.

In the integrated system a 'horizontal' cross-functional management structure can be established. See Figure 5.5. In this structure the three functions are integrated at the corporate level and then semi-integrated or fully integrated at implementation level as a 'semi-integrated' or 'fully integrated' system is adopted. The integrated approach provides for cross-functional expertise, open communication, shared information and co-ordinated practices. It encourages a holistic view of the organisation in which functions of management contribute to fulfilling not only their own requirements, but also those of the wider organisation. The approach brings together best practice across the range of management functions. As it promotes and disseminates insight from each function of management it has the inherent ability to create synergy and strengthen the whole.

The advantages of an integrated horizontal system structure are as follows:

- An integrated system communicates more effectively with, and usefully informs, the company.
- Integrated policy can better meet the company's holistic business and organisation needs.

- Best practice in discrete functional management areas can be shared with others.
- An integrated implementation plan can focus on vital aspects, such as risk assessment, across management functions simultaneously.
- Integration provides for more focused and streamlined working methods.
- A single feedback loop facilitates more rapid assimilation of lessons learnt and potential improvements to policy, procedures and practices.

It is not suggested that there are no disadvantages with horizontal structured systems. One of the prominent arguments in favour of vertically structured separate systems is the need to meet independent management system registration, or certification, schemes for quality, environment, and health and safety. However, the common elements of systems standards are such that the three management functions closely interrelate and, if needs be, the system can be integrated internally to fulfil the holistic needs of the company while also having sufficient clarity of differentiation to satisfy the external influences from particular management system standards. Although described subsequently, it is worth noting at this point that integrated management system assessment (IMSA) (BSI, 1999) is available to companies seeking dual and triple certification of multiple management systems. Likewise, national and international accredited certification organisations, such as the exemplary Davis Langdon Certification Services (DLCS), are coming to the management systems registration marketplace.

By seeking to integrate management systems, an organisation can identify wasteful activities through the rationalisation and simplification of the following key aspects:

- corporate functions of policy making, objectives setting, audit and review;
- management functions and working instructions;
- document structures.

Organisational opinion and conjecture can still surround the feasibility and cost-effectiveness of, for example, the following:

- simplification of system integration;
- practicality of operation;
- clarity of monitoring, control, audit and review;
- ease of certification.

Integrated management interpreted for practice

An interpretation of integrated systems has been suggested by the Construction Industry Research and Information Association (CIRIA) for use within the construction and engineering industries and is shown in Figure 5.6 (CIRIA, 2000). The focus is on the consideration of all those factors which impinge upon the organisation and its levels of management to bring together the wherewithal to provide all the necessary information in a single instruction to those undertaking

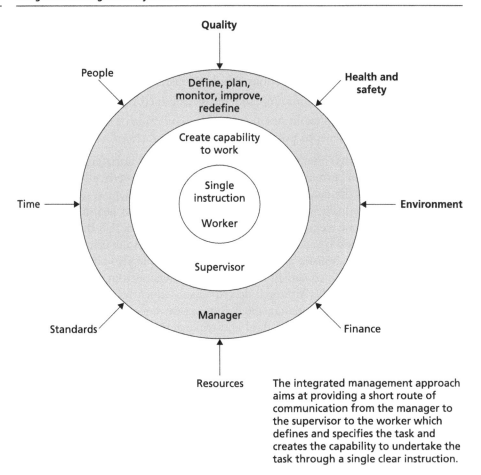

Figure 5.6 The integrated management approach

Source: Adapted from CIRIA (2000)

The integrated management approach aims at providing a short route of communication from the manager to the supervisor to the worker which defines and specifies the task and creates the capability to undertake the task through a single clear instruction.

the work. The instruction should embrace the performance requirements for the work in each function of quality, the environment and safety. Where integration takes place, the result should be one focused and easily understood instruction. Where there is no integration, a multitude of disparate instructions will impose complexity, confusion and miscommunication.

It should be accepted in viewing Figure 5.6 that construction and engineering are exceptionally complex in terms of information flows. Unlike manufacturing, for example, where work typically follows an assembly line of repeating operations, construction involves multiple resource inputs across a multi-task worksite. Instructions have to be transmitted to many members of the workforce in different work locations simultaneously. Furthermore, instructions need to be well co-ordinated if the many separately undertaken work tasks are to be fulfilled and not hinder other activities. Such complexity is not confined to construction and engineering and is seen in many other industries.

CIRIA's interpretation is useful as it exemplifies just how difficult it is to capture the essence of integration within a complex multi-variable work environment. While an ideal situation is to promote a single instruction output from integrated inputs, the reality is more complicated. If some of the impinging variables are excluded with a view to simplicity and embrace only quality, environment and safety, the amount of information that needs to be co-ordinated remains considerable. In reality, what integration does attempt to achieve is the minimisation of the amount of information arriving from the management functions. This can be done by integrating the key requirements rather than allowing those requirements to arrive separately at the workplace. The worker receives only the information needed to fulfil a given task to the required specification.

To make this work, supervisory tasks need to be co-ordinated effectively. There is little sense in carrying out the work if the required monitoring and control are not focused on the work. Therefore, ongoing instruction, checking to work specification and measuring to performance criteria must be conducted in line with the way the work was assigned. This requires integrated and simplified documentation and recording mechanisms. In fact, all management systems require clear and easy-to-use document structures and recording procedures if they are to operate efficiently and effectively. Integration can be problematic because of the amount of information that is required to specify work tasks in many situations. Notwithstanding, overlaps in procedures, work instructions, documents and record keeping present a good opportunity for simplifying information through integration. It is at operational level that integration may perhaps be most beneficial. It will be seen later how rationalising and integrating documentation and record keeping plays an important part in fostering management system integration.

Common difficulties experienced with management system implementation

The following bullet points highlight issues significant to organisational thinking and positioning when seeking to develop and implement management systems within construction. These reflect a synthesis of views from a number of small-product manufacturing and supply companies operating within the construction sector:

- the lack of common terminology for and understanding of management system standards and practice;
- the ambiguity presented by multiple systems standards to be met in a variety of functional management disciplines;
- the confusion arising from prioritising which standards to meet and which systems to develop to satisfy organisational needs;
- the enormity of the task surrounding the structure and development of documentation and the practicality of co-ordinating documentation with recording mechanisms;
- the magnitude of resources and costs required to address system development;

- the lack of cost–benefit information available to quantify the effort to be deployed;
- the upfront investment for developing and the long-term overhead commitment for maintaining management systems;
- the commercial consequences of not engaging in such developments.

Potential benefits

The above difficulties can be balanced with a perspective of potential benefits reported by the same group of respondents. They suggested that:

- one core policy could better drive the business;
- common goals throughout a business would focus attention on the important business matters;
- a process map linked to management structure and core processes would be beneficial;
- risk throughout the processes of a business would be better determined;
- there would be improved management throughout the organisation;
- costs of systems establishment and ongoing support would be reduced.

Quality, environmental, and health and safety management standards, specifications and systems

This section examines: the compatibilities between the standards applicable to quality, environment and safety, namely ISO 9001, ISO 14001 and OHSAS 18001; and the potential of and opportunities afforded by management system integration.

International and national standards and specifications for management systems

Details of the standards governing quality, environmental, and health and safety management systems were presented earlier in the respective parts of this book. Notwithstanding, a summary of each is presented here to alleviate the need to refer back to the preceding parts. It is also worth taking note at this point of how international management system standards originate and how these cascade to become national standards.

ISO, CEN and BSI

International standards are developed and presented by the International Organization for Standardization (ISO) – a worldwide federation of national standards bodies (ISO Member Bodies) – based in Geneva. Such standards are prepared by ISO technical committees, composed of representatives from the member bodies. These standards are translated into European Standards, or a European Normalisation (EN), by the European Committee for Standardization (CEN), based in Brussels, its members being the national standards bodies of participant countries throughout Europe. CEN members, of which the UK is one, are bound to comply with the CEN Internal Regulations, which give a European Standard the status of a national standard without alteration, such that in the UK it automatically becomes a British Standard (BS) presented by the British Standards Institution (BSI).

A British Standard is the official English-language version of an EN ISO standard and is identical with an ISO standard. A European Standard is presented in three official language versions – English, French and German – although a member of CEN can translate it into its own language, with that version notified to CEN. The home language version then has the same status as the official versions. In practice, the standards are commonly noted and referred to conveniently as ISO standards, simply deleting the EN and BS designations, which signify the acceptance and use of a standard within European CEN member countries and the UK.

The QMS standard

The international standard specifying the requirements for a QMS is ISO 9001 Quality management systems – Requirements, where 9001 represents the series number of the standard. The series number is followed by the calendar year signifying the year of publication. This has been omitted in the above presentation to maintain long-term currency of reading, but to illustrate the complete signifier for the version published in 2008 it would be presented as ISO 9001:2008.

ISO 9001 is one of a series of standards relating to quality. ISO 9000 is the classification, often referred to as a family, of standards which has been established to assist companies in all business sectors, of all types and sizes, to develop and implement effective QMS. The series of standards are (ISO, 2008):

- ISO 9000 – describes the fundamentals of a QMS and specifies the terminology used within the documentation for QMS;
- ISO 9001 – specifies the requirements for a QMS where an organisation needs to demonstrate its ability to provide products that fulfil customer and applicable regulatory requirements and aims to enhance customer satisfaction;
- ISO 9004 – provides guidelines that consider both the effectiveness and efficiency of the QMS, the aim being an improvement in the performance of the organisation and satisfaction of customers and other interested parties;
- ISO 19011 – provides guidance on auditing quality and environmental management systems.

Evolution of the quality management standard – procedure to process
ISO 9001 replaced the preceding 1994 versions of the ISO 9002 and ISO 9003 standards in the series. In noting this revision it is important to appreciate that the evolution of the ISO 9000 series has accompanied a changing philosophy and

general approach. The most important change from earlier versions has been the conscious move away from an ostensibly procedural approach to one which is business process orientated. There is now a much greater requirement for the QMS to focus on the business processes and to deliver outputs which clearly meet consumer needs and expectations. In addition, there is greater emphasis on management commitment and the responsibilities not only to make sure that the management system functions effectively, but also to accommodate the requirements of the corporate organisation, stakeholders, industry regulators and customers. Essentially, today's standards have perpetuated a much broader and holistic vision of quality management.

The EMS standard

The international standard specifying the requirements for an EMS is ISO 14001 Environmental management systems – Requirements with guidance for use (ISO, 2004). This standard was developed and presented by ISO and CEN in the same way as that described previously for QMS.

ISO 14001 is augmented in its family of standards by ISO 14004 Environmental management systems – General guidelines on principles, systems and support techniques.

The H&SMS standard (specification)

The Occupational Health and Safety Assessment Series (OHSAS) specification for occupational health and safety management systems, published by the BSI, was developed 'in response to urgent customer demand for a recognisable occupational health and safety management system standard against which their management systems can be assessed and certified' (BSI, 1999). This specification is: BS-OHSAS 18001 Occupational health and safety management systems – Specification.

Standards and specifications

The status of standards and specifications are, of course, different and should not be confused.

A standard is:

> A document, established by consensus and approved by a recognized body, that provides, for common and repeated use, rules, guidelines or characteristics for activities or their results, aimed at the achievement of the optimum degree of order in a given context.
>
> (British Standards Society, 1998)

A specification is:

> A document stating requirements.
>
> (British Standards Society, 1998)

Aims of standardisation

The aims of standardisation and the standards that are created are of immense value to an organisation (British Standards Society, 1998). These include:

- *improving the economy* – through removal of trade barriers, interchangeability and simplification;
- *consumer protection* – through setting standards for quality, environmental regulation, and rules for manufacture, testing and use;
- *safety and health* – through standards recognised by legislation;
- *improved communication* – through terminology, methodology, specification and codes of practice.

It is in the interests of all organisations, as providers or users of products and services, to support standardisation in the wider context as it can contribute to enhancing their industry sector marketplace and their commercial competitiveness. In addition, it can help promote greater clarity of business purpose, policy, organisation, systems and operations in the pursuit of quality, environmental, and health and safety management.

Basic structure of standards

An ISO management system standard is structured into main sections and subsections. These embrace the requirements for each component of the management system and each is numbered sequentially. In addition, annexes are used to provide information which corresponds to other management systems. A bibliography is provided to record sources of information used in its composition.

Compatibility of ISO 9001, ISO 14001 and OHSAS 18001

The British Standards Society (BSS), the UK member body of the International Federation for the Application of Standards (IFAN), published *Standards and quality management – An integrated approach* (1998). BSS supports compatibility and integration, commenting:

> There are three facets of management demanding integrated systems and documentation. They are quality, environmental, and health and safety management systems. The last decade has seen a growth in the recognition of quality management and the expansion of registration to ISO 9000. In the next decade one can expect a similar explosion in environmental and health and safety standards with varying performance requirements but needing parallel guidance and documentation.

To enhance compatibility between ISO 9001 and ISO 14001 the standards are aligned in their configuration. The quality systems standard does not include requirements that are specific to other management systems, such as EMSs, but it does provide sufficient compatibility to enable an organisation to align or integrate its own systems. OHSAS 18001 is compatible with both ISO 9001 and ISO 14001 'in order to facilitate the integration of quality, environmental and occupational health and safety management systems by organizations, should they wish to do so' (BSI, 1999). As a collective set, these standards and specifications provide an organisation with the requisite systems guidance and structure for documentation.

The nature of compatibility between ISO 9001 and ISO 14001 is presented in Table 5.1. Appropriate to the specific content of each standard, the sections and

Table 5.1 Similarity of elements in ISO 9001 and ISO 14001

ISO 9001			ISO 14001	
Quality management system (title only)	4	4		Environmental management system requirements
General requirements	4.1	4.1		General requirements
Documentation requirements (title only)	4.2			
General	4.2.1	4.4.4		Documentation
Quality manual	4.2.2			
Control of documents	4.2.3	4.4.5		Control of documents
Control of records	4.2.4	4.5.4		Control of records
Management responsibility (title only)	5			
Management commitment	5.1	4.2		Environmental policy
		4.4.1		Resources, roles, responsibility and authority
Customer focus	5.2	4.3.1		Environmental aspects
		4.3.2		Legal and other requirements
		4.6		Management review
Quality policy	5.3	4.2		Environmental policy
Planning (title only)	5.4	4.3		Planning
Quality objectives	5.4.1	4.3.3		Objectives, targets and programmes
Quality management system planning	5.4.2	4.3.3		Objectives, targets and programmes
Responsibility, authority and communication (title only)	5.5			
Responsibility and authority	5.5.1	4.4.1		Resources, roles, responsibility and authority
Management representative	5.5.2	4.4.1		Resources, roles, responsibility and authority
Internal communication	5.5.3	4.4.3		Communication
Management review (title only)	5.6			
General	5.6.1	4.6		Management review
Review input	5.6.2	4.6		Management review
Review output	5.6.3	4.6		Management review
Resource management (title only)	6			
Provision of resources	6.1	4.4.1		Resources, roles, responsibility and authority
Human resources (title only)	6.2			
General	6.2.1	4.4.2		Competence, training and awareness
Competence, awareness and training	6.2.2	4.4.2		Competence, training and awareness
Infrastructure	6.3	4.4.1		Resources, roles, responsibility and authority
Work environment	6.4			
Product realisation (title only)	7	4.4		Implementation and operation
Planning of product realisation	7.1	4.4.6		Operational control
Customer-related processes (title only)	7.2			
Determination of requirements related to the product	7.2.1	4.3.1		Environmental aspects
		4.3.2		Legal and other requirements
		4.4.6		Operational control
Review of requirements related to the product	7.2.2	4.3.1		Environmental aspects
		4.4.6		Operational control
Customer communication	7.2.3	4.4.3		Communication

Table 5.1 (*continued*)

ISO 9001		ISO 14001	
Design and development (title only)	7.3		
Design and development planning	7.3.1	4.4.6	Operational control
Design and development inputs	7.3.2	4.4.6	Operational control
Design and development outputs	7.3.3	4.4.6	Operational control
Design and development review	7.3.4	4.4.6	Operational control
Design and development verification	7.3.5	4.4.6	Operational control
Design and development validation	7.3.6	4.4.6	Operational control
Control of design and development changes	7.3.7	4.4.6	Operational control
Purchasing (title only)	7.4		
Purchasing process	7.4.1	4.4.6	Operational control
Purchasing information	7.4.2	4.4.6	Operational control
Verification of purchased product	7.4.3	4.4.6	Operational control
Production and service provision (title only)	7.5		
Control of production and service provision	7.5.1	4.4.6	Operational control
Validation of processes for production and service provision	7.5.2	4.4.6	Operational control
Identification and traceability	7.5.3		
Customer property	7.5.4		
Preservation of product	7.5.5	4.4.6	Operational control
Control of monitoring and measuring devices	7.6	4.5.1	Monitoring and measurement
Measurement, analysis and improvement (title only)	8	4.5	Checking
General	8.1	4.5.1	Monitoring and measurement
Monitoring and measurement (title only)	8.2		
Customer satisfaction	8.2.1		
Internal audit	8.2.2	4.5.5	
Monitoring and measurement of processes	8.2.3	4.5.1	Monitoring and measurement
		4.5.2	Evaluation of compliance
Monitoring and measurement of product	8.2.4	4.5.1	Monitoring and measurement
		4.5.2	Evaluation of compliance
Control of non-conforming product	8.3	4.4.7	Emergency preparedness and response
		4.5.3	Non-conformity, corrective action and preventative action
Analysis of data	8.4	4.5.1	Monitoring and measurement
Improvement (title only)	8.5		
Continual improvement	8.5.1	4.2	Environmental policy
		4.3.3	Objectives, targets and programme(s)
		4.6	Management review
Corrective action	8.5.2	4.5.3	Non-conformity, corrective action and preventative action
Preventative action	8.5.3	4.5.3	Non-conformity, corrective action and preventative action

numbering are individual. Where activities are alike or similar, the sections of both standards correspond and are shown accordingly. This allows an organisation to see where its activities in one system lie within the other system such that alignment and integration are facilitated. It is not an intention at this stage to describe and discuss the requirements of each section of these standards. This will be seen later where the QMS, EMS and H&SMS are presented, reviewed and evaluated in detail. Table 5.2 shows the similarities in structure and potential compatibility between the OHSAS 18001 specification and ISO 9001.

Similarities between ISO 9001, ISO 14001 and OHSAS 18001 are readily apparent in the structure of documentation required. As seen previously, the general documentation structure for any management system resembles a pyramid where documents evolve and cascade from company policies via a manual, procedures, plans and work instructions to forms used to maintain records. This structure is common to all three management systems as shown in Figure 5.7.

The latest derivations of the standards for both the QMS and EMS have shown that much attention has been given to making the standards easier to appreciate and interrelate in the context of business application. This can be seen in their:

- greater business focus;
- commitment to business improvement;
- focus on customer requirements;
- attention to the process model and its management;
- compatibility of specifications for management systems;
- encouragement of an integrated approach to systems development.

In addition, the standards have attempted to simplify the assimilation and understanding of their content by using:

- common terms and phraseology;
- consistent structure;
- similar layout.

The standards demonstrate that:

- there are considerable aspects of commonality and compatibility between ISO 9001, ISO 14001 and OHSAS 18001 to encourage integration of management systems if desired;
- the integration of management systems is a company-specific consideration and decision, based on familiarity and use of standards, and different circumstances and situations will determine different outcomes on the type of integration that is desirable or feasible within an organisation;
- separate management systems for quality, environment, and health and safety can be brought together successfully where standards are used as a basis for sensible integration focused on the company's business activities.

Table 5.2 Similarity of elements in OHSAS 18001 and ISO 9001

OHSAS 18001		ISO 9001	
		0	Introduction
		0.1	General
		0.2	Process approach
		0.3	Relationship with ISO 9004
		0.4	Compatibility with other management systems
Scope	1	1	Scope
		1.1	General
		1.2	Application
Reference publications	2	2	Normative reference
Definitions	3	3	Terms and definitions
OH&S management system elements	4	4	Quality management system
General requirements	4.1	4.1	General requirements
		5.5	Responsibility, authority and communication
		5.5.1	Responsibility and authority
OH&S policy	4.2	5.1	Management commitment
		5.3	Quality policy
		8.5	Improvement
Planning	4.3	5.4	Planning
Planning for hazard identification, risk assessment and risk control	4.3.1	5.2	Customer focus
		7.2.1	Determination of requirements related to the product
		7.2.2	Review of requirements related to the product
Legal and other requirements	4.3.2	5.2	Customer focus
		7.2.1	Determination of requirements related to the product
Objectives	4.3.3	5.4.1	Quality objectives
OH&S management programme(s)	4.3.4	5.4.2	Quality management system planning
		8.5.1	Continual improvement
Implementation and operation	4.4	7	Product realisation
		7.1	Planning of product realisation
Structure and responsibility	4.4.1	5	Management responsibility
		5.1	Management commitment
		5.5.1	Responsibility and authority
		5.5.2	Management representative
		6	Resource management
		6.1	Provision of resources
		6.2	Human resources
		6.2.1	General
		6.3	Infrastructure
		6.4	Work environment
Training, awareness and competence	4.4.2	6.2.2	Competence, awareness and training

Table 5.2 (continued)

OHSAS 18001		ISO 9001	
Consultation and communication	4.4.3	5.5.3	Internal communication
		7.2.3	Customer communication
Documentation	4.4.4	4.2	Documentation requirements
		4.2.1	General
		4.2.2	Quality manual
Document and data control	4.4.5	4.2.3	Control of documents
Operational control	4.4.6	7	Product realisation
		7.1	Planning of product realisation
		7.2	Customer-related processes
		7.2.1	Determination of requirements
		7.2.2	Review of requirements related to the product
		7.3	Design and development
		7.3.1	Design and development planning
		7.3.2	Design and development inputs
		7.3.3	Design and development outputs
		7.3.4	Design and development review
		7.3.5	Design and development verification
		7.3.6	Design and development validation
		7.3.7	Control of design and development changes
		7.4	Purchasing
		7.4.1	Purchasing process
		7.4.2	Purchasing information
		7.4.3	Verification of purchased product
		7.5	Production and service provision
		7.5.1	Control of production and service provision
		7.5.3	Identification and traceability
		7.5.4	Customer property
		7.5.5	Preservation of product
		7.5.2	Validation of processes for production and service provision
Emergency preparedness and response	4.4.7	8.3	Control of non-conforming product
Checking and corrective action	4.5	8	Measurement, analysis and improvement
Performance measurement and monitoring	4.5.1	7.6	Control of measuring and monitoring devices
		8.1	General
		8.2	Monitoring and measurement
		8.2.1	Customer satisfaction
		8.2.3	Monitoring and measurement of processes
		8.2.4	Monitoring and measurement of product
		8.4	Analysis of data

Table 5.2 (*continued*)

OHSAS 18001		ISO 9001	
Accidents, incidents, non-conformance and corrective and preventative action	4.5.2	8.3	Control of non-conforming product
		8.5.2	Corrective action
		8.5.3	Preventative action
Records and records management	4.5.3	4.2.4	Control of records
Audit	4.5.4	8.2.2	Internal audit
Management review	4.6	5.6	Management review
		5.6.1	General
		5.6.2	Review input
		5.6.3	Review output
Correspondence to ISO 14001 and ISO 9001	Annexes A and B	Annex A	Correspondence to ISO 14001
Bibliography			Bibliography

The positive aspects mentioned above should not be taken to imply that the consideration of management system standards and their potential integration is straightforward. Some less positive views are highlighted in the following case study.

Some views on management system compatibility for integration

Opinions from a group of small companies from an engineering sub-sector of the construction industry demonstrate that there are counter-views to the feasibility and use of the IMS. These are as follows:

- The structure of the standards suggests opportunities for integration but does not explicitly guide their development and implementation.

- There can be confusion over the priority placed by different standards in the context of business activities.

- The standards assist integration of the higher management-level considerations such as policies for quality, environment and safety, but could complicate activities for management involved at operational levels.

- The standards may not have equal standing in practice, with health and safety management considered differently from quality and environmental management because of its legislative importance.

A QMS focus

The quality management system is that part of the organization's management system that focuses on the achievement of results, in relation to the quality objectives, to satisfy the needs, expectations and requirements of interested parties, as appropriate. The quality objectives complement other objectives of the organisation such as those related to growth, funding, profitability, the environment and occupational health and safety. The various parts of an organization's management system might be integrated,

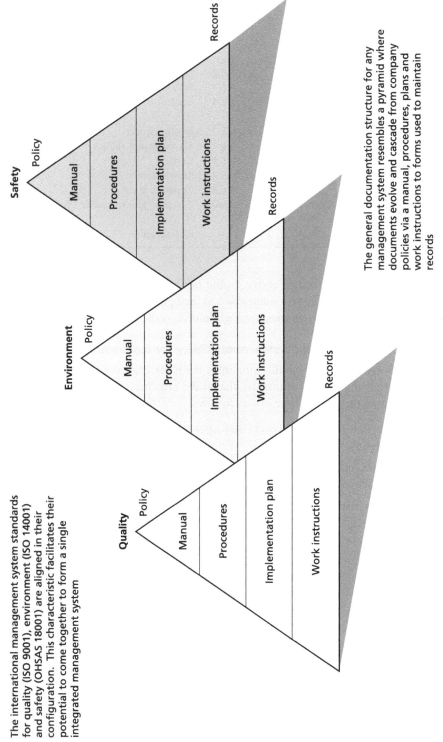

The international management system standards for quality (ISO 9001), environment (ISO 14001) and safety (OHSAS 18001) are aligned in their configuration. This characteristic facilitates their potential to come together to form a single integrated management system

The general documentation structure for any management system resembles a pyramid where documents evolve and cascade from company policies via a manual, procedures, plans and work instructions to forms used to maintain records

Figure 5.7 Common structure of management systems

together with the quality management system, into a single management system using common elements. This can facilitate planning, allocation of resources, definition of complementary objectives and evaluation of the overall effectiveness of the organization. The organization's management system can be assessed against the organization's management system requirements. The management system can also be audited against the requirements of International Standards such as ISO 9000 and ISO 14001. These management system audits can be carried out separately or in combination.

(ISO, 2000)

In practice, the majority of companies seeking to integrate systems, in part or in full, have used their ISO 9001 QMS as a basis for developing other compatible systems.

Other management standards and systems and their compatibility

IiP, EFQM, TQM and EMAS

In addition to the international standards ISO 9001, ISO 14001 and OHSAS 18001, other approaches to configuring management systems exist. Like the systems mentioned, these also provide companies with a formal means of self-assessment and improvement in key management functions.

Prominent business models for good management practice include the Investors in People (IiP) Standard and the European Foundation for Quality Management (EFQM) Business Excellence Model (BEM). Total quality management, or TQM, is another approach based on providing products and services through continuously improved management systems. TQM is:

a way of managing an organisation to ensure the satisfaction at every stage of the needs and expectation of both internal and external customers, that is stakeholders, consumers of its goods and services, employees and the community in which it operates, by means of every job, every process being carried out right, first time and every time.

(Latham, 1994)

In the area of environmental management there is the European Community's Eco-Management and Audit Scheme (EMAS). This presents the specification for any organisation in any sector of business seeking to establish an environmental protection system. Within the UK, a similar scheme, the Small Company Environmental and Energy Management Assistance Scheme (SCEEMAS), was to make EMAS accessible to small and medium enterprises (SMEs) but was abolished some years ago.

While the aforementioned standards, management approaches and schemes do not directly relate to ISO 9001, ISO 14001 and OHSAS 18001, they all share aspects of good management and business practice. Although it is not intended here to explore these, it is worth noting that common objectives and elements do pervade all approaches. For example, the Deming model of continual improvement (plan–do–check–act) is a fundamental element of all the systems, as is adherence to the process model of delivering products and services.

Document structure is also similar across the systems, with the pyramid hierarchy used extensively.

QUENSH

QUENSH is an approach that brings together the aspects of quality (QU), the environment (EN) and health and safety (SH), to present a method of managing the three disciplines within a single organisational management system: 'The purpose behind this collectivism is to bring together these three sciences which are also assessed and accredited separately into a single accreditation duly supported by the appropriate accreditation bodies in combination' (Quality Register, 2010). The focus of the approach is to ensure that all products and services are sourced, procured and managed in a safe and environmentally empathic way and applied to a set standard of performance or workmanship.

QUENSH is based on the fundamental premise that in organisations which focus on strategic and holistic management, each and every decision and activity is assessed in terms of its risk. Whereas it would be easy for any organisation inadvertently to ill consider risk in the plethora of its business processes and operational activities, QUENSH highlights risk assessment and management as an intrinsic element of a standards-based process-orientated model.

The QUENSH approach is based on the requirements of ISO 9001, ISO 14001 and BS-OHSAS 18001 and follows the same principles of development and implementation. It therefore provides an integrated system rooted in the culture of the company applying its concepts, and one which is fundamentally based on continuous quality improvement. A QUENSH-based management system will seek to ensure that construction work, together with its input products and services, meets the client's specification and satisfaction with regard to quality, environment and safety. Some principal contractors and larger subcontractors have adopted the QUENSH approach to good effect where intra-organisational, industry and client requirements can be met collectively and systematically.

QUENSH is used as a basis for contractual terms and conditions in the procurement arrangements of public bodies such as London Underground Limited (LUL), and by waste management and water supply companies in the private sector. The principal objectives in such applications are to seek to control risk through the supply chain, identify risks contained in particular work packages, and carefully consider and devise mitigation measures to combat such risks. QUENSH is noted as being a supplier-facing approach such that the client's requirements in relation to quality, environment and safety are correctly understood by providers and accommodated within appropriate method statements, site-specific plans and working arrangements.

A number of consulting practices working within the construction sector and other business sectors provide advice and assistance in the development of QUENSH-based management systems. The focus of their approach and therein the basis of their commercial services are to devise systems which can provide organisational management streamlining and efficiency gains together with operational improvements for the user. Some consultants have also adopted the QUENSH concept in naming their organisations, and it is not uncommon to see the acronym QUENSH, in a variety of forms and presentations, used in marketing a range of services addressing quality, environment and safety within the construction sector and other business sectors.

Systems

Almost all companies that use standards-based management systems have adopted one of three configurations, the principles of which were discussed in an earlier section of this book. These are as follows:

1 *Separate systems* – independent QMS, EMS and H&SMS with separate corporate policies and separate implementation plans.
2 *Semi-integrated system* – QMS + EMS, but independent H&SMS with separate corporate policies and single implementation plan.
3 *Fully integrated system* – QMS + EMS (described as an environmental quality management system, or EQS) + H&SMS with single corporate policy and single implementation plan.

In these approaches, all three management systems are accommodated to varying degrees of integration according to organisational circumstances and priority of need.

The above categorisations are simplifications of those systems used in practice. It is not always easy to discern the precise configuration that organisations use. In fact, many companies use hybrid versions which cannot be easily categorised. In addition, the terminology used throughout industry is such that ambiguity surrounds the attempt to place the systems used within categories clearly. What is more clear is that companies tend to support the use of separate management systems or commit to using a form of IMS to their own configuration as dictated by their organisational requirements. Such an approach is rightfully appropriate as companies must meet required standards, and for that matter legislation and regulation, but should do so within the context of the needs of their organisation and its business activities.

Application

The development and use of a standards-based management system is voluntary, and with a multitude of factors to consider it is likely that management systems will differ from one organisation to the next. A standard does not imply that all organisations will have the same approach towards management systems. What any standard does do is provide a set of recognised requirements which, if the organisation complies, will have a management system framework that is uniform with other organisations working to the same standard in the same functional management discipline.

Although there are many similarities in the three structures it must always be remembered that they are intended for completely different uses. Therefore, they must be used sensibly by organisations as tools to help understand and configure their own management systems to meet their unique circumstances and situations. They should not be applied blindly without careful reference to the company and its business. The standards are similar in structure such that where an organisation is using more than one system, then those systems have sufficient compatibility to be integrated in the interest of improving organisational performance, efficiency and effectiveness.

This section has conceptualised and contextualised the nature of management systems, defined and explained the IMS and its potential together with the implications of international management system standards. It has shown that there is sufficient correspondence and compatibility between quality, environmental, and health and safety management to enable a company to consider the integration of these important organisational management functions. The next step is for a company to consider management systems in the context of its business.

Management systems and the business

This section examines: why a company or organisation would seek to integrate its management systems; and the considerations that need to be made in relation to organisational and system framework and structure.

Fundamental considerations

A company may seek to integrate its management systems for three key reasons:

1 To create greater corporate value.
2 To enhance brand value and customer focus.
3 To remain competitive.

A company develops and implements effective management systems because it generates the desire from within the organisation to serve its marketplace and customers better. There are no government demands placing pressure for compliance. The company establishes its own systems to meet its own perceived business needs.

A company seeking to develop and implement a management system or integrating its existing systems might consider the following fundamental aspects:

• What improvement to the core business is sought from management system implementation?
• What type and amount of organisational change may be required to facilitate improvement?
• What external aspects influence the organisation and structure?
• What organisational structure is required to embed any approach?

- What management functions need to be deployed?
- What management systems are required to structure the management functions?
- What management systems might and can be integrated?

These can be simplified to the following key organisational aspects:

- Organisational culture
- Organisational structure
- People
- External influences
- Systems.

Organisational culture

For many organisations the basic consideration of potential, development and implementation of management systems will require a profound shift in organisational culture. There will be a move away from having systems just for the perceived need of having them to structure and resource given parts of the organisation, to having them because they serve the core business and add value to the company. Such change has mirrored the development in management thinking through theorists including Deming, Juran, Feigenbaum and Ishikawa, as described in Chapter 1 of this book. This requires a reorientation of management system perception and value. In addition, it will be necessary to distinguish the core processes from the support processes and structure any management system to best serve the policies and objectives of the whole organisation. This may seem obvious, yet many companies have management support systems which do not truly support the core processes and intrinsically perpetuate themselves at a vast proportion of organisational overhead costs.

There has been a profound change in the orientation and practice of management over recent decades. The more customer-focused and fiercely competitive marketplace in almost all industry and commercial sectors has demanded clear attention to performance improvement and added-value delivery. Changing and more demanding business environments have highlighted the need for dedicated corporate policies, objectives and strategies coupled with more efficient and effective directive and operational management designed to deliver products and services to optimum performance standards. Over the last fifty years there has been a shift in emphasis in customer need from just price to choice and unique specification. This has resulted in a business shift in response from efficiency to quality and bespoke provision meeting multi-dimensional needs and wants.

Organisational change has become synonymous with exercises in outsourcing, downsizing and business re-engineering. Such practice has been commonplace not only nationally but on an international footing as companies have transferred their business operations overseas. The basis for this has been to reduce capital and recurrent operating costs, married to providing better service and value for customers, although the reality for some has been customer-interface difficulties, lack of convenience and loss of service. Within many organisations the effects of change have often been radical, severe and not without workforce scepticism. A prominent issue has been the de-layering of management functions, removing levels of specialists often to be replaced by generalists. Conversely, re-layering has also taken place as the management of support processes has self-perpetuated, often at the expense of managing the core processes. All of these issues and others have taken their toll on many companies, which have responded to change in their various ways. One thing is clear: those companies which are sophisticated and forward thinking have made their changes carefully and proactively while others appear to have acted haphazardly.

There has been a profound culture shift from the inward-looking, or morpho-static, organisation with its traditional and often outdated business practices, to the outward-looking and dynamic, or morphogenic, organisation in which change is proactive and championed. Such change has influenced significantly the employment structure and skills valued by companies. In the 1970s, the bureau-cratic structures and strictly prescribed task assignment within many organisations valued educational basics such as literacy, numeracy, honesty and reliability. In the 1980s a greater focus was on employees, individual contribution and the drive for quality organisation culture where more employees were directly responsible for parts of the delivery of outcomes by empowered focused management teams. A 'less is more' philosophy has, in general, meant fewer people in smaller teams accomplishing more activity.

Dramatic changes in information technology and communications have influenced company structures, processes and resources. The need for functional specialists has reduced as generalists utilise technology in their wake. Accessible communications media and their power of persuasion, in particular in the inter-face with the general public, have emphasised celebrity, charisma and profile in many fields of endeavour. Industry, business and commerce have not been slow to capitalise on this as high-profile individuals have been recruited to executive and non-executive directorships of many large companies. Moreover, the attributes that they might bring to a company are mirrored by managers within the organisation.

With organisations becoming outward looking and dynamic, external influ-ences have become equally as important as intra-organisational aspects to struc-ture and human resources. Greater cohesiveness in management is needed to accommodate more demanding customer requirements within the marketplace. Competitiveness must be better understood to ensure the added value of prod-ucts and services. Increasingly stringent legislation, in particular in relation to the environment and health and safety, must be recognised and responded to. Such important factors require that companies are clear in the designation and assignment of enabling activities.

Within many companies the structure and organisation of a management system and its elements, or sub-systems that make up the system, are relatively simple in concept. This is, for example, where activity is based on a single and central corporate management location and perhaps a small number of production sites, as seen in manufacturing. For other companies, however, activities and resources may be geographically dispersed to remote locations where many interdisciplinary inputs are managed de-centrally in a project situation, as would be the case, for example, in marine engineering. Such factors differ according to type of company, and the nature of its business and the management systems configured should reflect this.

In many medium and large-size companies, the arrangement of management systems will require at least two broad tiers of management, one established to configure the corporate level of management and the second to facilitate the provision of its products or service – that is, the operations level of management. For many small organisations and, in particular, sole traders this would not be required as, intrinsic to their nature and structure, the corporate and operation levels are essentially one and the same – they are integrated in their basic form. The efficient and effective management of both the corporate and operations organisations are dependent on the synthesis of the interdisciplinary inputs. It is the functional management and interaction of the elements that produce the potential synergistic effects that lead to the successful delivery of the product or service and in so doing maintain and perpetuate the business of the company.

Organisational structure

Companies within the process engineering or manufacturing sectors will likely be based on a single and centralised corporate management infrastructure in one geographic location with one or a limited number of processing or production sites. In such situations, the structure of the management functions involved is well defined, clearly assigned and requires little flexibility of operation. Other companies, however, have decentralised management, operating through sub-organisations removed from the central corporate management infrastructure. Examples of this include retail chain stores. A more extreme example of decentralised management, structure and organisation can be seen in the construction industry, where contractors provide services to clients on a project basis with projects often being far removed from the company's corporate centre. Furthermore, construction projects are temporary and transient in nature, such that the integration of a wide range of interdisciplinary elements can be difficult. In the examples mentioned, there is a requirement for the company to maintain its corporate management centre to direct the holistic aspects of its business while detached sub-organisations provide the functional management to deliver the service or product in other locations as the marketplace demands. The effective structure of an organisation is, therefore, essential to arranging management functions and systems, which in turn are crucial to the successful provision of its outputs.

For a company to be managed efficiently and effectively, there must be clear policies, objectives, strategies, plans, procedures and, moreover, resources that support the core business processes. To translate these requirements and embed them in workable and effective arrangements, there must be a strong, clear and recognised pattern of structure to the organisation. Both internally and externally

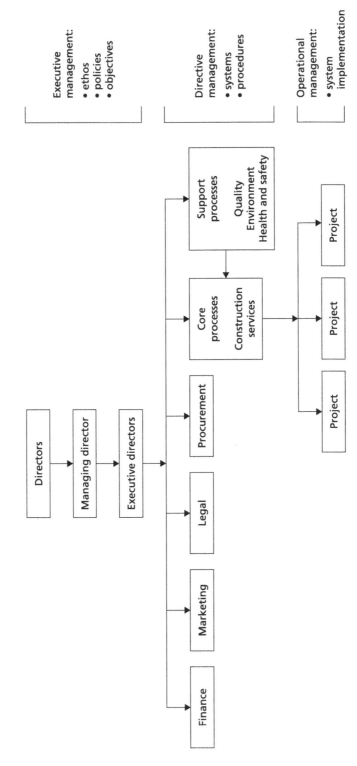

Figure 5.8 Organisational structure for a medium-sized contractor providing construction services to multiple projects

a company must establish its organisation to facilitate intra-organisational working, inter-organisational relationships and situational interaction with the external environment. This must allow the formal structure to become recognised and accepted and also facilitate the informal structures developed by its people. Only in this way will the organisation be sufficiently stable and healthy to drive and maintain its business.

Organisational structure should be sturdy to provide clear channels of communication, authority, responsibility and accountability but also be flexible to accommodate change and be adaptable to commercial, situational and environmental fluctuations. Organisations simply cannot be responsive if they are inward looking and static. All should be outward looking with a good perspective on their holistic activities in addition to the provision of their business outputs. A robust organisational structure to arrange functional management supported by well-configured management systems will allow a company to be fully aware of its customers' needs and operational surroundings while maintaining a healthy inward perspective.

All companies need an effective organisational structure to give form and focus to their activities. Principally, this has to be arranged around the core business processes. Specialisation, as mentioned earlier, forms the basis of a company's structure, with vertical specialisation influencing the typical hierarchical structure of authority while horizontal specialisation differentiates management functions within the various levels of the hierarchy. A company with strong vertical specialisation, where policies, objectives and procedures are rigidly documented and functions are strictly defined and deployed, is said to be a 'mechanistic' organisation. Companies with a flat structure and strong horizontal specialisation with fewer rigid procedures and greater informal control are said to be 'organic' organisations. In practice, most organisations reflect the characteristics of both mechanistic and organic forms of structure. This is because most utilise an overarching management structure through formal arrangements while also supporting the informal structures developed naturally through the social interaction of their employees. The importance of effective organisational structure, the positive arrangement of groups and the encouragement of the individual cannot be overemphasised. These aspects, propounded by the management theorist Deming (1960), form the core of understanding the practical aspects of how organisations operate successfully.

Figure 5.8 reflects an organisational structure for a medium-sized engineering contractor providing specialist services to projects in the wider construction sector. The diagram illustrates the link between the two tiers of management – the corporate level of management and the operations level of management – involved in delivering a service through the project. The purpose of this structure is to enable the company to channel managerial responsibilities from executive and higher managerial levels in the organisation to specialist functional managers who oversee service delivery at project level. This hierarchical structure establishes policies and objectives for the company at executive level which are translated into managerial systems and procedures by functional directive management for implementation of services at operations level, in this case within a project. The figure shows the core business process, which is the provision of specialist consultant inputs to an engineering project, together with the support processes of quality, environmental, and health and safety management. It can be seen clearly that the structure is important in establishing what management does at each level, together with shaping channels of communication, authority, responsibility and accountability.

People

It is frequently, sometimes conveniently, forgotten that management functions and systems do not simply structure themselves but that people are at the heart of their creation and perpetuation. Employees are the core resource of any company and make the organisation work. Obviously, business activities vary in the extent to which human inputs are required, but somewhere along the way all endeavour requires human input. The organisational structure is fundamental to arranging the human resource inputs of any company. The way in which an organisation brings structure to its activities and orchestrates its people is essential to creating the right working conditions to promote effectiveness. The key demand of structure in this regard is to create and promote teamwork.

Effective working by individuals within teams and the interaction of those teams are essential. Managers who are skilled and able and have the commitment to lead and motivate teamwork are vital. Good teamwork, through the effective structuring of individuals and groups, provides a sound basis for developing harmony and, in turn, synergy, both absolutely essential to the success of the company. In almost all industrial activity, and certainly in the construction industry, around one-half of all problems are associated with technological issues surrounding the delivery of processes while the other half result from difficulties in human interactions and relationships. People management is therefore fundamental and prerequisite. A company must foster effective human resource management, and this is founded in its organisational structure.

External influences

A company's organisational structure should be developed with respect to the company's situation in the business environment. Structure will be established in direct response to the core processes and its supporting processes, management functions and systems, but also a multitude of aspects that impinge directly and indirectly upon the company's nature and activities. The company must consider its intra-organisation, inter-organisation and wider environmental context. Furthermore, it must look at these in relation to its holistic activities in addition to the provision of its outputs. Organisational structure must reflect the activities of corporate management, which drives and maintains the business, and the activities at operations level that deliver the outputs, and must provide a robust and workable link between the two.

External influences on a company can impact at all organisational levels, on some more than others, and affect them in different ways and to varying degrees.

Management function	Influence	System type	Standards	Key driver	Performance indicator
Project planning	Internal	Company	Internal	Planning software	Completion schedules
Cost planning	Internal	Company	Internal	Cost modelling software	Completion budgets
Procurement of work	Internal	Company	Internal	Procurement databases	Successful contracts
Human resources	External	Standards based	Investors in People (IiP)	IiP standard	Certification
Quality	External	Standards based	ISO 9001	System specification	Certification
Environment	External	Standards based	ISO 14001	System specification	Certification
Health and safety	External	Standards based	OHSAS 18001	System guidelines	Recognised system

Figure 5.9 Management function and system interaction with the business environment

At corporate level, responsibilities will be directly affected by the effects and impact of the company's business. For example, the potential environmental effects of a chemical production process will impact on company thinking throughout all aspects of the organisational structure, management functions, management systems and operations. A perceived matter at corporate level can reflect in specific practice at worker level. Similarly, a particular influence can impact on a specific aspect of the organisation. For example, a formal approach to risk assessment is required by health and safety legislation, and this will impact on the legal and financial regulation of a company's business. Another example is the promotion of a company's quality of performance, which will be reflected in the organisational aspect of marketing. Figure 5.9 shows prominent influences on organisational structure.

Open systems

The intent of any management system including an integrated form for quality, environmental, and health and safety management is to provide an 'open system'. An open system is one which recognises, understands and interacts with its environment. In reality, all systems are open systems since by definition a closed system is one that has no interaction with its environment, and in practice a system's environment is too influential to be disregarded or simply ignored. By implication, an open system presents an intrinsic dilemma. A system is surrounded by an environment which is generally so important to it that both merge at their boundaries and exist symbiotically. However, this can reach a point where one cannot exist without the other, and this can become problematic. An example of this is a geographic area that relies exclusively on a particular business activity for its existence – motor manufacturing based in large cities is typical of this phenomenon. A severe downturn in the economy reduces spending on large capital purchases such as cars, and manufacturing is hit directly. Moreover, where there is geographic clustering of such business, cities, towns and entire communities can be severely affected through effects on the direct and supply chain businesses. A system, therefore, must co-exist and make use of its environment to assist the company without becoming wholly dependent upon it. A system must also be flexible such that the company retains the ability to accommodate changes brought about by its environment.

Companies use management systems to structure their approach to management functions in different ways. The requirement for functions to interact with other functions throughout the organisation is a prerequisite. The requirement for functions to interact with the wider environment is dependent upon the degree of influence that the environment has on the particular management function. For example, the function of quality management is standards based, configured to an international standard for system specification and where a performance indicator is management system certification with an approved external body. Conversely, cost planning is undertaken to a company-specific system with no applicable standard and no external performance indicator. The former system therefore has extensive interaction with the external environment whereas the latter system does not. The extent to which a management system is influenced by and interacts with its environment is shown in Figure 5.10.

Developing the framework for integrated management

The use of management systems generally and forms of integrated management specifically had been problematic until the widespread recognition of international management system standards. Such standards have led to a better understanding of system development frameworks for quality, environmental, and health and safety management throughout many sectors of industry and commerce. It was mentioned earlier that one trend in organisational culture and change has been a greater use of general managers in preference to specialists in some industry sectors. Standards-based management systems actually assist this as they tend to be more readily understood by non-cognate managers than may have been the case in the past. This is not to discount or undermine any dedicated function of management but rather to put matters in perspective. Modern business creates a greater requirement for informed generalists who can work across traditional system boundaries rather than those whose knowledge and skill base is limited

Aspect of organisation	Aspect of influence
Corporate responsibilities	• Ethics • Ethos • Policies • Organisational effects • Responsibilities • Accountability
Business strategy	• Business mission • Commitment to policy • Market awareness • Commercial positioning • Customer focus
Legal and financial regulation	• EU Directives • International standards • National legislation • Investment • Risk assessments • Liabilities • Insurances
Marketing and corporate image	• Public perception • Internal image • Company promotion • Corporate literature • Media interfacing and communications
Organisational management systems	• Organisational structure • Management systems and procedures • Performance indicators • Operational and project control • Business delivery improvement • Communication, information technology and knowledge
Human resources	• Staff deployment • Recruitment • Training and development • Incentives • Investment in people
Operations	• Time, cost and quality • Environment • Health, welfare and safety • Resources • Knowledge and skill base

Figure 5.10 Influences on organisational structure

to individual management functions. Also, knowledge and capabilities from one management function are used to develop and support others far more extensively than in the past.

Management frameworks and those persons who work within them need to be able to understand the wider business of the organisation and how this translates into activities that are carried out throughout the whole organisation.

A development framework for management system development and implementation requires managers who can appreciate company policies right through to supporting the operatives who put them into day-to-day practice. This has become more crucial today as companies structure their organisations into a number of sub-organisations, each with delegated responsibilities for delivering sections of the business whether they are product lines or discrete services.

Jonker and Klaver (1998) suggested a framework based on 'five different levels of integration ranging from abstract to practical', which are linked with the requirements for quality, environment and safety, as follows:

1 Policy integration
2 Conceptual integration
3 System integration
4 Normative integration
5 Pragmatic integration.

Policy integration

The company's policies with respect to quality, environment, and health and safety should, where possible, be brought together in a set of documents which complement each other. This may be a new all-embracing policy or modification of an existing policy. Furthermore, policies may be integrated themselves into one statement document, although it will be seen subsequently that this is hard to achieve totally effectively. Therefore, a set of policies following a common aim of the business is often prepared. This can be useful to perceptions outside the company where separate perspectives of quality, environment and safety are influenced by multiple standards and, in particular, dedicated legislation and regulations governing business activities.

Conceptual integration

This requires a prominent, or dominant, model, for example a QMS, around which the IMS can be developed. The use of a core management system is essential as, without this, the integrated system will not be integrated at all and simply be a haphazard collection of procedures. In addition, the core element from each individual system should be identified, which can be used as a vehicle to help bring otherwise disparate processes together. The common element in almost all management systems is the identification, assessment and mitigation of risk. Therefore, risk management plays a fundamental and important role in all system developments and implementations.

System integration

Any management system must be translated from policies into formal procedures and working instructions through documentation. Integration should ensure that fundamental elements of a good systems management approach are adopted – policies, objectives, strategies, structure, organisation, resources – all essential to effective implementation. Monitoring, measurement, assessment, audit, review and improvement must also feature in providing an open system with feedback loops.

Normative integration

Integration requires that systems which would otherwise be separate are brought together under a common nomenclature and format such that they merge seamlessly and operate to common principles and consistent practices. Standards-based systems for quality, environment, and health and safety facilitate this.

Pragmatic integration

For system integration and normative integration to succeed, both management and the workforce must be aware of, understand and be fully committed to the holistic aims of the organisation. There must be clear communication throughout the company of the importance of the core business processes and the support processes, with everyone understanding their appropriate role and function. It is in this last facet that most problems with management system implementation occur.

Comments on framework development

On introducing any management system or integrating existing systems, current activities need to be maintained while the new arrangements are introduced and embedded. A key feature of the framework must be the ability to design systems around the organisational needs while also appreciating the company's external environment and customer demands. If systems are introduced randomly and without careful planning they can quickly become burdensome and unsupported. A prominent difficulty is that systems are sometimes introduced randomly and without appropriate planning as a company moves quickly to meet changing situations such as, for example, new legislation. The result of this is frequently an overlap of individual management systems, duplication of procedures and disrupted practice. In fact, rather than integration, wider organisational disintegration occurs. It is often best and more cost effective to use an existing system as a framework for developing new or integrated systems. This has been the case with the QMS. The framework for integration outlined certainly encourages this.

An acceptance of new or adapted systems by management and the workforce is crucial to successful development and implementation. The separate and dedicated documentation that accompanies traditional systems is well understood. One system is often enough for people to handle at one time, so matters can become exacerbated when a multiplicity of management functions are encompassed in a single but more wide-ranging system. There may be a perception of extra work, duplication of activities, unnecessary tasks, uncertainty and even a lack of basic understanding of what the changes are seeking to achieve. Establishing any system is, therefore, a delicate process of reviewing the framework of existing activities, modifying rather than drastically changing accepted working practices and gaining understanding and support throughout the organisation. For this reason pragmatic integration is most important. The expected advantages of a system must be explained to all concerned. Moving towards an IMS presents an opportunity to reconfigure the fundamental basis and performance of the company, but it must be done with great care and respect for the people in the organisation.

Developing the structure for an IMS

There are strong fundamental links between the framework for the development of an IMS and the structure of the organisation and its management functions. This is true, in particular, from the perspective of how the company views the arrangement of its core processes, its support processes and its assurance processes. Within many organisations the term 'system' is sometimes misunderstood. It is not uncommon for management, administration or a paperwork regime to be confused with a system. It is often said within organisations that there is a system for this or that when, in fact, what is really meant is that there is merely a paperwork exercise conducted within the remit of an administrative function. The procedure may appear to function appropriately but it is not focused on anything more than completing set paperwork within the administrative process. It has no remit outside of the set procedure, does not impact on other functions and does not produce synergy with other organisational activities to the benefit of the wider or whole organisation.

Traditionally, a company focuses on delivering its business through carrying out core processes. The functions of management needed to support the core processes are structured into what are, conventionally, termed systems. Each system provides support processes to enable the core processes to be completed. Because the company has to provide support processes through a number of different management functions, it establishes a separate system for each. This is shown in Figure 5.11.

The principal disadvantages with this situation are that the systems may be:

- developed separately and have different structures;
- using different operating characteristics;
- functioning independently;
- focusing only on their own identities;
- perpetuating themselves rather than the core processes;
- losing sight of the core processes;
- not supporting the core processes effectively.

With an IMS the ways in which the company sees its core processes and support processes are modified. The core processes form and become 'the system' around which all support 'inputs', or sub-systems, are directed. This fundamental modification is highly significant because the focus is now on a single, or 'parent', or sometimes called the 'umbrella' management system embracing the management functions needed to deliver the core business.

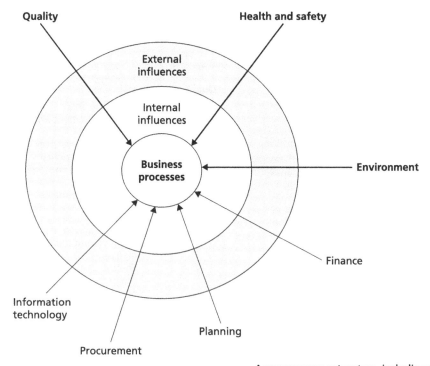

Any management system, including those shown, must support and add value to the core business processes and must operate to the requirements of those internal and external influences upon the organisation

Figure 5.11 Inputs by separate management systems to core business processes

The inputs are support processes based on explicit management functions. In the case of quality, environment, and health and safety, the support processes are, coincidently, assurance processes. This is because those particular processes are connected with intra-organisational and external management system approval, or certification, schemes, accredited by third-party, or external, bodies. The main system, focusing on the core processes of the business, whether they are the provision of products or of services, must be assisted effectively through the networking of those management functions delivered through the sub-system inputs. To achieve this, the structure of those inputs and their characteristics of operation must correspond. This does not imply that they must be identical in every respect, but that there should be much in common, they should work in similar ways, and they must be capable of working together.

Figure 5.12 shows the configuration of an organisation when viewed as a network of sub-system inputs supporting a single, or parent, management system. The principal advantages of this situation are that the inputs reflecting the management functions:

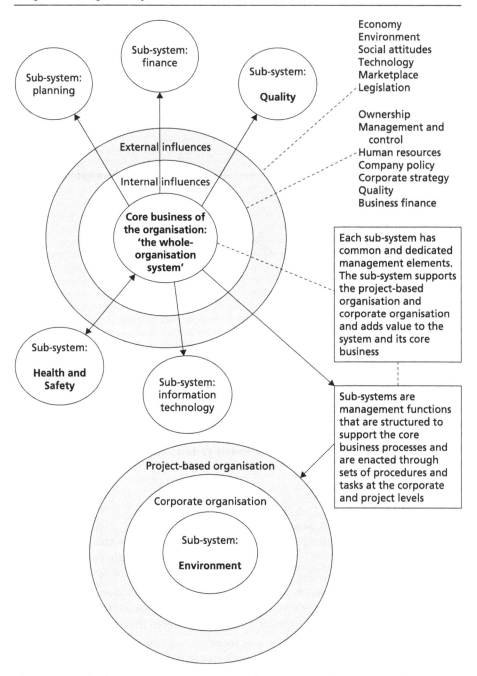

Economy
Environment
Social attitudes
Technology
Marketplace
Legislation

Ownership
Management and
   control
Human resources
Company policy
Corporate strategy
Quality
Business finance

Each sub-system has common and dedicated management elements. The sub-system supports the project-based organisation and corporate organisation and adds value to the system and its core business

Sub-systems are management functions that are structured to support the core business processes and are enacted through sets of procedures and tasks at the corporate and project levels

Sub-system: finance

Sub-system: planning

Sub-system: Quality

External influences

Internal influences

Core business of the organisation: 'the whole-organisation system'

Sub-system: Health and Safety

Sub-system: information technology

Project-based organisation

Corporate organisation

Sub-system: Environment

Figure 5.12 Single management system with supporting inputs, or sub-systems

- are co-ordinated;
- have common structure;
- have unified operating characteristics;
- have common elements of procedure (and documentation);
- have working compatibility;
- are integral to the single management system;
- have common focus and purpose;
- serve the corporate organisation and operational organisations (site of product, service or project) simultaneously and equally;
- are added to at any time with additional sub-systems to cover other management functions.

It can be seen from the structure that the various sub-systems reflect dedicated management functions, the combined network of these being essential to the operation of the core business processes. Where a company carries out its core processes in one physical location, then the single management system and its sub-systems will require only one tier of management – that which manages the company and its operations together. This would occur in, for example, a small manufacturing company based in one premises. However, where a company's corporate organisation is distanced from its operations, whether they are the manufacture of products or provision of services, then the company will require two tiers of management: one to focus on the management functions as carried out within the corporate organisation, and one to focus on the management functions as conducted at the operations sites. An example of this would be chain retailing, where a corporate head office devolves production, distribution and sales of its goods to a number of remote facilities. This is one of the key reasons why companies implement corporate strategies and operations strategies for their businesses.

The single management system is an open system in that it interacts with its environment and is influenced by many factors: economy, marketplace and legislation are prominent examples. It is also subject to influences from within the organisation including culture and ownership aspects. As the corporate dimensions of the system are cascaded to operations, the sub-system, guided by company policies, strategies and objectives, will be translated into management and supervisory procedures, working instructions and day-to-day tasks. The sub-systems have elements dedicated to their specific management functions but these are also common from one sub-system to the next. These facets (policy, organisation and planning, risk assessment, implementation, auditing, review and more) promote correspondence between the sub-systems which allows them to do more than would be the case with separate systems. As the support processes provided by the sub-systems are strongly directed to serving the core processes, they have the capability to create synergy and add value to the wider organisation and the business.

A prominent criticism of management system integration is the perceived difficulty with registration of the system by certification bodies. While any standard addresses only its own functional management area, it does not discourage the integration of elements where it would be sensible and feasible to do so. In fact, certification bodies encourage integration as combined assessment is more efficient for the certification body and more efficient and cost effective for the company submitting the system for certification.

A key area in which sub-systems can be integrated effectively is in the documentation used. It was seen earlier how the pyramid structure of management systems allows documents to overlap in implementation. In the single management system, its sub-systems are subject to a regime of corresponding and compatible monitoring checklists and performance records. In this way, duplication of tasks can be removed, paperwork can be reduced and information can be collected, collated and stored to the same methods.

The effectiveness of a sub-system lies much in its ability and desire to communicate across its boundaries to other sub-systems and also in the sharing of information of common value throughout the whole network of sub-systems. These abilities are assisted through the horizontal structure of the sub-systems rather than the traditional vertical structure, as explained previously in the chapter. In addition, the application of standards-based specifications for management systems, in the case of quality, environment, and health and safety, encourages the integration of corresponding and compatible elements. An IMS can at best remove or at least break down some of the barriers that impede organisational communication and practices. Similarly, implementation plans, management procedures and working instructions can be configured to correspond across the sub-systems. Moreover, the three areas have much in common conceptually and contextually. Each sub-system:

- uses a well-defined framework;
- is formalised in approach;
- has corresponding characteristics;
- is proactive in nature;
- advocates a holistic philosophy.

Quality, environmental, and health and safety management are three management functions that have a most significant part to play in promoting integration because between them they embrace a large proportion of organisational activities in most companies. If any company is to meet its commercial challenges then it must be able to call on all its management resources in the sure knowledge that they are working very efficiently and also effectively in the spirit and interest of the business. Continuous improvement is a paramount requirement placed on almost all companies by demanding clients and customers. Companies need to achieve, and to demonstrate, improved performance year on year and show that the standards achieved meet national and global benchmarks.

The first major step in fulfilling these demands is the development of a framework within which management integration can be facilitated. The second step is the development of organisational structure where a single management system, clearly focused and directed on the core business processes, is effectively supported by well-configured functional management sub-systems. For any company, the next steps are to develop, implement, optimise and certificate its IMS.

Development and implementation of an IMS

This section examines: the establishment of an IMS by a company or organisation following the requirements of the international standards for quality, environmental, and safety management systems, the process approach to systematic control and a core business focus; the elements of developmental structure and key enablers; the stages of system development; and the key elements of an effective IMS.

Systematic control

Each of the international management system standards ISO 9001, ISO 14001 and OHSAS 18001 adopts the Deming principle of systematic control as an underpinning philosophy to management system development (Griffith *et al.*, 2000; Griffith and Watson, 2004). The Deming principle is based on the methodology known as plan–do–check–act, or PDCA. This has been briefly described in all the earlier parts of this book and is reaffirmed briefly as follows:

- *Plan* – identify customer needs and expectations to establish objectives and organisational processes to deliver business outputs in accordance with stated policies.
- *Do* – implement the developed organisational processes.
- *Check* – monitor and measure organisational processes for their performance against policy, objectives and other requirements, analyse and review.
- *Act* – continually improve the performance of the processes.

The Deming principle of management control is sometimes referred to as the Deming dynamic control cycle or continual improvement loop. It features, in some form, in many books on management theory and practice. PDCA is important to management system development, first because it is prominent in management system standards and, second, because companies across all sectors of industry, business and commerce can frame and structure their systems using PDCA principles throughout the various stages of development and implementation involved. Figure 5.13 reaffirms the Deming principle of control.

The process approach

International standards promote the implementation of a process approach to systems development. Process thinking propounds that the reasons underlying

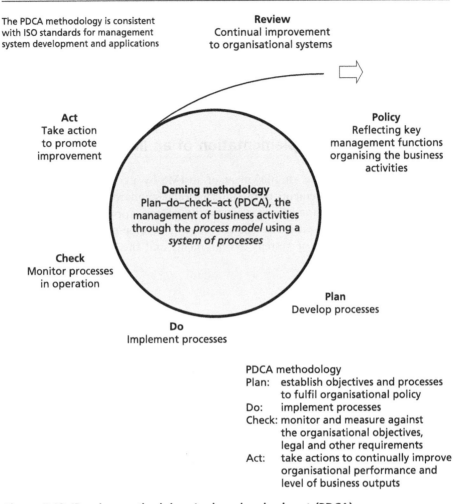

The PDCA methodology is consistent
with ISO standards for management
system development and applications

Review
Continual improvement
to organisational systems

Act
Take action
to promote
improvement

Policy
Reflecting key
management functions
organising the business
activities

Deming methodology
Plan–do–check–act (PDCA), the
management of business activities
through the *process model* using a
system of processes

Check
Monitor processes
in operation

Plan
Develop processes

Do
Implement processes

PDCA methodology
Plan: establish objectives and processes
        to fulfil organisational policy
Do: implement processes
Check: monitor and measure against
        the organisational objectives,
        legal and other requirements
Act: take actions to continually improve
        organisational performance and
        level of business outputs

Figure 5.13 'Deming methodology': plan–do–check–act (PDCA)

the success or failure in performance lie in the business processes and their management. For effective processes and management functions, a company needs the capability to manage the multitude of interrelated activities that convert inputs to outputs. This involves focusing activities on the core processes and support processes which deliver the product or service. These dimensions were introduced earlier. The process approach is encouraged by standards to enhance customer satisfaction through meeting customer requirements more effectively. This requires a good understanding of those management functions and systems which reflect customer requirements in the translation of inputs, through the conversion process, into outputs that satisfy customer needs. The principal method to achieve this is the close and continuous control of any individual process, afforded by the PDCA cycle, together with control of the interaction between multiple processes.

The process approach emphasises the importance of specific system facets, as follows:

- to understand and meet specific requirements;
- to consider core, support and assurance processes in terms of added value;
- to determine the performance of processes and levels of effectiveness;
- to pursue continual improvement of processes based on evidence of performance.

The model of processed-based management systems is frequently depicted in the style of illustration which the BSI presents to explain its model of a process-based quality management system. This is shown in Figure 5.14. The emphasis of the model is the provision of systematic mechanisms to ensure the continual improvement of the management systems responsible for overseeing the core business process. Product realisation, or for that matter the effective provision of services, becomes susceptible to a cycle of understanding and response to client or

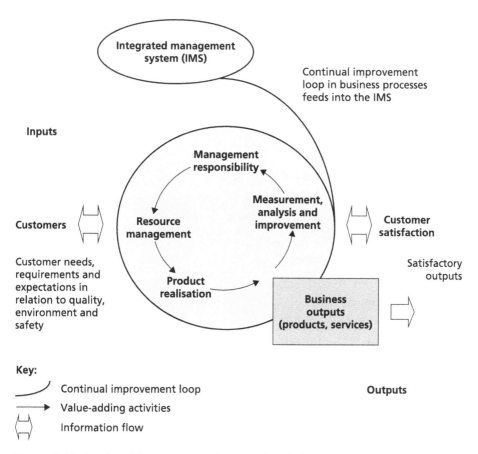

Figure 5.14 Continual improvement loop and its link to the integrated management system (IMS)

Source: Adapted from ISO 9001

customer needs, the conversion of inputs into outputs that the customer expects and, moreover, the pursuit of improvement in the whole process with a view to adding real value not only to the processes but also to the holistic business.

Stakeholder satisfaction

For any company to achieve success it must ensure that the core business achieves its corporate objectives. In addition to meeting the triple-bottom-line requirements of time, cost and quality, the business must be carried out safely for all concerned and with safeguards for its environment in the widest sense. Business objectives are, in the main, the same as those demanded by a company's stakeholders. Commensurate with the process model and the Deming principle of control, the way in which the objectives are to be met is fundamental to the arrangement of processes and the systems of management used. Stakeholders represent a number of interested groups, and each has many requirements within which quality, environment and safety aspects need to be embraced. These are as follows:

- *Shareholders:*
  - legal and regulatory compliance;
  - profitable business;
  - return on investment;
  - maintenance of turnover;
  - growth of the business;
  - fulfilment of policies and objectives;
  - minimisation of commercial risks;
  - image of the business and its outputs.
- *Clients and customers:*
  - provision to specification;
  - fitness for purpose;
  - value for money;
  - sustainable supply;
  - environmentally considerate;
  - safe products and services.
- *Employees:*
  - reward for good performance;
  - job security and prospects;
  - job satisfaction;
  - safe working procedures;
  - healthy working environment.

- *Community:*
  - employment potential;
  - profitable and sustainable local business development;
  - environmentally sound facilities and infrastructure;
  - safe and secure living environs.

Organisational principles

At the outset of management system development, a company should adopt a number of key principles in attempting to meet the range of requirements set by stakeholders. These must focus on many different organisational dimensions, but in the context of quality, environmental, and health and safety management they are as follows:

- An organisation's business activities and operations must be carried out in accordance with the legal requirements, industry regulations and business sector best practice that impact on quality, environment, and health and safety.

- An organisation must plan for, monitor, measure and assess the potential and actual impacts of its business activities on the quality of its products and services, the environment, and health and safety of anyone who is involved or comes into contact with business activities.

- An organisation must ensure that its policies and objectives for quality, environment, and health and safety reflect its commitment to these aspects, and that its activities meet these policies and objectives.

- An organisation must work in collaboration with governmental authorities, statutory bodies, industry and trade associations and the public to ensure that measures are implemented which maximise quality, environmental safeguards and the health and safety of all involved with its business activities.

- An organisation must ensure that its management and workforce are aware of and have commitment to its policies and objectives together with the fulfilment of policies and objectives for quality, environment, and health and safety.

- An organisation must ensure that management and workers have appropriate knowledge and training, skills and the necessary competence to safeguard quality, environment, and health and safety in the undertaking of work activities.

- An organisation must ensure that clients, customers, workers and the public have sufficient communication with and information on its business activities in respect of quality, environment, and health and safety such that they can assume informed responsibility.

- An organisation must ensure that it has the mechanisms in place not only to maintain continual improvement of its business activities, but also to be sufficiently proactive to mitigate and rapidly reactive to minimise instances of poor quality, environmental, and health and safety performance.

The process of integration

Preparatory business review

In considering management systems and their potential integration a company should ask some key questions in relation to its business activities and their organisation, as follows:

- Where is the company now?
- Where does the company want to get to?
- How is it going to get there?

The appropriate starting point to address these questions is for the company to carry out a preparatory business review. Reviews of this type are denoted by a variety of terms including: preparatory systems review; preparatory integration review; and business process review. All do the same broad task: to review the business activities, their organisation and operations in the context of the way the company manages quality, environment, and health and safety. Although the review encompasses the whole organisation, it is conducted predominantly at executive level. The review should include all processes involved in the provision of a company's products and services in the light of stakeholder requirements, and associate this with the requirements for systematic control within the PDCA methodology. An organisation's preparatory review will focus on the core, support and assurance processes. It should identify those aspects which impact upon the processes together with the requirements for development, implementation and maintenance of the management functions involved. Here, the focus is quality, environment, and health and safety, but many other dimensions of a company's activities may require review and would therefore be incorporated. The review should include the following aspects:

- *Identification:*
  - legislation;
  - regulations;
  - stakeholder requirements;
  - risks, hazards and business impacts;
  - management system standards;
  - gaps in systems;
  - overlaps and duplication in system procedures;
  - costs of processes and systems;
  - change management requirements.
- *Potentials:*
  - rationalisation of policies;
  - refocusing of objectives;
  - redirection of responsibilities;

- simplification of planning;
- streamlining of procedures;
- clarity in communication;
- removal of duplication and overlaps;
- reduction in paperwork and bureaucracy;
- minimisation of risk;
- optimisation of costs.
- *Plan:*
  - select key personnel;
  - determine objectives and goals;
  - prioritise key activities;
  - establish organisation;
  - plan implementation;
  - identify awareness and training needs.
- *Do:*
  - communicate policy, objectives and goals to employees;
  - implement awareness and training plans;
  - deploy resources;
  - implement system;
  - commit budget;
  - establish performance measurement.
- *Check:*
  - monitor activities;
  - audit performance;
  - identify potential improvements.
- *Act:*
  - implement improvements;
  - evaluate performance;
  - review experiences;
  - apply knowledge to future policy and objectives.

Uniform thinking

To be prescriptive on the development and implementation of management systems and their integration would be wholly inappropriate. The establishment of any system is the strategic decision of the company. The design of the system is influenced by a variety of organisational factors including business activities, policies, objectives, resources, processes, size, structure, situation and circumstances. There is no singular uniform or best approach. A company has to make up its own mind how it will approach management system integration based upon its assessment of all the factors identified in its preparatory review. Notwithstanding, international management system standards present a basis for uniform thinking in the development and implementation of a system. Furthermore, they provide a company with the elements of effective management systems around which they can configure their own approach to best suit their requirements. This is important as it ensures, as far as it is possible, that the system will be sufficiently comprehensive and robust in operation. This provides an assurance to the company that the management of processes is appropriate while also conveying confidence to the purchaser of products and services that they meet recognised performance standards. Quality, environmental, and health and safety management are at the forefront of business thinking by companies together with the business perception of clients and customers. This is because these three management functions embrace such a large proportion of company activities. If the management system controlling these functions is effective then the vast majority of the organisation is probably being managed successfully.

Documents and records

Before presenting options for integration it is pertinent to point out that management functions and, in particular, the documentation used are subject to specific requirements within legislation in some sectors of business. Requirements include, for example, the need for a risk assessment to be formal and documented, plans and programmes to be submitted as part of contractual arrangements between customer and supplier, and post-contract records to be kept in the form of files retained by clients. Such requirements feature prominently in connection with the EMS and H&SMS. This is so, to a large extent, because the focus of both is towards compliance with legislation and regulation. In the case of quality, its focus is more towards meeting customer demands through delivering outputs to specification. In all cases, however, there are overriding needs to meet customer care and responsibility and to perpetuate a continual improvement philosophy, and this necessitates the upkeep of records. However, in systems terms, documents are not the same as records. The development and implementation of a management system should, therefore, be considered around the PDCA methodology, the processes of the business and, to a large extent, the requirements or desire to maintain records of performance.

Predominant system

A company will, more often than not, integrate its management systems based on the predominant system in use at the time. This is usually, although not exclusively, a QMS. This may be an ISO 9001 standards-based system or an intra-organisational framework of quality planning and management. The important attribute is that the existing system should have sufficient structure such that it can be used as the established system and be the focus of integration for the other management functions.

Options for integration

A company can seek to integrate its management systems in a number of ways, although it is debatable as to whether all of the options are really true systems integration, as follows.

Merging

Where a company has more than one formal system, such as a QMS or an EMS, then it can merge the two systems and integrate other systems as they are introduced. This approach is also referred to as 'integration through compatibility'. The focus of this method is to merge documentation where it supports the same process. The main disadvantage with this approach is that the two systems really remain separate because they are implemented as two systems sharing co-operative documentation. Unless the management function labels of quality and environment are removed from the documentation, then the systems never truly merge or integrate at all. In fact, this approach can create much confusion in the use of documentation. When the identifying labels are removed it cannot always be ensured that both quality and environment have been given appropriate attention in application. The benefit is that at least some processes and procedures share documentation, which is a step in the right direction.

Conversion

Where a company has a formal and certificated QMS, then it can add the processes and procedures necessary to accommodate an EMS, H&SMS and other management functions as required. It achieves this by sharing common processes and procedures throughout all management functions and adding those specific ones required for particular functions such as environment or safety. This approach is also referred to as 'integration through alignment'. The main disadvantage with this approach is that the effectiveness of the resulting system is highly dependent on the efficacy of the original QMS. The benefits are that it is relatively simple and the common processes do become integrated.

System engineering

Where a company has an existing system such as a QMS, or indeed no system at all, it can use a 'system engineering' approach. This develops and implements a management system designed around the holistic approach to systems development. The disadvantage of this approach is that the system can be quite complex due to the multiplicity of management functions. This has an impact on the structure, organisation and resources. The great benefit of this approach is that it provides integration of systems based on processes which serve the wider interests of the organisation. It was seen earlier that integration is not about management functions that merely sit alongside each other, but that they must come together to create harmony and synergy.

IMS model

The business process model

The most successful way to achieve the establishment of an IMS is to adopt an overarching system engineering approach. An appropriate model for systems development and implementation is the business process model (BPM). This uses the ISO 9001 standards-based QMS as its core system and integrates the ISO 14001 EMS and OHSAS 18001 H&SMS.

The model is structured around the core policy, objectives and processes of the company's business activities, hence the term business process model. Rather than allowing the QMS to dominate system development, it places the business at its core and structures the three management functions – quality, environment, and health and safety – as formal support and assurance processes. The model, in fact, is denoted by its business orientation rather than any management function title. The approach reflects the structural requirements for providing key system elements (policy, organisation, planning, implementation and review), the requirements for formalised documentation (policy, plan, procedures, instructions and records) and management system certification requirements. The model is shown in Figure 5.15.

As the orientation of the system is the business core processes, the principal system elements are structured around a company 'business management plan'. The elements in the plan are developed in accordance with the elements of good practice contained in the QMS, which will meet ISO 9001. The elements of both the EMS and the H&SMS will also be contained in the business management plan. These elements will also comply with their respective standards. Of course, a company may have a requirement to integrate any number or any combination of management systems, and the model can accommodate this. Sometimes a company may wish to integrate two of its systems but not a third, and again the model enables this.

The integration occurs throughout the entire organisation from policy development through to record keeping during implementation, but is most evident in the way that documentation is structured and the methods by which procedures are conducted. Instead of operating management functions separately, they come together and are carried out simultaneously using integrated documents and integrated procedures.

To structure the system throughout the entire organisation of the company is far from straightforward. As explained previously, the company may need to put in place a regime which operates beyond its corporate organisation boundary into sometimes a number of sites in remote locations. Using a business process model facilitates this, as the focus is always placed on the business activities, the products and services, wherever they may be located and however many there are.

Elements of developmental structure and key enablers

The company will need to put in place a comprehensively developed, highly focused developmental structure and follow this with key enablers to ensure that the

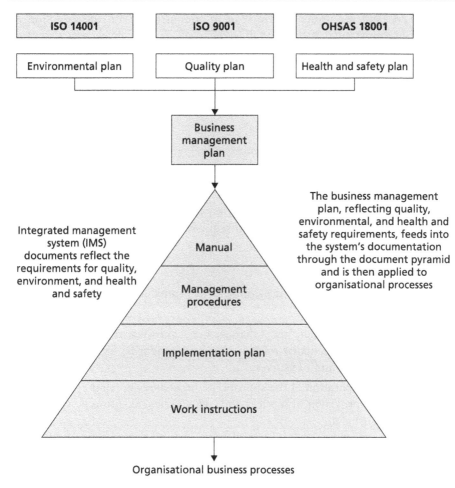

Figure 5.15 Business process model supporting IMS development

system is appropriate and has a high degree of effectiveness. Moreover, the developmental structure must be consistent in its perspective throughout the entire organisation if holistic needs are to be realised. The main elements are:

- vision – company mission and values;
- client and customer requirements – product/service specifications;
- business assessment – preparatory review;
- organisation – physical configuration of the company;
- business planning – definition of goals;
- strategic planning – objectives and targets;
- methods implementation – programme and operations;
- critical performance measurement – measures of efficiency and effectiveness;
- management review – including auditing;
- process and system improvement – including feedback to policy.

If the main elements presented are read in conjunction with the grouping of elements in any standards-based management system, it will be seen that they broadly correspond. Developing an approach around these main elements will place the company in an appropriate position to establish an effective system which will meet the requirements of management system standards.

The key enablers for the above elements are:

- change management;
- leadership management;
- organisational management;
- human resource management.

These aspects have been touched on previously. The way in which the elements and enablers are applied will be seen subsequently as the key elements of system development are presented.

Stages of system development

There are four broad stages of development and implementation involved in an IMS:

1 *Conformance* – ensuring that management sub-systems conform to all appropriate requirements of relevant systems standards, legislation and the company's policies and objectives.

2 *Procedures* – preparation of the appropriate procedures in relation to products and processes, provision of services and, where relevant, projects, which perpetuate the business.

3 *Documentation* – preparation of appropriate documents which define, describe and explain the business processes and their management.

4 *Implementation* – deployment of procedures as embraced by documentation, the maintenance of performance records, review and improvement.

It is clear that the four stages of system development and implementation are more complicated than appears in the enumerated headings. In fact, there are a multitude of system elements and sub-elements which need to be assembled. Most systems adopt key elements for simplicity, yet within each there are many aspects to be considered. Likewise, management system standards require the inclusion of many aspects. As a consequence a system may be highly detailed and complex. These four stages will be expanded upon subsequently when system documentation, plans and records are explained.

Key elements of an IMS

Any management system comprises a number of fundamental and essential elements. These form the broad structure of a system within which sub-elements also form important aspects of the total system. A company may use slightly different terminology to define and describe the elements of its system but essentially they follow a conventional nomenclature. For the purpose of this section

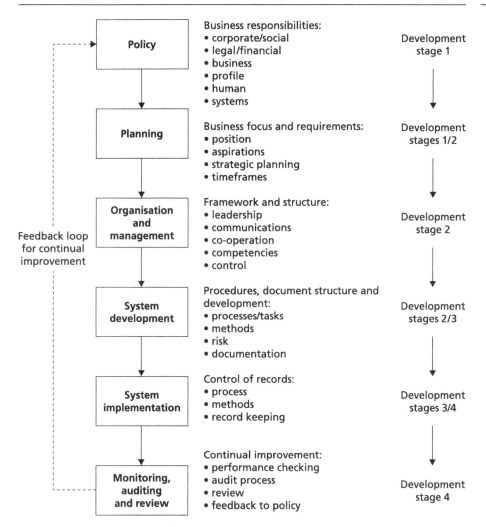

Figure 5.16 Key elements of an IMS

it will be assumed that a company wishes to develop and implement a system which integrates quality, environment, and health and safety. In practice, a company may establish any combination of management functions. The key elements, illustrated in Figure 5.16, are as follows:

- Policy: business responsibilities – development stage 1.
- Planning: business focus and requirements – development stages 1/2.
- Organisation and management: framework and structure – development stage 2.
- System development: procedures, document structure/development – development stages 2/3.

- System implementation: control of records – development stages 3/4.
- Monitoring, auditing and review: continual improvement – development stage 4.

These key elements form the fundamental structure of the IMS. These same elements can be used as the basis for the development of other support systems. As the elements are related to stages of system development it can be seen that there is a natural overlap and interaction of the elements. The key element of risk assessment, subsumed within planning, is not always overtly displayed in all system configurations. This would depend on the explicit nature of the management support processes involved. Risk assessment would always feature in a system which incorporates quality, environment, and health and safety, since assessing the mitigation of risk is intrinsic to the core values of each management function. The key elements correspond broadly with main sections of each of the management system standards ISO 9001, ISO 14001 and OHSAS 18001. Inevitably, the different foci of each of the standards mean that particular elements will differ but the underlying themes and intentions are present. Within system documentation the elements and sub-elements will be strictly defined and described in an explicit context. The following sections are more descriptive to provide explanations of the relevant system thinking.

Policy: business responsibilities

Company policies on quality, environment, and health and safety influence virtually all organisational activities and decisions taken. They have a strong bearing on the selection of resources, methods of working and delivery of products and services. Furthermore, the policies develop and establish a fundamental supporting culture within which all organisational activities are accomplished. Equally, policies serve in fulfilling an organisation's responsibility towards business legislation, in satisfying regulations which govern its activities and, in a wider remit, they allow the company to address public accountabilities.

For an IMS a policy is 'a set of statements which reflect the company's intentions in relation to the management of quality, environment, and health and safety'. These may be individually focused statements brought together under an umbrella statement or a broader and more generalised all-inclusive statement. Whichever is used, the policy should first define the company's philosophy and thinking on quality, environment, and health and safety in the context of its business activities and marketplace, and, second, be clearly presented in the form of a company policy statement. It is not always easy to present an integrated policy statement that holds currency. Changes in company activities or amendments to legislation mean that policy statements require adaptability. Also the complexity of bringing three functional management areas together may inhibit the clarity of an all-embracing statement and, therefore, individual policies may be preferred.

There are a multitude of influences and organisational responsibilities which need to be considered in determining company policy for an IMS. These are as follows:

- Corporate and social responsibilities:
  - philosophy and culture;
  - business ethics;
  - effects from business activities;
  - public accountability;
  - stakeholder satisfaction.
- Legal and financial responsibilities:
  - EU Directives;
  - international standards;
  - national and local legislation;
  - industry sector regulation;
  - investment;
  - risk and loss;
  - insurances;
  - budgetary control.
- Business responsibilities:
  - vision, mission and aims;
  - policy commitment;
  - objectives and targets;
  - industry and sector marketplace awareness;
  - commercial positioning;
  - customer focusing.
- Company profile responsibilities:
  - public image;
  - intra-organisational perspective;
  - product and service promotion;
  - company literature;
  - media interfacing.
- Human resource responsibilities:
  - recruitment and retention;
  - staff deployment;
  - training and staff development;
  - incentives;
  - welfare.
- Management system responsibilities:
  - organisational structure and processes;
  - procedures;

- resources;
- performance measurement;
- continual improvement (process and system).
- Communications and information technology responsibilities:
  - intra-organisational information;
  - collection, synthesis and analysis of data;
  - use of information technology.
- Operations responsibilities:
  - quality of products and services;
  - safeguard of environment;
  - safe and healthy working procedures;
  - emergency practices;
  - contingencies provision.

To embrace all of these responsibilities in an appropriate way a company must establish a policy which is:

- relevant to the core processes of the business;
- clear and unambiguous and reflects holistic vision;
- established by executive management;
- committed to throughout the organisation and its activities;
- transferable into a workable management system;
- supportive of practical procedures and instructions;
- transparent to sector interests, regulatory authorities and certification bodies;
- amenable to performance measurement, audit, review and continual improvement.

The policy of a company is a core statement of its business intentions. It is the pronouncement against which outside parties perceive and often judge the performance of the company. If an organisation is perceived as not fulfilling its policies then many, if not all, aspects of its activities may be called into question. Therefore, stating a clear policy, meeting the policy and, above all else, being seen to meet the policy are of fundamental and significant importance. Policy making and statement are wholly the responsibility of corporate executive management. It is often said that 'policy comes from the top', and this is true. Policy reflects corporate philosophy and business responsibility, and therefore it must emerge from this senior level. That is not to say that policy making and statement are considered draconian. Rather, effective policy will permeate throughout the entire organisation and assist managers and employees to appreciate their role and duties within the organisation. Clear policy also helps in developing structure, processes and procedures, which again is advantageous to the workforce.

Policy will be determined by the characteristics of the company and will be made as such. Notwithstanding, irrespective of company nature and orientation a policy should be formalised in written form. An IMS policy should reflect the company's position in respect of its activities as they impinge upon quality, environment, and health and safety. The policy statement should be a formal written declaration of intent covering the scope of each management function as it applies to controlling the core processes. The statement should be:

- clear and amenable to subsequent performance evaluation;
- published with company identity;
- transparent to internal and external scrutiny and judgement;
- structured for robustness while maintaining flexibility to accommodate fluctuating business circumstance.

The statement of policy forms an initial and important part of management system documentation. It will be included as an early section in the IMS manual and set the tone for subsequent system documentation. The philosophy and culture that policy can set throughout the activities of an organisation should not be underestimated. In many ways the policy generates and maintains a mindset among management and workforce that their work matters and that it is part of a combination of practices that maintains a company standard. Policy making is as much about engaging the commitment of employees as it is about stating the company's intentions to the wider community. A watchword on policy formulation is that some organisations over-detail their policy and in so doing can lose focus and clarity. A policy needs to be relevant but overarching and authoritative, with detail left to other system documentation. A robust policy helps establish the company's vision, culture, dedication and commitment to the business, without which it would likely flounder. Policy is, therefore, a prerequisite to developing and implementing an IMS.

Statement of policy

The following paragraph, taken from a principal contractor's dual-system environmental and safety IMS, illustrates the nature of a company's policy statement. As all organisations are different, the statement should be taken only as one example of approach. It can be seen in this example that the attempt was made to provide a statement embracing both health and safety and environmental policy. Brevity is exemplified within the statement, yet it can be seen that there are appropriate references to the company management manual and supplementary documents which contain the details of the management system.

Environmental Quality Policy

The continuing policy of the Company is to provide our services in such a manner as to ensure, in so far as is reasonably practicable, the health, welfare and safety of its employees and the general public together with maintaining safeguard of the natural environment and all within it relevant to the delivery of quality services by the Company.

The Company commits to its responsibilities for complying with current legislation and regulations and to cooperate with those responsible for their monitoring and enforcement.

The Board of Directors have appointed an executive member to oversee all Company activities relevant to quality, environment, and safety. The implementation of policy in respect of these functions is contained in the Company Management Manual and supplementary documents as required by our service provision. In addition, all our employees have a duty to take practicable care in the execution of their duties relevant to safety, the environment and the provision of quality.

This Company Policy will be reviewed as required based on Company experience of the provision of our services.

Signed on behalf of the Company ....   Date ....

Importance of policy statements

An interesting example of policy statement development was highlighted during a research study undertaken in the Far East. At first glance it would appear somewhat animated and perhaps unconvincing, yet it underlines the importance of communicating company policy both within the organisation and to external onlookers:

> A large construction and engineering practice based on the twenty-fourth floor of a high-rise office complex is accessed by elevator from the ground-floor entrance. Upon arrival at floor 24 the elevator doors open to reveal the company lobby. On the wall directly opposite the elevator doors are hung the organisation's ISO 9000 QMS Certificate, ISO 14001 EMS Certificate and company policies on quality and environment. These are displayed in oversized gold-coloured frames. Every employee entering the office complex together with visitors, clients and customers see this prominent display reflecting the company's high standard and commitment to these important management functions. Upon discussing this with the chief executive officer it emerged that: 'the company had taken the decision not to hide the certificates and policies away but to hang them in such a place that the first thing one sees when entering the practice is their commitment to quality and environment and the last thing one sees when leaving is a reinforcement of the same'. Discussion with visitors revealed that far from being perceived as tawdry it was seen to reinforce customer confidence in the total service of the practice and also employee attachment to the workplace.

In this way it can be seen that policy statements do much more than express the business intent of the company. Policy can perpetuate ethos, culture and commitment while also generating positive external perspectives about the organisation and how it goes about its business. All of these aspects and more can only be good for any company.

Planning: business focus and requirements

Effective planning is a prerequisite to the development of a successful IMS. Planning commences long before system development and it was explained earlier that a company will engage in a *preparatory business review*. At that stage, the company will ask fundamental questions of its business and organisation to ascertain: its orientation, configuration and position in the marketplace; where it wishes to place itself in the future; and how it intends to achieve its ambitions. In undertaking the review it will determine how it intends to frame and structure the core processes which deliver its products and or services together with the required support and assurance processes. It is these which come together to form the management system.

Planning not only considers the structural components of a system, but also considers the human resources required to make the management system work effectively. It therefore looks at how a management team will be formed and how management leadership will be provided within the structure. A crucial aspect of planning is to determine the current position of the organisation in relation to process and system development, a process initiated at the preparatory business review stage, and what will be required to move the exercise forward – an aspect referred to, conventionally, as gap analysis. A concluding aspect of planning is to consider an appropriate timeframe for system development and implementation, as often an extensive period is required.

Management and the team

Management commitment is absolutely essential to the establishment of the management system. The drive and efforts of management will be needed from the preparatory business review right through to certification of the system. Few business endeavours are successful without an appropriate team of people, their ability to work together and effective leadership. Selecting the system leader and sub-systems leaders and forming a team are crucial to success. The nature of the team and its leadership will vary from company to company according to its needs and size. A small organisation may require only a small core team while a large and complex organisation may require a larger core team with many persons in support.

An essential facet of the establishment of the team and its leadership is that the company must ensure that roles, duties, responsibilities and, most importantly, authority are well defined and clearly stated. Potential ambiguities and conflicts of interest must be avoided. It was discussed earlier that the potential for conflict is high among different management functions, a characteristic which integrated management seeks to alleviate. The system leader should have access to corporate management. This is important as it is the leader's responsibility to ensure that company policy, aims, objectives, goals and targets are communicated to sub-system managers and the workforce, and that appropriate procedures and working practices are employed. This is also important in communicating to the workforce that the company is committed to its policies and business responsibilities.

Part of deploying management and forming a system team is the provision of training. The company should ensure that management and workforce are suitably aware, knowledgeable and skilled to implement the system. Although this may appear obvious, the complexity of implementing an integrated quality, environmental, and health and safety management system requires a knowledge base both broad and deep. Again, constant change in business circumstance, regulation and standards will require a programme of training to ensure current and leading-edge knowledge. There may be a requirement for external inputs to the training process, and the company should plan for this. An example would be education and training in the processes of management system auditing and certification.

Gap analysis

An important part of developing and implementing a management system is the examination of the core processes of the business with a view to identifying where gaps exist in their management. More fundamentally, it may involve the determination of difficulties with parts of the processes themselves. This will usually manifest in non-conformance to specified processes or procedures. In terms of an IMS, it will involve an analysis of core, support and assurance processes to see where new or amended procedures will be needed to ensure that quality, environment, and health and safety are maintained in all products or services delivered. Gaps will also need to be identified between the company's management sub-systems and those requirements specified by management system standards.

An IMS differs from individual management systems in this regard. Part of gap analysis will involve the determination of compatibility between the sub-systems activities, quality, environment, and health and safety, and how these can be brought together in integrated documents and records. There may be natural overlap in some aspects which facilitate easy integration while other aspects may require specific documentation to be created.

There is an important link between potential gaps in business processes, documentation and the team who implement the system. It is essential that gap analysis involves those persons at the sharp end of the processes as it is they who can more readily identify where difficulties have arisen and how they might be alleviated.

Corrective actions

Once gaps in processes and their management have been identified, *corrective actions* need to be taken. Any non-conformance to specified standards of performance must be rectified by modifying parts of or possibly an entire process. Where appropriate, it is important that modifications are made in accordance with the specifications of management system standards. As a system comprises documentation, gaps in processes or procedures will also need to be reflected in new or amended documents and record mechanisms. Training was identified earlier as important to the planning process and is maintained here, as any corrective actions taken to bridge gaps in processes may well require system personnel to be updated in terms of their knowledge or skills.

Timeframes

As the development and implementation of an IMS can be complex and time consuming, it is important for the company to set a timeframe for its establishment.

The timeframe is an important management tool for communication about system development and its relationship to other organisational activities. A timeframe for the development and implementation of a system from the preparatory business review through to certification could be anything up to three years, and so maintaining control of time is essential. As time invariably means money to a company, cost budgets for development will also need to be prepared. Planning will therefore involve the preparation of a master development programme incorporating key milestone targets together with a detailed cost budget, resource schedules and contingency plan. The last is, of course, important should the process experience delay.

Organisation and management – framework and structure

An IMS is heavily influenced by the organisational structure established by the company. It was seen earlier that the adoption of an integrated approach requires that the company, to some extent, challenges convention. It needs to adopt a framework and structure that allow management functions to work as effective support and assurance processes, or as sub-systems, within a single system focused on the core business processes. As explained earlier, this may be far from straightforward and not without difficulty. The crucially important aspect of organisation in this regard is that it must facilitate the undertaking of multiple management functions concurrently. It requires the clear definition and formatting of quality, environmental, and health and safety management in an integrated form within the organisational structure. If this position is ambiguous it will produce organisational uncertainty and confusion in the minds of employees, which will ultimately permeate to procedures and documentation and lead to partial or total system failure. Of equal importance to establishing physical structure to the company, organisation allows employees to place and understand their position in the company. Organisation depicts their role, duties and responsibilities and their relationship with other persons. It also provides the wherewithal for key elements to be established which are essential to the implementation of any management system, namely:

- communication;
- co-operation;
- competencies;
- control.

Communication

A company, its organisation and its management functions receive a multitude of communications and information. This can be intra-organisational from individuals and groups operating within the company through formal and informal routes. It can also be inter-organisational from other organisations involved with the company's business, from external parties such as regulatory bodies or from the business environs, for example from the public. Communications and information also flow out from the organisation, for example to stakeholders in the business. The ways in which a company handles communications and the information needed to upkeep its business are paramount. In supporting the core business processes, an IMS is required to channel, assimilate, synthesise and evaluate communication and information efficiently and effectively. It is also required to handle the three sources of information highlighted above. Due to the complexity of handling information concerning two or three management functions simultaneously, channels of communication and information and their clarity are most important. The formal organisation of a company necessitates that communication will be up, down and across the organisation. Structuring communications and information flows is essential if they are to function appropriately and effectively. Without an overarching regime, all parts of the company will be transmitting and receiving in an ad hoc way such that communication and information become dysfunctional.

Co-operation
Co-operation by individuals and groups both within and outside the company is important to business success. Company organisation must be fashioned in such a way that co-operation is encouraged. This entails structuring the formal organisation so that those management functions constituting the support and assurance processes underpin the core processes. This requires the formalisation of channels for co-operation so that synergy can perpetuate rather than disharmony. Within large companies in particular, the numbers of personnel can make the organisation of co-operation far from easy. Inevitably, there will be individuals and groups who function alongside or outside the requirement for free and easy co-operation. This is often borne out of inadequate organisational structure, where ambiguity of role can lead to dissatisfaction. Company organisation should specify where and why co-operation is required and how the contribution of individuals and groups is valued by the company. Involvement through carefully conceived organisation is vital if those policies which lie at the heart of the company's business activities are to be fulfilled. Although vision, policy and objectives are executive driven, the shared effort and commitment of all employees are needed for them to be delivered through the procedures implemented. This aspect is particularly important for support and assurance processes. There might be a temptation to favour core process resources to the detriment of support process resources, and therefore management should ensure that equality of purpose and value is given.

Competencies
It is axiomatic that effective organisation is based on competent personnel. However, the knowledge and abilities of managers and workforce should never be taken for granted. An IMS requires a greater range and amount of detailed knowledge than that required for any individual management function. However, a manager leading an integrated system may require less detailed knowledge if supported by highly knowledgeable staff, but may require a greater range of interpersonal abilities to communicate with, co-ordinate and manage system

personnel. Company organisation must provide for knowledge building and skills training. For a management system to function effectively, it requires not just a robust system but employees who appreciate and understand what that system seeks to achieve. The requirement is to maintain competent personnel who have awareness, knowledge, skills, capabilities and a healthy attitude to the company's activities. While much will be natural in the individual characteristics of the managers and workforce, some will require organisational support. In the case of quality, environment, and health and safety there is probably far too much information needed for any one person to be naturally gifted with all the information or to remain updated of developments in the management function. Therefore, training will be required. The training required may be organisational, for example education concerning the implementation of policy or performance targets. It may be job related, for example awareness of personnel duties and responsibilities. It may also be individual, for example instruction in new working techniques. Effective training is vital to maintaining a competent workforce, and a competent workforce is essential to underpinning the implementation of an effective management system.

Control

Like any management system, integrated management must exert control of its support and assurance processes, which in turn control the core business processes. A company must provide organisation within which control mechanisms are central to its philosophy. This is especially important in the management functions of quality, environment, and health and safety. Control is paramount to mitigating the potentially damaging effects from business activities. Efficient and effective control is concerned with planning, monitoring and review. The development of plans is needed which reflect company policy and objectives in the functional management areas. Consistent and vigilant monitoring of processes and outputs is required to ensure that plans are implemented effectively. Review of activities and performance is needed to learn from difficulties experienced, and actions taken such that continual improvement to process and system is ensured. An important aspect of control is the assignment of responsibilities and setting of performance expectations. The configuration of company organisation should have clear reference to individuals and groups and their responsibilities, and also specify performance levels. This can be contained within job descriptions and monitored and updated as needed through periodic performance review and staff development programmes. This is an area which can be problematic in dual and triple management functions because of the range of responsibility and sheer amount of technical information. Moreover, the frequent change and evolution of management system standards and legislation concerned with quality, environment, and health and safety only exacerbate the issue. This makes frequent review of activities, responsibilities and performance more important, and compels a strong link to continuous training and education by the company.

An absolutely essential aspect of organisation has been left until last, that of managerial leadership. A company's organisation must be clear in terms of its

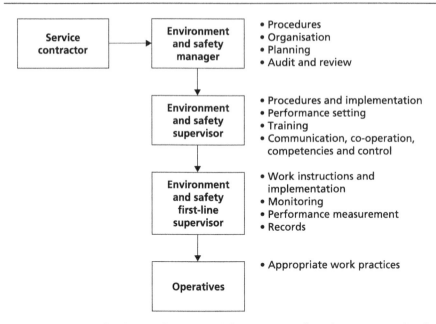

Figure 5.17 Organisation and personnel for integrated environment and safety approach

leadership. It should have a strong and committed leader at each key point in the organisation to distil policy, objectives, goals and targets, to ensure that workable procedures are maintained, and to support, motivate and help train employees. Furthermore, leaders must be effective communicators and have the ability to handle the multitude of information which the company generates. Effective leadership is concerned with eliciting participation, co-operation, effort and commitment and with ensuring that the right conditions exist and are maintained to enable employees to deliver to their potential. It is sometimes said that work should be fun, but certainly a contented workforce are likely to perform better, and leadership therefore is about facilitating this.

Figure 5.17 shows the organisational structure for an integrated environment and safety management approach by a large service contractor operating in the engineering sector. While this industry sector is influenced by specific legal requirements and regulations, this is not central to the example. The focus of attention is towards the organisation put in place to develop and implement environmental and safety policy, procedures and working practices throughout the various levels of management, supervision and operations. It can be seen that the system flows from the corporate organisation to implementation, and maintains its focus throughout. To ensure this, management leadership is provided at each level in the system.

System development: procedures, document structure and development

An IMS seeks to ensure that the core business processes are supported in delivering quality outputs using healthy and safe working methods while mitigating potentially adverse environmental effects. Business processes are characterised by the ways in which they are undertaken and management systems by the ways in which they are explained by documents and records.

There are, therefore, three requirements to be developed:

1 Procedures for carrying out a process, parts of a process and sequences of processes.
2 Documents for defining, describing and explaining procedures.
3 Performance records for procedures carried out.

It was mentioned earlier and should be noted now that a *document* is different to a *record*.

Documents:
Describe a management system's operation – what procedures are used and how they are undertaken.

Documents are often termed controlled documents, as within a management system development, circulation, use and update are strictly controlled. *Version control* is adopted so that when updated or new versions are introduced the previous versions are withdrawn.

Records:
Provide evidence that the management system operates according to the documents.

In examining the core processes of the business, their requirements for quality, environmental, and health and safety management will be considered. These will form the management procedures for the support and assurance processes. It is these procedures that need to come together to form the IMS to manage the core processes. An IMS will come to fruition when the procedures operate across the management functions and throughout the core processes seamlessly, harmoniously and with synergy. To maintain continuity with the logic of management system development, procedures and documentation are explained next while records will be covered under the heading 'System implementation'.

Procedures

A procedure describes the methods of working to be used when carrying out a process. A process may comprise a number of individual work tasks, and they need to be identified and described in terms of what procedures are to be used to complete them. A process may be simple and short in duration or can be complex and time consuming. Moreover, sequences of processes are often required to deliver any product or service. In the case of a product, a process embraces not only the technical content and materials element as inputs, but also the application of human effort. This requires a socio-technical understanding of how any process in undertaken.

Processes and tasks

To develop procedures for undertaking any process, a process needs to be broken down into tasks, these being the types and sequence of activities that combine to carry out the process. An analytical method referred to as *work breakdown structure*, or *WBS*, is often used. This approach breaks down a process into its constituents, which can be examined in detail and the impacts of the method of working, materials, equipment and human inputs can be determined. From this analysis, a procedure, or set of procedures, can be developed which will complete the process effectively and with optimum efficiency. The procedures derived from the analysis of each process are recorded in a document commonly termed a *method statement*.

Procedures require a two-level analysis. The first is the development of general management procedures applicable throughout the execution of all processes, and the second is specific procedures to accommodate particular characteristics of a process, or special tasks that need to be conducted within the process, or tasks specific to work location. This may also apply where services rather than products are the outputs. Generic procedures may apply throughout the company to describe its services but specific procedures may apply when those services are delivered in a particular contractual or project situation. An example of this is seen in the petrochemical industry where a corporate organisation has in place a set of management procedures covering, for example, environmental and safety aspects of all its worldwide activities and augments these with sub-sets of specific environmental and safety procedures applicable to its remote production sites.

To differentiate the general from the specific, a company may designate its generic procedures as *management procedures* while specific procedures are designated *working instructions*. In addition, where a service is provided in a project situation, procedures may also be subject to *site rules*. As the term implies, these are site specific and may involve, for example, compliance with explicit regulations.

Procedures and management functions

When developing procedures, the management functions of quality, environment, and health and safety need to be considered. Tasks which involve a potential for compromise in any of the three management areas need to be examined. While quality is fundamentally concerned with complying with customer requirements for product or service performance, environment, and health and safety are concerned with compliance with legislation. Together, these aspects should ensure, for all processes, that employees:

- carry out tasks to required performance levels in a safe and healthy manner using methods which do not adversely impact upon the environment;

- are aware of potential compromise to performance, any hazard to the environment and any danger to their safety and health;

- implement measures in the course of their tasks to eliminate or reduce the risk of compromise to performance, hazard to the environment and danger to their safety and health.

Risk assessment

All three management functions have a key aspect in common, that of *risk assessment*. Risk assessment is an analytical technique used widely in industry, business and commerce. It involves three key activities: (1) hazard identification; (2) evaluation of risk; and (3) prevention and protection measures. A sound basic framework for risk assessment during the development of procedures will involve the following key steps:

- Identification of the tasks to be carried out, the risks associated with those tasks, which employees will carry out the tasks and what equipment and resource inputs will be used.
- Determination of who might be placed in a compromised situation with regard to quality, environment, or health and safety.
- Evaluation of potential risk to alleviate compromise or, where this is unavoidable, minimise the effects of compromise.
- Record the outcomes from risk assessment to assist company awareness and in establishing appropriate measures of control.
- Review experiences for future risk assessment of processes and procedures.

These steps are consistent with the PDCA methodology and commitment to continual improvement of processes and management systems.

Method statements

Method statements which assume an integrated format can incorporate management procedures, working instructions and any site rules for quality, environment, and health and safety and link them directly to processes and tasks. Method statements assist in planning the methods and resources necessary for undertaking processes and are also used in monitoring and control, as activities are ongoing. Specific documents in the style of a pro forma are commonly used as method statements.

A method statement should:

- describe the processes and tasks;
- identify the physical location of the work (pertinent in large/busy environments);
- specify the management/supervisory arrangements (some tasks may need specific arrangements, for example safety regimes or environmental mitigation measures);
- state the inputs to the process/task (materials, equipment, people);
- identify potential risk/compromise to the task;
- specify precautions to be taken in carrying out the tasks.

Written procedures

As procedures are developed from the analysis of processes and tasks they are formalised in written form and brought together in a compendium of procedures. These descriptions and explanations of activities form the core of system documents and record keeping. Documents and records are presented in a hierarchy of documentation. This will be seen in the next section of this chapter. It is extremely important that the company develops a clear and consistent structure of documents and record-keeping mechanisms. A coherent and transparent structure and document trail will be required for effective system implementation and also in satisfying the requirements for system auditing and certification.

Document structure and development

Document hierarchy

Consistent with management system standards, documents and records assume a *document hierarchy*, frequently referred to as the *document pyramid*. Introduced earlier, the document hierarchy forms a pyramid shape as there is one document at the top – the policy – supported by an increasing number of documents throughout the lower levels of the pyramid to implementation at the base. This is shown in Figure 5.18.

Document levels

Convention suggests that there are three levels of system documents: manuals; procedures; and work instructions. However, such documents, while meeting the requirements of many applications, would likely be generic in orientation, and

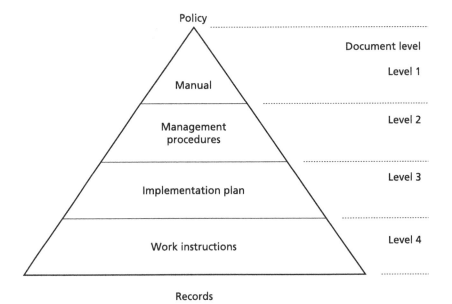

Figure 5.18 The document pyramid showing the four levels of management system documentation

there needs to be some reference to situation-specific use, hence the inclusion of an implementation plan. Therefore, a document hierarchy for an IMS, like other systems, tends to have four levels:

1 Manual

2 Procedures

3 Implementation plan

4 Work instructions.

In some systems levels 1 and 2 may be embraced within the manual while in other systems the two are kept separate. Level 3 may or may not be required in some systems, with application depending upon the specifics of application. Level 4 also includes record-keeping mechanisms in almost all systems.

Level 1
The manual is a documented introduction to the management system and contains the following:

- Management policy (and rationale).
- Organisation structure (in outline).
- Management responsibilities.
- Procedure references (document trail to level 2).

Level 2
The procedures describe processes and how they are to be managed and contain the following:

- Definition of terms (within the description of processes).
- Description of the processes.
- Process responsibilities (of managers).
- Implementation plan/working instructions references (document trail to levels 3/4).

Level 3
The implementation plan describes the characteristics of a specific application of the procedures, for example a service provided within a project scenario, and contains the following:

- Definition of terms.
- Management plan (specific to situation).
- Special processes (where relevant).
- Programme (specific to situation).
- Special procedures (where relevant to special processes).

- Work instructions references (document trail to level 4, where working instructions are affected by the specifics of the situation, such as special procedures).

Level 4

The work instructions describe tasks and how they are to be carried out, and contain the following:

- Definition of terms (within the description of tasks).
- References to standards and specifications (performance criteria).
- Training and instruction requirements (for specific tasks).
- Work instructions (written guidance/checklists on how to carry out the tasks).
- Record-keeping references (document trail to written records).

Document organisation

The organisation of system documents is essential to the effective structure and implementation of a management system. The amount of documentation in the system manual, sets of procedures, implementation plans, work instructions and records will be considerable, and all of these must be controlled. Document organisation is as much about recognising what is not in the system as what is in the system. Only that documentation which is truly part of the system should be included. For example, if certain environmental management documentation does not impact upon safeguarding the environment in the course of undertaking the processes then it does not need to be included. Conversely, quality checks of a product realisation process would be essential.

Document organisation should have the following:

- A formal structure.
- A development process.
- Ownership of documents.
- Identification of documents to activity.
- A master file and master list of all documents.

A useful starting point in organising system documents is to draw up a *document organisation chart*. This allows the company to refer to documents in relation to the organisational structure, the business processes and management responsibilities. Hierarchy forms a common thread between the organisation of a company and system documentation. Therefore, establishing the document hierarchy in the form of a chart allows document levels to be clarified and numbering to be established. It also provides a further opportunity to identify gaps in processes and procedures that may have been overlooked in earlier gap analysis such that corrective actions can be developed.

Documentation organisation involves relationships and correspondence with the inputs of third parties to the company's processes and procedures. This will be looked at subsequently in supply chain management. There may be an impact on the ways in which the company operates as a result of contributions from designers, consultants, suppliers and sub-contractors, and where applicable these need to identified and evaluated in the context of the organisation's activities. Third parties may implement their own management systems and have different ways of dealing with quality, environment, and health and safety, and these should be examined with a view to ensuring consistency and compatibility of approach. Their documentation might be different, and again this may require some degree of scrutiny.

In addition to system-specific documents, a company will likely have other documentation which may be pertinent to the operation of the management system. Where this is the case, such documents should be drawn into the system structure or alternatively remain outside the structure but with reference made to it. Examples of this include company handbooks, common to most organisations, software manuals for computer systems, and operating instructions for equipment.

Functional breakdown of documents

It is normal to use the document organisation chart as a broad base for structuring documents and then to break this down into smaller elements for more detailed attention. This approach is termed *functional breakdown* of documents. Breakdown can be by company divisions, departments or sections but it is conventional to structure documents by management function, for example human resources, training, finance, legal, marketing. In the context here the pertinent functions are, of course, quality, environment, and health and safety. The advantage of functional breakdown is that there are direct and strong links between: (1) the requirements specified by management system standards; (2) the processes used by companies to structure the delivery of their products and services; (3) the functions of management as they control organisational activities; and (4) the structure of system documentation.

Further advantages are that functional breakdown is clear and well understood by employees, and therefore they have a good sense of why things are done in a certain way and can see the part that they play in them. Also, documents structured in this way are arranged in broad groups and the boundaries between different functions are clear. Of course, an IMS must bring these functions together. The advantage afforded by functional breakdown is that, broadly, the management functions for quality, environment, and health and safety have similar elements which assist integration. This similarity pervades the links mentioned above and again this is helpful.

Document development process

Any document requires writing at conception or rewriting when being updated, and this must be controlled. Documents, therefore, should be conceived as a strict *document development process*. A practical process for document development contains the following stages:

1 *Need identification* – the requirement for a document is identified, sanctioned by the system document co-ordinator, assigned a document identification number to signify its place in the set of documents and responsibility given to the document owner, the manager whose functional area the document is relevant to.

2 *Preliminary planning* – the document owner considers the document required based on the purpose (need within process and procedures), the users and resources.

3 *Initial draft* – the document owner writes the initial draft of the document, usually in consultation with the functional management team, and submits it to the system document co-ordinator.

4 *Final draft* – the system document co-ordinator checks and edits the document for operational efficacy and submits it to the system authoriser. Where this is a change to a document, this will be signified by a document change notice to the current document.

5 *Authorisation* – the final draft is authorised by company management by signing off the document, if new, or the change notice, if applicable.

6 *Distribution* – the document co-ordinator circulates the document to copy holders for application within the management system. A copy is filed in a master document file and the master document file list is updated.

Document requirements

Management procedures, work instructions and record-keeping mechanisms will be required for each and all processes involved in delivering a business product or service. As a rule, these will focus on satisfying those requirements specified within the management system standards. If for example a QMS was being developed in isolation, then the procedures and work instructions would focus on complying with the ISO 9001 specification. Each element contained within the standard would be reflected in the system documents. The standard specifies what needs to be complied with in terms of the elements but leaves the company to decide how best it chooses to meet these given the nature of its business. Therefore, a company has flexibility in meeting the requirements for a system within the standard as long as it addresses all of the requirements.

Integration of management functions and documents

Where a single management system is involved, system requirements are reasonably straightforward. Where dual or triple systems are involved, it is more complicated. It would be unrealistic to suggest that a company can implement a completely seamless single system to embrace multiple management functions. What an IMS achieves is that, where it is possible to integrate quality, environment, and health and safety management functions within a system, then it does so. Some elements will not be amenable to integration and where this occurs then those elements are allowed to operate within the system as individual but related, co-ordinated and systematically controlled elements.

It was explained earlier that a *semi-integrated* management system and a *fully integrated* management system seek to bring elements of functional management together where possible. In the semi-integrated approach separate policies and procedure manuals were apparent but with integrated implementation plans and work instructions. In the fully integrated approach policies, procedures, implementation plans, work instructions and records were all brought together. As processes and tasks are examined through work breakdown structure and as procedures are developed to manage them, the three support process management functions – quality, environment and safety – should be considered and where appropriate built in to the procedure. The focus should be on including these not as separate functions around the processes and tasks but as integrated functions. For example, in developing procedures to carry out 'monitoring and measurement', the requirements of ISO 9001, ISO 14001 and OHSAS 18001 should be consulted and an all-embracing procedure developed for that element. Likewise, work instructions and record-keeping mechanisms should be developed in this way. The documentation associated with each procedure, work instruction and record can be integrated accordingly. See Figure 5.19. It is here that integration can be achieved, so leading to organisational benefit through reducing duplication of activities and paperwork.

It should be remembered that integration is concerned with much more than integrating text in documents – it is about consciously undertaking an activity with thought for multiple requirements. Whenever a procedure is carried out and records of its undertaking are gathered, there is a fundamental need to think in terms of quality, environment and safety. Only when this is done will integration be achieved in terms of procedures and practice. This has as much to do with

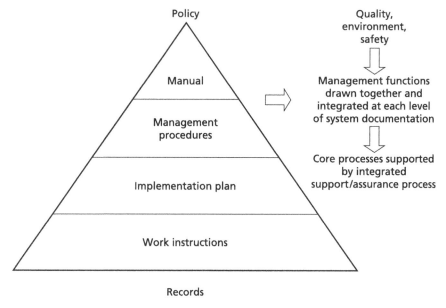

Figure 5.19 Document hierarchy for an IMS

the efforts and commitment of functional managers and the workforce as it does with system documentation structure and development. Effective system implementation is therefore fundamental to success.

System implementation: control of records

As processes and their associated tasks are completed, records should be gathered and maintained to ensure that there is evidence of conformance to specified system requirements. An effective management system will have in place robust mechanisms to facilitate this. As with document organisation, the maintenance of records should be a formal process. A practical process for records development and implementation contains the following stages:

1 *Need identification* – requirements for records identified, responsibilities defined, content of the record determined, ownership of record established and how versions will be controlled.

2 *Acquisition* – establish what is required for each record, how evidence will be collected, how record will be maintained and responsibilities for information collection established.

3 *Collection* – implement process to collect evidence from activities when carried out.

4 *Indexing* – process of acknowledging receipt, filing and indexing for storage, retrieval and disposal when new/updated versions are introduced.

5 *Filing* – reports are filed in predetermined organisational locations for storage (hard copy and electronic).

6 *Storage* – provided centrally by the corporate organisation or possibly on-site storage where operations take place remotely.

7 *Disposition* – discarding of records as they are replaced by updated versions and disposal of same to avoid potential record ambiguity.

The fundamental purpose of records is to demonstrate with objective evidence, both quantitative and qualitative, that activities conformed to documented procedures, ensuring that they were carried out appropriately and to the requirements of management system standards. A record-keeping mechanism will be required for each procedure, so when procedures are being developed it is as well to consider how records relate to the procedure documents. If appropriate methods have been developed then there will be a direct and strong link between the procedure manual, implementation plan, work instructions and record-keeping mechanisms.

Auditing and review: continual improvement

Definition

An audit is:

The detailed evaluation of the management system and documentation.

Purpose

A company needs to have in place suitable measures to provide management with information on the conformance with and effectiveness of the management system. This is provided by the development and implementation of a system auditing plan and a structured auditing process. The output of the process supports the development of corrective actions to any shortcomings or difficulties in the way in which core and support processes are undertaken. Auditing provides the feedback loop from system implementation to corporate management to facilitate a review which informs update and revision to company policies and objectives. Auditing and review are therefore linked and are both essential to the maintenance of an efficient and effective IMS.

A management system audit is internal to the company and may be conducted by company personnel or, where required, by external consultants. The audit should focus on the implementation of the management system yet will also consider the dynamics of the organisation and the behavioural aspects of system managers and their impacts on system operation. This is necessary because auditing will evaluate the effectiveness of actions taken by the organisation to address identified difficulties with managing company processes.

Methods

Audits are normally carried out annually and are formal in nature. The core business processes and support processes together with all elements of the management system will be looked at and evaluated. In respect of controlled documents, all will be scrutinised to a schedule and procedure which ensures that all are treated in the same way. For system records, all will be scrutinised with an emphasis on evaluating the effectiveness of actions taken when process or system difficulties were encountered together with how processes and procedures were revised to implement improvements. Audits may be carried out by *longitudinal analysis* or *cross-sectional analysis*. These are complementary methods and often used in combination. The former involves an in-depth examination of a specific element within the management system, for example the planning aspects of a process. The latter involves the detailed examination of a particular activity or task within an element, for example the input of an adviser to planning the specific element. Together, these methods enable the company to carry out both macro and micro analyses of the processes it uses to deliver its products or services.

Organisational aspects

An effective audit will involve the evaluation of aspects throughout the company linking practice back to policy making. It will therefore look at:

- current company policy, aims and objectives;
- corporate and operational organisation and resources;
- responsibilities of managers;
- core processes;

- support processes (management functions and the management system within which they are deployed);
- control mechanisms (system implementation) to ensure compliance with policy and objectives;
- record-keeping methods (including actions and revisions to process and system);
- feedback on performance to inform company policy making (review).

Audit questions

A company should ask a range of questions of its management system to ensure comprehensive coverage. Such questions might include:

- What is to be included in the audit?
- Can the company undertake the audit or are consultants required?
- How might the audit be conducted and who will be involved?
- How will the audit check performance and determine compliance with requirements?
- How will results from the audit be published and communicated within the organisation?
- How will identified actions be implemented and subsequently monitored?
- How will the results from the audit inform management review?

Types of audit

There are three types of audit that may be carried out to evaluate the company's management system, characterised by the party conducting it:

1 *First-party audit* – performed by the company's own auditor.

2 *Second-party audit* – performed by a client or customer.

3 *Third-party audit* – performed by an independent auditor.

Focus

The auditor will focus on three aspects of the controlled documents for the system: standards; operations; and records, as follows:

- *Standards* – do the documents conform to the management system standards?
- *Operations* – do the operations conform to the documents?
- *Records* – do the records evidence conformance to the documents?

Audit report

When information has been collected, collated and analysed, a detailed audit report will be compiled which will form the basis of review by company management.

Review

The purpose of review is for the company to consider the findings from the audit. An essential aspect of review is the clear identification of elements and tasks within the company's processes and management system which require remedial action, determining how and by whom such action will be implemented. Executive and senior management should be at the forefront of review where reflection on company activities will focus on the most recent organisational performance and its correspondence with policies and objectives.

Review should ensure:

- development and improvement to policy in respect of quality, environment, and health and safety;
- continuation of effective organisation and resources to support existing and future policies;
- reinforcement of performance standards, monitoring, reporting, record keeping and audit to underpin policy.

As effective review is based on comprehensive audit and audit is reliant on evidence, review should be founded on the objective consideration of factual information. Performance indication is fundamental to this. Review should:

- determine the level of compliance by quality, environmental and safety management with set performance standards;
- identify activities where performance has been inadequate;
- consider the effectiveness of remedial actions taken;
- identify where actions still need to be taken;
- relate performance back to organisational policy and objectives;
- evaluate the efficacy of processes and their constituents;
- consider the effectiveness of system managers.

The scope of a company's review is primarily internal. Notwithstanding, review can also facilitate external benchmarking with other organisations. It is not uncommon for information on quality, environmental, and health and safety performance to feature in company reports and publicity materials. In addition, information is required to be filed where there have been environmental and health and safety transgressions, and such information will be accessible to corporate management. Review, although usually an annual event, should be one which takes place on an ongoing basis through interim reviews. Information from these can be collated, synthesised, analysed and fed into the detailed annual review. In this way, the management review can assume useful breadth and depth and contribute greater insight to organisational activity than would be the case with the annual review.

A framework for IMS development and implementation by principal contractors

This section examines: a framework for establishing an IMS by a principal contracting company or other organisations operating within the construction industry; system certification; and the integrated system supply chain. It is based on a synthesis of case studies.

IMS case study example

A framework for IMS based on case studies

Based on research and professional practice outlined earlier in this book and the IMS concepts, principles and practices described and discussed extensively in this chapter, this section presents a framework for IMS establishment. The framework is based on a synthesis of case studies of IMS development and implementation provided by five principal contracting companies. These companies are major organisations in terms of annual turnover of work, number of employees and scope of operations. Several operate in the European and international marketplaces, and all five can be said to be at the leading edge of management system utilisation. The framework presented is a 'composite' based on those characteristics and attributes of the most effective IMS in operation at this time. What the framework achieves is the presentation of a generic configuration for the IMS, adopting the business process model, and positioning this within the operating environment of construction and in particular the remit of the principal contracting organisation. It is a framework which, while orientated towards the activities of the principal contractor, can be applied to many other organisations operating within the construction industry.

It is emphasised that the IMS framework presented does not purport to be prescriptive or the best possible system that can be configured. A principal contractor, sub-contractor or any other organisation that chooses to adopt a management system approach must establish a system to satisfy its own unique structure, organisation, business objectives and modes of operation. Moreover, the organisation must do all of this within the context of its own business environment, circumstance and marketplace.

The framework for IMS development and implementation is shown in Figure 5.20. The framework is described and followed by discussion of the key considerations and decisions which informed its arrangement. As the framework uses the business process model as a skeleton, it should be noted that headed references to quality, environment and safety are not used, but rather that 'business' heads up the nomenclature adopted. Business management policy and objectives are identified, determined and confirmed by undertaking organisational and process mapping. Policy, goals, objectives and targets embracing quality, environment and safety as they affect business activities are brought together through an integrated policy statement. Quality, environmental, and safety management functions are encompassed within a single integrated business management manual.

As with all management systems, company-wide standing procedures and work instructions are contained within the manual and accompanying sets of documents. A QMS may be used as the base system, and in the systems reviewed for this composite framework that was the case. A prerequisite propounded

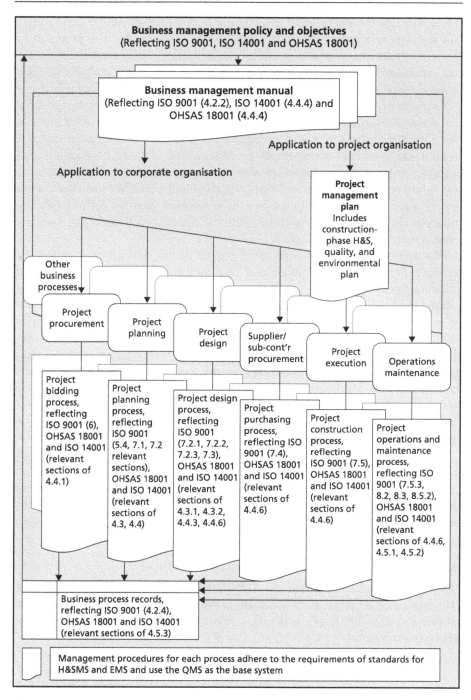

Figure 5.20 Framework for IMS development and implementation

Source: Adapted from Griffith and Bhutto (2008)

throughout this book is the need to integrate the different systems into one unified and seamless system but also to integrate the system with the core processes of the business of the organisation. Within construction, the principal contractor has a number of core processes to manage, ranging from procurement through to maintenance of the completed product – a building, structure or facility.

To accommodate these processes the business management manual is translated into a single project management plan (PMP). The PMP considers each process, identifies and assesses risk and configures the necessary management controls to tailor the nature of management procedures and working instructions. For example, project procurement will need procedures appropriate to contractor selection while project execution will need site rules and procedures to check work performance on site. It can be seen in the framework that for each of the core processes the management function meets the relevant sections of the management system standards applicable.

This approach provides a structured regime of key performance indicators, feedback to organisational policy and continual improvement, all precepts of efficient, effective and successful management systems. The core processes are augmented by other business processes which also serve to support product, service or, in this case, project delivery in specific organisational areas, for example human resource management (HRM) or legal affairs. During project execution, work-checking procedures are integrated and simplified to minimise duplication of tasks and paperwork. Administrative forms, or pro formas, will be used to initiate, monitor, check and record operational activities for quality, environment and safety simultaneously, and these also provide business process records and documents for project files. Information from the business process records is channelled to system and organisational feedback and review.

A crucial characteristic of this framework is that it intrinsically facilitates horizontal integration. With this established, information is shared across and throughout the core processes of the principal contractor's business and throughout the three management functions. It also brings sharply into focus the positioning of the management systems within the organisation as key management, assurance and business support systems focused on and serving the core business.

Developing and implementing the framework

Developing and implementing the IMS framework involves five key phases: (1) organisational mapping; (2) statement of vision, policy and objectives; (3) commitment to IMS-BPM; (4) structure; and (5) management system evolution. These are as follows:

1 *Organisational mapping* – to configure functional management in direct relation to processes within the context of the holistic perspective of the organisation.

2 *Statement of vision, policy and objectives* – to be clearly defined, communicated and accessible to all employees.

3 *Commitment to IMS-BPM* – to deliver 'integration' rather than a piecemeal approach.

4 *Structure* – to ensure homogeneous and consistent application across management functions and throughout the whole organisation.

5 *Evolution* – to ensure continual improvement to the management system and organisational betterment.

Main system elements

Key elements

The key elements applicable to the development and implementation of any management system need to be considered. These are: policy; organisation; risk assessment; planning; implementation; and audit and review. These may, however, be modified to reflect the particular requirements of an IMS. The BSI propounds the use of PAS 99 (Publicly Available Specification 99) when an organisation is seeking to integrate its management systems. PAS 99 is the world's first IMS requirements specification based on the six common requirements of ISO Guide 72 – the standard for writing management system standards. The six common requirements are: (1) policy; (2) planning; (3) implementation and operation; (4) performance assessment; (5) improvement; and (6) management review. Each management system application has its own specific requirements, as has been seen. However, the six common requirements feature in all systems and can therefore be used as a sound basis for systems integration. This approach is shown in Figure 5.21.

It is not proposed to present a detailed discussion of these elements as these have featured in Chapters 2, 3 and 4 and their development and application have been well covered. Notwithstanding, the important dimension to the IMS is that each element seeks to integrate the quality, environmental and safety management functions into one seamless system and integrates this with the management of the processes which deliver the business outputs. Examples of this are an integrated company policy which embraces quality, environment and safety into one corporate statement of business intentions and the application of a risk assessment approach which, for any business process, identifies and evaluates risk to quality, environment and safety simultaneously. A likely reality of such integration is that there may be a trade-off in absolute precision of management across the three integrated functions. Overall, however, the integrated management of the three functions should, across the board, be more efficient and effective. The truth is that much relies on the manager's ability to treat each management function appropriately and fairly when applying and administering the system in practice.

System documents

Documentation of the IMS broadly follows the structure also explained previously with the four levels of documentation:

1 System manual

2 Procedures

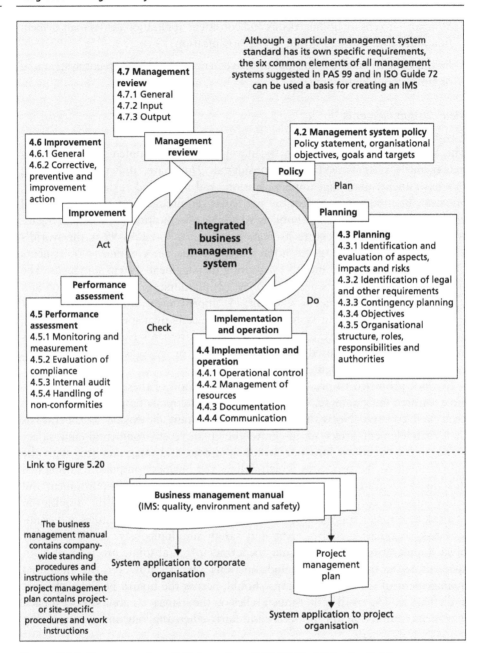

Figure 5.21 Elements of an IMS based on PAS 99 (Publicly Available Specification 99) for management systems

3 Implementation plan

4 Work instructions.

The difference from the framework shown in Figure 5.20 is that an IMS best follows the business process model and as such the document hierarchy suggests the following:

1 *A business management manual* – describing the framework and structure of the organisation and the arrangement of company standing management procedures and work instructions applied to the corporate organisation and reflecting a unified business management policy and set of objectives.

2 *A project management plan* – describing the application of the management system to the organisation's business products or services and, in the case of construction, the application to projects. This plan is project specific and will, obviously, vary from project to project yet contains the main thrusts of quality, environment and safety as designed within the organisation's business policy, objectives and management approach.

The framework has the benefit of providing sufficient structure to support a cohesive and strong approach throughout the whole organisation yet has sufficient flexibility to accommodate the specific requirements of different project applications. The project management plan also has the flexibility to accommodate as few or as many processes as designated by the business and may be changed as organisational processes change and evolve. This means that the parent IMS applied throughout the corporate organisation provides a uniform, consistent and continuous lead in business management while the project management plan enables the organisation to adapt to the requirements of the project and in so doing better respond to the needs of the customer.

The particular demands of an IMS mean that the project management plan must embrace the managerial requirements of quality, environment and safety simultaneously and must do so within each and every process that constitutes the business of the organisation. With this in mind, Figure 5.20 shows the relevant sections of each management system standard which must be accommodated by the process.

Administrative forms

Where the IMS differs substantially from a single management system is in the way in which administrative forms or pro forma templates are designed and used. In an IMS the forms must, first, focus on the process being managed and, second, consider the balanced requirements for quality, environmental and safety administration in one management procedure. This is essential in eliminating the duplication of tasks and documents raised by separate systems. It may mean that more time is needed to administer any particular task or process, since the manager needs to check and record activity for three management functions rather than one, but it does mean that all three functions are covered in a single observation event.

Again, it is not intended to present and describe IMS-related administrative forms as these are similar to those shown previously, albeit with modifications to reflect dual or triple applications rather than a single application. An organisation using administrative forms for application to, say, a QMS will have little difficulty in modifying the forms for applications with wider scope. In addition, and as explained previously, management consultancies can provide system templates through their commercial services to industry, and indeed some system documentation is available from website sources.

Experiences of IMS application

The following points reflect the experiences of IMS applications by principal contractors:

- IMS demands the encouragement, commitment and active support of executive management. This was achieved by appointing an IMS director who oversaw all aspects of system management in connection with the core business processes. In reality, this job turned out to be one of trouble-shooter as the role quickly became one of solving problems and difficulties as the system was being established.

- Executive management must delegate direct responsibility to process and system managers with support given from the system director as and when needed. This was felt appropriate to establish ownership of the IMS quickly and unequivocally.

- The IMS should focus on the core business and how this impinges upon the customer rather than becoming bogged down in the system itself. As such, procedures and documents were kept to a minimum with triple-function checking and records used.

- Management needs to target rapid and visible gains to the core processes from IMS to convince employees of the system's benefits in operation.

- IMS operation should encourage simple checklist supervision and brief but salient record keeping to win over system supervisors.

- The triple-function procedures can lead to ambiguous early practices which need to be rectified through more on-the-job awareness training and instruction.

Key success factors

Companies using the IMS report the following factors as highly influential in its establishment:

- *Organisational culture* – instilling a trusting and co-operative workforce is vital to embedding the system.

- *Involvement* – bottom-up involvement from grassroots level in system development is essential, as is inviting contribution and feedback to management.

- *Resources* – trained and capable managers, supervisors and workforce are essential and, as such, investment in training and system ownership should be a priority.

- *Flexibility* – the system should be allowed considerable flexibility in performance upon system establishment, incrementally becoming more demanding as familiarity with its operation is developed.
- *Shared commitment* – management must develop a blame-free culture where learning and improvement are preferred to difficulty and blame.

Key requirements

The key requirements of a workable and effective IMS, as highlighted by practice, are as follows:

- a robust and consistent policy statement which embraces all organisational aspects and operations;
- procedures for hazard identification, risk assessment, mitigation measures and controls;
- documented measurable goals, objectives and plans;
- a clear management structure and assignment of duties and responsibilities;
- assurance of a competent management base and skilled workforce;
- appropriate education and training mechanisms;
- clear organisational processes and efficient operating procedures for their undertaking;
- arrangements for communication, liaison and engagement with organisational and business stakeholders;
- a process of document control, revision and retention;
- sensible and active performance indicators;
- a process to detect non-conformances to system norms;
- a process to respond to and normalise system non-conformance;
- an approach which identifies learning opportunities and best system practices and shares improvements;
- a monitoring and audit facility;
- regular reporting of system operation and organisational performance;
- thorough management review;
- active and continual system enhancement;
- continual organisational and business improvement.

Corporate social responsibility

Respondent organisations highlight that a positive way to engender and embed the use of an IMS within an organisation's day-to-day operations is to link directly the ethos of the management system with the organisation's perspective of corporate social responsibility (CSR). Most companies uphold a set of corporate social responsibilities in key areas of their business activities. For example, statements in respect of corporate governance, community responsibilities, environmental safeguards, and health and safety are often published in company public-relations documents and annual statements of business. This is an ideal vehicle for the support of management systems, in particular the IMS, as the system can be used as a galvanising medium to co-ordinate system use with the core business processes. A statement of commitment to IMS application can feature covertly throughout a set of CSRs or can stand overtly as a management philosophy and approach as a CSR in its own right.

Optimising IMS effectiveness

People

While the effectiveness of an IMS is highly dependent upon the correct choice and appropriate configuration of the organisational structure and a framework for management, the greater influence is undoubtedly the vision, drive and commitment of management staff and the workforce. A management system does not create itself or run on its own – it must be carefully crafted and applied sensibly. This has everything to do with the people who design it, those who operate it and those whose work takes place in and around the system.

An IMS must be supported by all within an organisation and chiefly at two managerial levels: corporate level; and operations level.

To ensure that an integrated system has the opportunity to work with optimal effectiveness there must be actions taken at corporate level, the key actions being:

- demonstrable commitment to the organisational policy and strategy stated within the system's documentation;
- clear statements of organisational aims and objectives circulated throughout the whole company;
- employee understanding and ownership of the system through involvement in its design, development and implementation;
- setting of goals against which performance is transparent and appropriate;
- availability of adequate resources to facilitate system operation and maintenance;
- appropriate education, training and support for management and the workforce in system operation;
- ongoing evaluation and review of system implementation to enhance employee experience and expertise.

At operations level an IMS will be supported by a range of actions, the key ones being:

- distribution of system guidelines to all staff;
- briefing operations staff on system implementation;
- practice notes on system procedures;
- provision of checklists for implementing and monitoring particular procedures;
- training in the implementation of system procedures;
- use of self-audit and review mechanisms for system managers;

- guidance notes on emergency procedures and actions;
- reference to corporate management and other means of support for the system.

It is absolutely essential that an organisation carefully considers its management system and applies it to its activities with purpose, common sense and the utmost regard for those who implement it and must work with it. For the IMS or for that matter any management system to work and have optimum effectiveness it needs to invest not only in the system infrastructure but also in people. Moreover, it needs to do this within a morphogenic culture which facilitates continuous organisational development and improvement.

Organisational balance

A dilemma facing core business process support functions is the vexed question of balancing the appropriate amount of integration against the degree of autonomy, and the level of bureaucracy against the allowance for independence. There is no one right or best way of doing this. The organisation must do what appears best for the organisation at the time, appreciating that it might not get it right, but be willing to change and adapt its system until it does get it right. Optimising any system is about being open-minded to the renewal of organisational strategy and the continuous transformation of operations in the interest of the company, its business and its customers.

IMS certification

There is, currently, no international standard specification for the IMS or a certification scheme for such systems. QMSs, EMSs and H&SMSs are specified by their individual standards and specification series ISO 9001, ISO 14001 and OHSAS 18001 respectively. The certification of these management systems assesses the individual system against the individual standards. Notwithstanding, an organisation can pursue single, dual or triple certification of its management systems by applying for certification concurrently. The BSI has 'Integrated Management System Assessment' (IMSA) which allows companies to gain certification of their management systems in any combination through one submission.

In IMSA the individual management systems must satisfy the requirements of each of the standards, although there is full recognition of the common, overlapping and interactive elements of the systems. The integration of company policies, documentation, operational activities, audits and management reviews is recognised by IMSA, and if the IMS meets the various standards then certification is granted for each management system. In this way, an IMS can be implemented to fulfil the precise requirements of the company while conforming to the convention of the certification process.

Such integration is advantageous as it reduces certification costs due to the combined assessment audit and saves time, cost and effort by the organisation through single rather than multiple tasks involved in intra-organisational preparation for audit. While individual system certification is currently the norm, it is likely in the future that an IMS standard will be developed which incorporates the existing standards for quality, environment and safety. This would, likely, lead to the establishment of a framework standard embracing generic management system principles and specifications against which any type and number of management systems could be assessed. Correspondingly, certification schemes would follow, which would have the capacity to assess an IMS comprising any combination of management functions.

IMS supply chain

The quality, environmental and safety performance of a company's suppliers can have a direct bearing on the business performance of the company. For this reason a very important aspect of business is the management of the supply of products and services, or management of the *IMS supply chain*. A fundamental aim of supply chain management is to increase added value to business activities through the effectiveness of the organisation's links to its suppliers. While achieving this it also seeks to reduce costs, resources and non-added-value activities associated with the links in the chain. The supply chain is usually described as a cycle in which, for products, inputs arrive at the organisation, are converted into goods and are then distributed to customers. Inputs may come to the company in a variety of forms from different suppliers before being consumed. Management must have in place the wherewithal to manage this chain through establishing interrelations with suppliers. The chain extends to managing the outputs through to customers.

Where an organisation has a management system for ensuring the quality of its outputs with environmental consideration in a safe and healthy way, this good work can be let down if other parts of the supply chain compromise performance. A company must therefore ensure that it has a mechanism to determine the efficacy of the supply chain organisations for compatibility with its own levels of performance. Moreover, it must ensure that only those suppliers which can maintain performance to its requirements form part of the supply chain.

Within the scope of the IMS, companies use quality supply chain management (QSCM), environmental supply chain management (ESCM) and safety supply chain management (SSCM). Risk assessment forms a core element of each. The assessment of correspondence between the principal contractor's management system and, say, that of a materials supplier can be determined. Where a supplier has a management system along the same lines as the principal contractor, then there should be a good degree of confidence in the supplier's ability to perform and manage its inputs to at least the same standard as the contractor. Where a supplier has no system, or a poorly structured system, then the principal contractor might be rightly sceptical of the supplier's abilities.

A final thought

The Institution of Occupational Safety and Health (IOSH) concludes that:

> Our considered view is that an IMS should be the preferred option for many, but not all, organisations. A well planned IMS should be more efficient, and capable of taking

optimal decisions in the face of a range of uncertainties. The process of integration presents distinctive challenges for different organisations. Organisations that are most likely to integrate their systems successfully already enjoy multiple channels of communication founded on trust, respect for the expertise of colleagues, experience and confidence in the management of change.

(IOSH, 2010)

Management system integration: key points, overview and references

This section presents: a summary of the key points from the collective sections of Chapter 5; an overview of Chapter 5; and a list of references used in the compilation of Chapter 5.

Key points

The key points are arranged under a number of sub-headings as follows.

Management systems generally
There are characteristics displayed by almost all management systems:

- Quality management systems (QMSs) meeting the requirements of ISO 9001 have been widely implemented by many companies across all sectors of business, industry and commerce.

- There is increasing emphasis being placed on the implementation by companies of environmental management systems (EMSs) meeting the requirements of ISO 14001.

- There is also a growing support for the formalisation of health and safety management systems (H&SMSs) meeting the requirements of OHSAS 18001.

- The QMS is supported primarily by companies which seek to comply with the specified requirements of their client and customer base while the EMS and H&SMS seek primarily to comply with national legislation and industry regulation.

- The QMS, EMS and H&SMS to international management system requirements are well recognised, used and accepted within the construction industry, with the QMS in particular having been implemented widely and successfully for over fifty years.

- Management systems have a formal framework and set of procedures to establish structure and consistency of application.

- Management structures are traditionally vertical and separate for each management function system, leading to major difficulties in organisational communication and co-ordination.

- Documentation requirements of systems are often considered by those who implement systems as over-formal, onerous and bureaucratic.
- System operations staff do not always fully understand management systems such that maximum benefits may not be achieved.
- Effective systems implementation is reliant upon intra-organisational cultural change and a paradigm shift in outlook by management and workforce.
- There is acceptance of a necessary degree of formality in systems developments but this is often considered a burden with excessive time spent completing laborious and repetitive paperwork.

Management integration

Key aspects of management integration include the following:

- Management system standards for quality, environment, and health and safety have sufficient correspondence to enable conventionally separate systems to be integrated.
- Standards and systems for management have much in common with the process model of business activity.
- There are three processes involved in the business of most organisations: core processes; support processes; and assurance processes.
- Management systems for quality, environment, and health and safety are both support and assurance processes which facilitate and can influence a company's core business.
- There are three broad arrangements for management systems: separate; semi-integrated; and fully integrated.
- Management system integration requires the establishment of horizontal cross-functional management.
- Integrated management has much in common with systems theory, where holism and the creation of synergy are key attributes.
- The structure of an integrated management system is based on the Deming methodology: plan–do–check–act (PDCA).
- The QMS, being the oldest of the three support process management functions, is appropriate as the base system for IMS establishment.
- Wide acceptance and recognition of the management system are attainable through formal certification.

Integrated management system

An integrated management system (IMS) provides an organisation with a different perspective on how to see, configure and carry out its business. An integrated management system:

- implements business process management within a structured and formal framework of quality, environmental and safety management;

- allows the organisation to remain focused on customers and other business stakeholders;
- deploys clear policies, strategies and goals throughout the management of the organisation, with a holistic perspective;
- empowers staff and workforce to pursue effectiveness and improvements to the core processes;
- focuses on the contribution of support and assurance processes to deliver the core business processes efficiently and effectively;
- mobilises and motivates employees towards business-critical issues;
- stimulates creativity and promotes innovation to make real improvements to the business;
- measures, audits and evaluates outputs and their impact on the organisation's business bottom line;
- can be applied not just to quality, environment and safety but to other designated functions of management using the same framework and, further, the system can be configured as part of a set of corporate social responsibilities (CSRs);
- has been applied successfully by construction-related organisations which operate both dual and triple integrated systems across the quality, environmental, and safety management disciplines.

Overview

The evolution of standards-based management systems in the fields of quality, environment, and health and safety has reached a point where forward-thinking and dynamic companies in many business sectors including construction have begun to question the convention of using separate management systems. The tried and tested management system standards of the International Organization for Standardization (ISO) – 9001 (quality), 14001 (environment) and 18001 (safety) – have recognised the similarity of system applications with their common elements of development and practice – so much so that modern management systems are tending towards integration where the more effective attributes of system use can be shared across and throughout multiple management functions.

Chapter 5 has explored the way in which a company considers the management functions and systems associated with seeking to optimise its business activities. The chapter has described how the integrated perspective and use of standards-based systems can be applied to developing company policy, organisation, resources and procedures. An integrated management system (IMS) is an innovative way of handling the multitude of management functions and procedures associated with the delivery of services and projects. The IMS can be used to structure a company's business activities at both the corporate level and project level, has the potential to deliver greater organisational effectiveness and efficiency, and can improve project performance.

References

Arifin, K., Kadaruddin, A., Azahan, A. Jamaluddin, M.J. and Rosman, I. (2009). Implementation of integrated management system in Malaysia: the level of organization's understanding and awareness. *European Journal of Scientific Research*, **31**, (2), 188–195.

BCC Research (2010). *Market Research*. www.bccresearch.com

British Standards Society (1998). *PD 3548:1998 Standards and quality management – An integrated approach*. British Standards Society, London.

BSI (1999). *Occupational Health and Safety Assessment Series [OHSAS] OHSAS 18001: Occupational health and safety management systems – Specifications*. British Standards Institution, London.

CIRIA (2000). www.ciria.org

Deming, W.E. (1960). *Sample Designs in Business Research*. Wiley, New York.

Griffith, A. (1999). Developing an integrated quality, safety and environmental management system. *Construction Papers*, No. 108, *Construction Information Quarterly*, **1**, (3), 6–18.

Griffith, A. and Bhutto, K. (2008). Contractors' experiences of integrated management systems. *Management, Procurement and Law*, **161** (MP3), 93–98.

Griffith, A. and Watson, P. (2004). *Construction Management: Principles and Practice*. Palgrave, Basingstoke.

Griffith, A., Stephenson, P. and Watson, P. (2000). *Management Systems for Construction*. Addison Wesley Longman, Harlow.

IOSH (2010). *The Institution of Occupational Safety and Health's Policy Statement on the Integration of Management Systems for Occupational Safety and Health (OSH), Environmental Performance and Quality (the 'Integration Policy')*. www.iosh.co.uk

ISO (2000). *ISO 9001:2000 Quality management systems – Requirements*. International Organization for Standardization, Geneva.

ISO (2004). *ISO 14001:2004 Environmental management systems – Requirements with guidance for use*. International Organization for Standardization, Geneva.

ISO (2008). BS EN ISO 9001:2008 *Quality management systems – Requirements*. International Organization for Standardization, Geneva.

Jonker, J. and Klaver, J. (1998). Integration: a methodological perspective. *Quality World*, August, 14–19.

JQA (2010). *Assessment and Registration of Management Systems*. Japan Quality Assurance Organization, Tokyo. www.jqa.jp

Latham, M. Sir (1994). *Constructing the Team*. HMSO, London.

Quality Register (2010). *QA Register*. www.quality-register.co.uk

APPENDIX I

Management system standards: applicable versions

The following standards are current at the time of writing this book:

Quality management systems:

ISO (2008). *ISO 9001:2008 Quality management systems – Requirements.* International Organization for Standardization, Geneva.

Environmental management systems:

ISO (2004). *ISO 14001:2004 Environmental management systems – Requirements with guidance for use.* International Organization for Standardization, Geneva.

Health and safety management systems:

BSI (2007). *Occupational Health and Safety Assessment Series (OHSAS) 18001.* British Standards Institution, London.

Statutory instruments and regulations

Preamble

The following lists relate to prominent statutory instruments and regulations in the UK which can have influence on the structure and arrangements made for management systems in their application to the construction industry and to construction projects. The lists are in no way exhaustive as many statutory instruments and regulations are in existence.

Note

Some aspects of the work of Her Majesty's Stationery Office (HMSO) are reflected in the work of the Office of Public Sector Information (OPSI). The reader is reminded of this when searching for HMSO information and website material.

Environment

DEFRA (2010). *The Environmental Permitting (England and Wales) Regulations 2010.* The Department for Environment, Food and Rural Affairs (DEFRA). www.defra.gov.uk

Environment Agency (EA) (1974). *The Control of Pollution Act 1974.* HMSO, London.

Environment Agency (EA) (1979). *The Ancient Monuments and Archaeological Areas Act 1979.* HMSO, London.

Environment Agency (EA) (1981). *The Wildlife and Countryside Act 1981.* HMSO, London.

Environment Agency (EA) (1990). *The Environmental Protection Act 1990.* HMSO, London.

Environment Agency (EA) (1990). *The Planning (Listed Buildings and Conservation Areas) Act 1990.* HMSO, London.

Environment Agency (EA) (1990). *The Town and Country Planning Act 1990.* HMSO, London.

Environment Agency (EA) (1993). *The Clean Air Act 1993.* HMSO, London.

Environment Agency (EA) (1995). *The Environment Act 1995.* HMSO, London.

OPSI (2010). *The Site Waste Management Plans Regulations 2008.* Office of Public Sector Information (OPSI). www.opsi.gov.uk

Health and safety

Health and Safety Executive (HSE) (1974). *The Health and Safety at Work etc. Act 1974.* HSE, London.

Health and Safety Executive (HSE) (1992). *The Manual Handling Operations Regulations 1992.* HSE, London.

Health and Safety Executive (HSE) (1992). *The Personal Protective Equipment at Work Regulations 1992.* HSE, London.

Health and Safety Executive (HSE) (1992). *The Provision and Use of Work Equipment Regulations 1992.* HSE, London.

Health and Safety Executive (HSE) (1993). *The Chemicals (Hazard Information and Packaging) Regulations 1993.* HSE, London.

Health and Safety Executive (HSE) (1995). *The Reporting of Injuries, Diseases and Dangerous Occurrences Regulations 1995 (RIDDOR).* HSE, London.

Health and Safety Executive (HSE) (1999). *The Control of Substances Hazardous to Health Regulations 1999 (COSHH).* HSE, London.

Health and Safety Executive (HSE) (1999). *The Management of Health and Safety at Work Regulations 1999.* HSE, London.

Health and Safety Executive (HSE) (2005). *The Control of Noise at Work Regulations 2005.* HSE, London.

Health and Safety Executive (HSE) (2005). *The Work at Height Regulations 2005.* HSE, London.

Health and Safety Executive (HSE) (2006). *The Control of Asbestos at Work Regulations 2006.* HSE, London.

Health and Safety Executive (HSE) (2007). *The Construction (Design and Management) Regulations 2007.* HSE, London.

Statutory instruments and regulations of other countries

As a member of the European Union (EU), the UK has to comply with EU legislation. European Directives set minimum standards for legislation at EU level, with the modes of implementation at the discretion of the national governments of member countries. EU legislation becomes UK law through Acts of Parliament and Enabling Acts, which in turn lead to national regulations. European member countries have their own legislation, as indeed do countries worldwide. The reader is directed to specific national legislation and regulations where relevant.

Contacts for further information

The following organisations are excellent contacts for any organisation seeking to establish quality, environmental, safety, and integrated management and associated systems. They can provide information to construction and other industry sector organisations on: standards; systems; assessment and certification; industry practices; research; training; and publications. Some provide professional consultancy services. Enquiries should be made to the relevant websites, where detailed information is available.

British Standards Institution (BSI)
389 Chiswick High Road
London
W4 4AL
UK
cservices@bsigroup.com

Bureau Veritas
Head Office
Great Guildford House
30 Great Guildford Street
London
SE1 0ES
UK
www.bureauveritas.co.uk

Construction Industry Research and Information Association (CIRIA)
Classic House
174–180 Old Street
London
EC1V 9BP
UK
enquiries@ciria.org

International Organization for Standardization (ISO)
1, Ch. de la voie-Creuse
Case postale 56
CH-1211
Geneva 20
Switzerland
www.iso.org

SafeScope Ltd
Felaw Maltings
44 Felaw Street
Ipswich
IP2 8SJ
UK
contact@safescope.co.uk

Wolters Kluwer (UK) Ltd (Croner)
145 London Road
Kingston Upon Thames
Surrey
KT2 6SR
info@croner.co.uk

APPENDIX IV

Integrated Management Systems for Construction website

Web-based material to support *Integrated Management Systems for Construction: Quality, environment and safety* is available on the Pearson Education website. This includes introductory presentations for academics to support the five chapters of this book, and self-assessment questions for students and practitioners.

Index

Printed and bound by CPI Group (UK) Ltd, Croydon, CR0 4YY

23/10/2024

01778254-0002